Martin Pfeifer

Automated Model Generation and Observer Design for Interconnected Systems

A Port-Hamiltonian Approach

Karlsruher Beiträge zur
Regelungs- und Steuerungstechnik
Karlsruher Institut für Technologie

Band 13

Automated Model Generation and Observer Design for Interconnected Systems

A Port-Hamiltonian Approach

by
Martin Pfeifer

KIT Scientific Publishing

Karlsruher Institut für Technologie
Institut für Regelungs- und Steuerungssysteme

Automated Model Generation and Observer Design
for Interconnected Systems: A Port-Hamiltonian Approach

Zur Erlangung des akademischen Grades eines Doktor-Ingenieurs
von der KIT-Fakultät für Elektrotechnik und Informationstechnik des
Karlsruher Instituts für Technologie (KIT) genehmigte Dissertation

von Martin Pfeifer, M.Sc.

Tag der mündlichen Prüfung: 11. Juni 2021
Referent: Prof. Dr.-Ing. Sören Hohmann
Korreferent: Univ.-Prof. Dr.techn. Andreas Kugi

Impressum

KIT Scientific Publishing

Karlsruher Institut für Technologie (KIT)
KIT Scientific Publishing
Straße am Forum 2
D-76131 Karlsruhe

KIT Scientific Publishing is a registered trademark
of Karlsruhe Institute of Technology.
Reprint using the book cover is not allowed.

www.ksp.kit.edu

Print on Demand 2022 – Gedruckt auf FSC-zertifiziertem Papier

ISSN 2511-6312
ISBN 978-3-7315-1135-9
DOI 10.5445/KSP/1000138970

When the physics is well
known and well understood
the only way to advance
knowledge and technology
further is by way of
MATHEMATICS,
with its help if it already
exists, with new research
otherwise.

Rudolf Emil Kálmán (1930 – 2016)

Abstract

Interconnected systems are a major pillar of modern society. In recent years, interconnected systems have gained an unprecedented extent. For such systems, existing approaches for deriving models and designing observers reach their limits. In consequence, models and observers are developed under severe simplifications of the systems' physics which leads to a poor applicability and underperformance. This thesis provides remedy by *automating* the processes of deriving models and designing observers. To this end, we develop automatable modeling and observation methods in the domain-unifying framework of port-Hamiltonian systems (PHSs). These methods are the first to allow for an automated model generation and observer design in a wide class of interconnected systems.

The methods and algorithms are implemented in a software prototype named AMOTO. AMOTO is applied to the automated model generation and observer design in two case studies. Numeric simulations verify the validity of the models and observers obtained from AMOTO. Moreover, they outperform the models and observers resulting from state of the art techniques which verifies the practical usefulness of the approach. Therewith, the methods, algorithms, and tools from this thesis can help to solve the upcoming challenges in the interconnected systems that constantly surround us.

Kurzfassung

Vernetzte Systeme stellen einen unverzichtbaren Teil moderner Gesellschaften dar. Mit dem Ausrollen neuer Kommunikationstechnologien und in Folge der fortgeschrittenen Nutzung von Synergiepotenzialen entstanden in den letzten Jahren vernetzte Systeme ungeahnten Ausmaßes. Aufgrund der Komplexität dieser Systeme, gelangen bestehende Modellierungs- und Beobachterentwurfsmethoden an ihre Grenzen. Modelle und Beobachter können deshalb häufig nur unter erheblichen Vereinfachungen entwickelt werden. Die vorliegende Dissertation schafft Abhilfe. Leitgedanke ist es, die Vorgänge der Modellerzeugung und des Beobachterentwurfs zu *automatisieren*. Hierzu werden in dieser Arbeit automatisierbare Modellierungs- und Beobachtermethoden auf Basis der Port-Hamiltonschen Systemtheorie entwickelt.

Diese Methoden sind in einem Software-Prototyp namens AMOTO implementiert. In zwei Fallstudien wird AMOTO jeweils zur automatisierten Modellherleitung und zum automatisierten Beobachterentwurf eingesetzt. Computersimulationen weisen in beiden Fallstudien die Funktionstüchtigkeit der erzeugten Modelle und Beobachter nach und zeigen, dass diese genauere Ergebnisse liefern, als Modelle und Beobachter, die mit Methoden des bisherigen Stands der Technik entwickelt wurden. Dies unterstreicht die praktische Nutzbarkeit des vorgestellten Ansatzes. Es zeigt sich ferner, dass der Ansatz auf eine große Klasse vernetzter Systeme anwendbar ist. Somit leisten die Methoden, Algorithmen und Werkzeuge aus dieser Arbeit einen wichtigen Beitrag zur Bewältigung zukünftiger Herausforderungen in vernetzen Systemen.

Contents

Chapter 1

Introduction

1.1 Motivation

Interconnected systems are a major pillar of our modern society. Well-known examples of interconnected systems are the energy systems that constantly surround us. Over the last few years, the rollout of powerful communication networks, the pursuit of efficiency, and increased performance requirements have led to the advent of interconnected systems of unforeseen extent. In the context of energy systems, large-scale multi-carrier energy systems become reality (O'Malley et al. [2020]). Other examples that penetrate into practice are interconnected industrial systems (Gao et al. [2019]), cooperative swarms of drones (Tahir et al. [2019]), adaptive mechanic structures (Warsewa et al. [2020]), and automotive power networks for hybrid and electric vehicles (Mantilla-Pérez et al. [2020]).

From a system theoretic point of view, such interconnected systems have demanding properties. First, they comprise a large number of interdependent system variables—a quality that Duindam et al. [2009] call the "curse of dimensionality". Second, many interconnected systems cover multiple physical domains as, e.g., the electric, mechanic, hydraulic, and thermal domain. Third, such systems consist of many subsystems each of which contributes to the overall system behavior. Hence, subsystems can hardly be neglected or abstracted. Fourth, in many cases, interconnected systems feature nonlinear dynamics. Due to these reasons, the literature refers to interconnected systems often as "complex physical systems" (see, e.g., van der Schaft and Jeltsema [2014]).

Models are inevitable for the development of control systems in interconnected systems. A model in form of a set of ordinary differential equations (ODEs) is the basis for many simulations and the natural starting point for a model-based control system design. The modeling of interconnected systems, however, is a very challenging task. This can be understood by two reasons:

1. The multi-physical nature of many interconnected systems disallows for an individual person to oversee the entire system. Hence, the model derivation involves experts from multiple disciplines. Naturally, different experts speak different technical languages, apply different methods, and use different notations.
2. The curse of dimensionality leads to extensive equations which makes it impossible to develop models by hand. Even a partially computer-aided development as an alternative leads to solutions in which the engineer is still forced to handle mathematical expressions of enormous size.

It is important to note that the curse of dimensionality goes beyond the system modeling. In fact, it complicates all steps in a model-based control system development. In this thesis, we will elaborate this issue particularly for the design of observers. In interconnected systems, observers are of considerable interest as they are essential for the supervision of a system and often required for the practical implementation of a control law.

Due to the above reasons, the development of models and observers for interconnected systems is cost-intensive, cumbersome, and prone to error. In consequence, models and observers are frequently developed under severe simplifications in which the systems' physics are only roughly approximated, e.g., by neglecting important dynamics or nonlinearities. In a wide operating range, however, these approximations are inadmissible which limits the practical applicability of the obtained models and observers.

To avoid such unjustified simplifications, there are different model derivation and observer design techniques in the literature. These approaches are outlined in the next section.

1.2 Literature Context

In the literature, there have been multiple computer-aided methods to handle the complexity of interconnected systems in the model derivation and observer design. This section presents a brief overview of these methods. A particular literature review on methods for the modeling and observation in the framework of port-Hamiltonian systems (PHSs) can be found in sections 3.1 and 4.1, respectively.

1.2.1 Model Derivation Methods

Existing methods for a computer-aided derivation of ODE models can be classified into two groups, viz. data-based methods and physical-based methods. Data-based methods compute a model based on data originating from the system or a virtual duplicate. In contrast, in physical-based methods, a model is derived from knowledge about the physical relations that govern the system. The focus of this thesis—and therewith of this subsection—is on physical-based methods.

In analytical mechanics, there is a long tradition in the algorithmic derivation of ODE models. This tradition traces back to the work of Joseph-Louis Lagrange and Sir William Rowan Hamilton and the formalisms named after them. The Lagrange and Hamilton formalisms and extensions thereof allow for a unified energy- and power-based modeling of multi-domain systems (see Jeltsema and Scherpen [2009]). Despite their prominence, the methods from analytical mechanics have received only limited attention in literature for the computer-aided model derivation in interconnected systems. Two exceptions are the papers of Leu and Hemati [1986] and Bachovchin and Ilić [2015]. In these papers, the authors present methods for deriving the dynamic equations of robotic manipulators and electric power systems, respectively, based on the Lagrange formalism.

Network-based modeling offers another approach to the computer-aided derivation of ODE models for interconnected systems. The idea of a network-based modeling is to describe the system under consideration by means of a graph. In contrast to the methods based on analytical mechanics, there exists extensive literature on the modeling of interconnected systems with network-based methods. It is neither in the scope of this section nor the intention to go into the details of all publications in this field. Instead, the pioneering works, some survey papers, and a few recent results are highlighted.

The history of network-based modeling ranges back to the work of Kirchhoff [1847] on electric network theory. The first network-based methods for an ODE modeling of electric circuits were published in the mid 1960s by Brayton and Moser [1964] and Chua and Rohrer [1965]. Roberson and Wittenburg [1968] propose a formalism for the derivation of the dynamic equations of a system of rigid bodies. The interconnection topology is described by a graph which is analytically characterized by its incidence matrix. In the mid 1980s, Cederbaum [1984] surveyed important applications of network-theory to the modeling, analysis, and synthesis of electric circuits. A method for the computer-aided generation of the equations of motion of planar multibody systems with open and closed kinematic chains was presented by McPhee [1998]. The method integrates network modeling methods and orthogonal projection methods. Schlacher et al. [1998] combine a network-based modeling approach with the Lagrangian framework. Based hereon, the authors propose a method for the computer-aided modeling of mechatronic systems.

The research activities in the field of a network-based modeling for mechanic, electric, and mechatronic systems in the years between 2000 and 2016 have been reviewed by Garziad and Saka [2017]. Dörfler et al. [2018] survey some recent results how network- and graph-theory informs the modeling, analysis, and design of electric systems.

Network-based models have an excellent scalability which makes them appealing for the treatment of interconnected systems. The main limitation of a network-based modeling approach is the assumption that the vertices and edges of the system-describing graph represent some pre-defined system components. This assumption limits the reusability and transferability of the models to other systems, in particular across different physical domains.

1.2.2 Observer Design Methods

As early as 1966, Luenberger presented a “simple algorithm for computing the observer” [Luenberger, 1966]. The development of algorithms for the practical observer design is closely linked to the development of the observer theory as a whole. Needless to say that a complete review of all algorithmic observer design methods is outside the scope of this subsection. Instead, it is focused on approaches which are particularly appealing for a computer-aided observer design in interconnected systems.

Well-known methods for the computer-aided design of observers date back to Ackermann [1972] and Kautsky et al. [1985]. The former is implemented in Wolfram Mathematica and can be accessed via the command `EstimatorGains`; the latter is behind the MATLAB command `place`. The innovation of the following decades were strongly influenced by the widespread availability of technical computing systems and an increasing computational power. These developments fostered the advent of optimization-based observer design techniques as proposed by Franceschini et al. [1994], Howell and Hedrick [2002], Ichihara [2007], and Shoukry et al. [2018]. Approaches particularly based on the solution of linear matrix inequalities (LMIs) were presented, e.g., by Cho and Rajamani [1997], Wang et al. [2003], Moreno [2004], and Chen and Saif [2006]. Cheng et al. [1994] and Raghavan and Hedrick [1994] derive an observer design strategy which is based on the numeric solution of a Riccati equation. Similar, Edwards and Spurgeon [1994] presented an algorithm for the model-based design of sliding mode observers which requires the solution of a Lyapunov equation. Syrmos [1993] provided a computationally efficient observer design framework based on the transmission zeros of the system. An approach based on orthogonal transformations was proposed by Laila et al. [2011]. Schweers [2017] presented a method for a computer-aided design of state estimators. The methods from Schweers [2017] rely on data-based models of multi-domain systems.

The subsequent section briefly summarizes the insights of the literature review and states the objectives of this thesis.

1.3 Research Objectives

The modeling of interconnected systems is a cumbersome and time-consuming task which may involve experts from different engineering domains. The model derivation using the power- and energy-based methods from analytical mechanics is a promising approach to overcome these problems (cf. Jeltsema and Scherpen [2009]). However, a computer-aided model generation with these methods has received only limited attention in the literature and is, as for now, restricted to a rather small class of interconnected systems.

On the other hand, many network-based modeling methods allow for an efficient model derivation. However, these approaches are also limited to particular classes of systems as, e.g., electric networks, power systems, mechatronic systems, or multi-body systems.

A unifying computer-aided modeling method which can be applied to a large class of interconnected systems is lacking in the main literature.

For the computer-aided design of observers for interconnected systems there exist powerful methods in the literature. A limitation, however, is that these approaches are based on a purely numerical model of the system. In such a model, there is hardly any information about the physical structure of the system to be observed. On the one hand, this limits the reusability of the resulting observers, e.g., in case of variations of the physical parameters. On the other hand, the lack of interpretability hampers the exploitation of physical system properties for the observer design.

Physical-oriented methods for a computer-aided observer design for interconnected systems have been rarely reported in the literature so far.

The thesis at hand addresses the above research gaps. The idea is to consistently *automate* the physical-based model derivation and observer design for interconnected systems. Something that is, to the best of our knowledge, neither present in the literature nor existing in practice. To this end, this thesis follows a port-Hamiltonian approach.

Port-Hamiltonian systems theory combines the Lagrangian/Hamiltonian-based modeling framework with the network-based modeling framework: on the one hand, PHSs rely on the Lagrangian/Hamiltonian paradigm in which energy serves as a unifying conserved quantity in different physical domains; on the other hand, PHSs are build upon the port-based modeling concept which emphasizes their affinity to the network-based paradigm. The combination of these two frameworks establishes the advantages of a port-Hamiltonian approach, four of which are:

i. its scalability to very large interconnected systems;
ii. its inherent ability to treat multi-domain systems;
iii. the deep physical insight provided by port-Hamiltonian models;
iv. the incorporation of nonlinearities while retaining underlying conservation laws.

For these reasons, port-Hamiltonian systems theory is the ideal methodological framework for this thesis. The research objectives—restated in the light of this methodological approach—are:

(O1) *to develop methods for an automated generation of port-Hamiltonian models for a wide class of interconnected systems,*

(O2) *to derive methods for an automated design of observers based on the port-Hamiltonian models obtained from the techniques from O1.*

1.4 Structure and Notation

To ensure orientation and readability, the structure and notation of the remainder of this thesis are presented here.

Structure Chapter 2 outlines the terminology applied in this thesis and provides the necessary basics in port-Hamiltonian systems theory. The following three chapters contain the main contributions of this work. Chapter 3 is devoted to the development of methods and algorithms for an automated model generation. Methods and algorithms for an automated observer design are presented in Chapter 4. Chapter 5 provides a proof of principle and demonstrates the practical usefulness of the automated model generation and observer design. Finally, the main conclusions of this work are drawn together in Chapter 6.

Notation Sets, groups, and spaces are written in blackboard bold. The cardinality of a set $\mathbb{M}$ is denoted as $|\mathbb{M}|$. The sets $\mathbb{N}$, $\mathbb{R}$, and $\mathbb{C}$ are the sets of natural, real, and complex numbers, respectively. For the dimension of a vector space $\mathbb{X}$ we write $\dim(\mathbb{X})$. The group of $n \times n$ orthogonal matrices is given by $\mathbb{O}(n)$. Vectors and matrices are written in bold font. Let $\boldsymbol{A} \in \mathbb{R}^{m \times n}$ be a matrix with m rows and n columns. For the transpose of $\boldsymbol{A}$ we write $\boldsymbol{A}^\top$. Now let $m = n$. The inverse of $\boldsymbol{A}$ is denoted by $\boldsymbol{A}^{-1}$ (if it exists). $\mathrm{Spec}(\boldsymbol{A})$ denotes the spectrum of $\boldsymbol{A}$, i.e., its set of eigenvalues. $\boldsymbol{A} \succ 0$ and $\boldsymbol{A} \succeq 0$ mean that $\boldsymbol{A}$ is positive-definite and positive semi-definite, respectively. A diagonal matrix is denoted by $\mathrm{diag}(\cdot)$; likewise, $\mathrm{blkdiag}(\cdot)$ is a block diagonal matrix of matrices. Now let $\boldsymbol{x} \in \mathbb{R}^m$ be a (column) vector. The Euclidean norm of $\boldsymbol{x}$ is denoted as $\|\boldsymbol{x}\|$. For the kernel, image, and rank of the linear map $\boldsymbol{x} \mapsto \boldsymbol{A}\boldsymbol{x}$ we write $\ker(\boldsymbol{A})$, $\mathrm{im}(\boldsymbol{A})$, and $\mathrm{rank}(\boldsymbol{A})$, respectively.

Throughout this thesis, when using the expressions "we" or "us", the author presumes that the readers agree to what he is saying. In particular, these types of expressions are not intended to imply any personalization of the academic discourse.

Chapter 2

Fundamentals

The aim of this thesis is to develop automatable port-Hamiltonian methods for the model derivation and the observer design in interconnected systems. This chapter provides the fundamentals towards this endeavor. To enable a clear and concise presentation of the technical contents, Section 2.1 outlines the terminology used throughout this work. Afterwards, Section 2.2 provides selected fundamentals in port-Hamiltonian systems theory which are necessary to derive automatable methods in chapters 3 and 4.

2.1 Terminology

This section is devoted to the definition of some terms of key importance for this work as, e.g., interconnected system, automated model generation, and automated observer design.

2.1.1 Interconnected System

Our notion of an interconnected system is guided by the terminology from Willems [2007]. To approach a formal definition of an interconnected system, we first introduce the notion of an *open graph*. The following definition is condensed from the elaboration of van der Schaft and Maschke [2013].

Term 2.1 (Open graph)

An open graph is a directed graph $\mathscr{G} = (\mathbb{V}, \mathbb{B})$ with vertices $\mathbb{V}$ and edges $\mathbb{B}$. The set $\mathbb{V}_\mathrm{B} \subset \mathbb{V}$ is the subset of boundary vertices which represent the interfaces between the graph system and its environment. The set $\mathbb{V}_\mathrm{I} := \mathbb{V} \setminus \mathbb{V}_\mathrm{B}$ contains the inner vertices of the open graph.

Based on the concept of an open graph, we may now introduce the concept of a *networked system*.

Term 2.2 (Networked system)

A networked system Σ is a system which can be described by an open graph $\mathscr{G} = (\mathbb{V}, \mathbb{B})$ with $\mathbb{V} = \mathbb{V}_\mathrm{I} \cup \mathbb{V}_\mathrm{B}$ and $\mathbb{B} = \mathbb{B}_\mathrm{I} \cup \mathbb{B}_\mathrm{B}$. The elements of $\mathbb{V}_\mathrm{I}$ represent multiple, possibly heterogeneous subsystems of the networked system. The interactions between the subsystems are expressed by the set of inner edges $\mathbb{B}_\mathrm{I} := \{(u,v) \in \mathbb{B} \mid u, v \in \mathbb{V}_\mathrm{I}\}$. The subsystems may interact with the system environment via a set of boundary vertices $\mathbb{V}_\mathrm{B}$. The boundary vertices are connected to the inner vertices by a set of boundary edges $\mathbb{B}_\mathrm{B} := \{(u,v) \in \mathbb{B} \mid u \in \mathbb{V}_\mathrm{I}, v \in \mathbb{V}_\mathrm{B}\}$.

Figure 2.1 illustrates the open graph representation of an exemplary networked system with nine inner vertices, three boundary vertices, twelve inner edges, and four boundary edges.

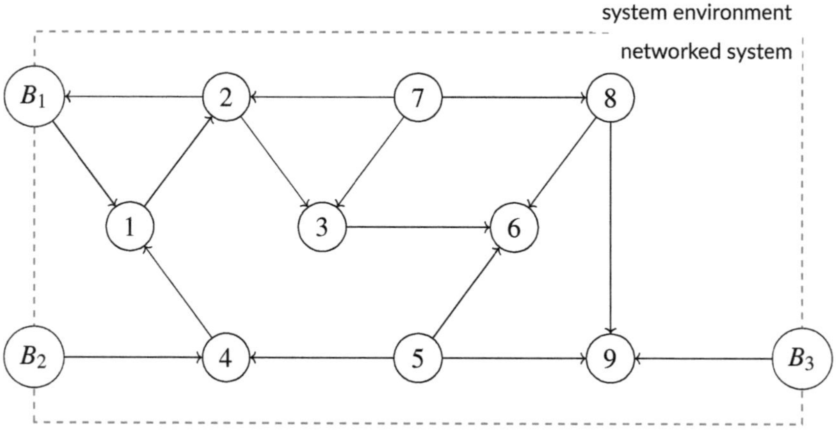

Figure 2.1: Example of a networked system with nine inner vertices $(1, \ldots, 9)$, three boundary vertices (B_1, B_2, B_3), twelve inner edges, and four boundary edges

A networked system allows for any kind of interaction between the vertices of the open graph, e.g., by an exchange of matter, information, or energy. This thesis will focus on networked systems in which the interactions are established by an exchange of energy. The exchange of energy between vertices requires the presence of a physical interconnection between them. This motivates the notion of an *interconnected system* [Willems, 2007].

Term 2.3 (Interconnected system)

An interconnected system Σ as a networked system in which the interactions between the vertices are established through an exchange of energy.

In this thesis, we consider continuous-time, finite-dimensional, deterministic interconnected systems. Some well-known examples of such systems are electric circuits (Kugi [2001]), power systems (Dörfler et al. [2018]), gas networks (Pfeifer et al. [2018]), district heating networks (Merkert et al. [2019]), interconnected industrial systems (Gao et al. [2019]), and battery systems (Kupper [2019]). Other examples are cooperative swarms of drones (Tahir et al. [2019]), automotive power networks (Mantilla-Pérez et al. [2020]), large-scale mechatronic systems (Deroo [2016]), and adaptive mechanic structures (Warsewa et al. [2020]).

2.1.2 Automated Model Generation

In order to approach the notion of an automated model generation let us first clarify the meaning of a *mathematical model*. The following elaboration is inspired from Wellstead [1979]:

Term 2.4 (Mathematical model)

Consider a system Σ. A mathematical model is an objective-specific description of Σ in the form of mathematical functions and equations deduced from available information about the system.

Next, the term *mathematical model* is particularized to a *physical-based model.*

Term 2.5 (Physical-based model)

Consider a system Σ. A physical-based model is a mathematical model of Σ deduced from the physical relations that govern the system.

Unless explicitly stated otherwise, in this work the term *model* denotes a physical-based model. We aim at explicit ODE models in a state-space representation.

Explicit state-space models are particularly appealing for the design, analysis, and simulation of control systems. The following definition combines the elaborations from Vidyasagar [1993, p. 1] and van der Schaft [2017, p. 10]:

Term 2.6 (Explicit state-space model)

Consider a system Σ. An explicit state-space model of Σ is a model of the form

$$\dot{\boldsymbol{x}}(t) = \boldsymbol{f}(\boldsymbol{x}(t), \boldsymbol{u}(t), t), \tag{2.1a}$$

$$\boldsymbol{y}(t) = \boldsymbol{g}(\boldsymbol{x}(t), \boldsymbol{u}(t), t), \tag{2.1b}$$

for all $t \in \mathbb{R}_{\geq 0}$, where $\boldsymbol{u}(t) \in \mathbb{U} \subseteq \mathbb{R}^p$, $\boldsymbol{x}(t) \in \mathbb{X} \subseteq \mathbb{R}^n$, $\boldsymbol{y}(t) \in \mathbb{Y} \subseteq \mathbb{R}^q$ and sufficiently smooth mappings $\boldsymbol{f}\colon \mathbb{X} \times \mathbb{U} \times \mathbb{R}_{\geq 0} \to \mathbb{X}$, $\boldsymbol{g}\colon \mathbb{X} \times \mathbb{U} \times \mathbb{R}_{\geq 0} \to \mathbb{Y}$. The quantities $\boldsymbol{u}(t)$, $\boldsymbol{x}(t)$, and $\boldsymbol{y}(t)$ are referred to as input, state, and output of the system, respectively; $\mathbb{U}$, $\mathbb{X}$, and $\mathbb{Y}$ are denoted as input-space, state-space, and output-space, respectively.

Remark 2.7 (Notation). In the remainder of this thesis, the time-dependence "(t)" of vectors is omitted in the notation.

For interconnected systems, one distinguishes between global models and local models. A global model of an interconnected system Σ is a model comprising the entire open graph $\mathscr{G} = (\mathbb{V}, \mathbb{B})$. In contrast, a local model is defined only on a connected subgraph $\tilde{\mathscr{G}} \subset \mathscr{G}$. Moreover, we distinguish between a symbolic and a numeric model. In a symbolic model, the functions $\boldsymbol{f}$ and $\boldsymbol{g}$ from (2.1) depend (besides on $\boldsymbol{x}$, $\boldsymbol{u}$, and t) on the physical system parameters in symbolic form as, e.g., a resistance R, a mass m, etc.; in contrast, in a numeric model, the physical parameters are numerically specified, e.g., by information from data sheets or parameter identification techniques. Hence, such a model is—except for the variables $\boldsymbol{u}$, $\boldsymbol{x}$, and $\boldsymbol{y}$—completely numerically determined.

Now we have everything prepared to introduce a term of key importance for this thesis, viz. *automated model generation*.

Term 2.8 (Automated model generation)

Automated model generation describes the computer-aided derivation of a physical-based explicit state-space model with minimal human assistance.

The modeling of interconnected systems may easily involve hundreds of equations. In Chapter 5, we examine two examples for this. For a human developer, dealing with such a number of equations is cumbersome and prone to error. Hence, this work particularly aims at automating the formalization step of the model derivation, i.e., the "equation deriving part".

2.1.3 Automated Observer Design

Observer theory is an important topic in systems and control theory. The observer concept traces back to Luenberger [1964]. The following conception of an *observer* is taken from Trumpf et al. [2011]:

> **Term 2.9 (Observer)**
>
> *Given a system Σ whose variables can be partitioned into a set of known or measured variables and a set of unknown variables to be reconstructed. An observer for Σ is a deterministic dynamic system which produces reconstructions of the unknown variables on the basis of the known or measured variables.*

We aim at "asymptotic observers" (Trumpf [2013]), i.e., observers where the reconstructions asymptotically converge towards the values of the unknown variables. If the Euclidean norm of the reconstruction error converges exponentially to zero, the observer is called exponentially convergent (cf. Khalil [2002, Def. 4.5]). Moreover, we distinguish between two observer architectures, viz. a centralized and a distributed observer architecture (cf. Kupper [2019, Def. 2.5]). Given an interconnected system Σ, in a centralized architecture, there is a single observer which receives information from all measurement sources to calculate reconstructions of all unknown variables from Σ. In contrast, in a distributed architecture, there is a dedicated observer for each subsystem of Σ. In each observer, the available measurement information and the variables to be reconstructed are local with respect to the subsystem.

Next, let us clarify the terms *observer design* and *model-based observer design*.

> **Term 2.10 (Observer design)**
>
> *Observer design refers to the process of specifying the parameters of an observer.*

Term 2.11 (Model-based observer design)

Model-based observer design is an observer design which is based on a mathematical model of the system under consideration.

This thesis focuses on an observer design on the basis of physical-based explicit state-space models. Such an observer design can rely either on global or local model knowledge (cf. Deroo [2016]). The latter is of particular interest for the design of a distributed observer.

A model-based observer design can rely either on a symbolic or a numeric model. Accordingly, these two cases are denoted as symbolic and numeric observer design, respectively. The former leads to an observer in which the observer parameters depend on the physical parameters in symbolic form; the latter results in observer parameters that are completely numerically specified. A symbolic observer design has two advantages over a numeric observer design. First, a symbolic design provides more insight into the physical background of the observer parameters. In Chapter 4, it will be shown that this insight allows us to use the physical properties of the system beneficially for the observer design. Second, a symbolic observer design enables the reusability of the resulting observer, e.g., in case of parameter variations. A drawback of a symbolic observer design is that it may lead to computationally expensive operations on large symbolic expressions. Both, symbolic and numeric observer designs, are in the scope of this thesis.

Next, let us introduce a notion of major importance for this work, viz. *automated observer design.*

Term 2.12 (Automated observer design)

Automated observer design refers to a computer-aided model-based observer design with minimal human assistance.

The following example motivates the practical usefulness of an automated observer design:

Example 2.13:

Power systems are well-known examples of interconnected systems. State estimation techniques are crucial for the operation of a power system. The state estimators applied in nowadays control centers are static. Recently, an IEEE taskforce pointed out that in consequence of the energy transition, a static state estimation is increasingly inadequate and "should be reassessed and enhanced with new monitoring tools, such as dynamic state estimation" [Zhao et al., 2019]. However, the extent of a power system with hundred and more inputs, states, and outputs significantly hampers the design of a dynamic state estimator. An automated observer design can overcome this problem—as will be seen in Section 5.2.

2.2 Port-Hamiltonian Systems

This section provides some fundamentals in port-Hamiltonian systems theory which are necessary for the development of automatable modeling and observer design methods in chapters 3 and 4, respectively.

In Subsection 2.2.1, we briefly recapitulate the notion of generalized power variables. Afterwards, Subsection 2.2.2 introduces the concept of a Dirac structure. The focus of this thesis is on explicit state-space models in form of input-state-output PHSs. This particular class of PHSs is introduced in Subsection 2.2.3.

2.2.1 Generalized Power Variables

PHSs follow the power- and energy-based concept of the Lagrangian/Hamiltonian modeling paradigm. A thorough introduction into this concept can be found, e.g., in Breedveld [2009], Jeltsema and Scherpen [2009], and in the ageless book authored by Wellstead [1979].

The core idea of an energy-based modeling is to use energy as universally, domain-independent conserved quantity. The interaction between two elements, subsystems, systems, etc. is coupled to an exchange of energy, i.e., $E = \int P \mathrm{d}t$. The power P is described by two time-dependent, generalized power variables, viz. a flow $\boldsymbol{f} \in \mathbb{F}$ and an effort $\boldsymbol{e} \in \mathbb{E}$ with $\mathbb{F}$ an abstract finite-dimensional vector space and $\mathbb{E}$ its dual vector space $\mathbb{E} := \mathbb{F}^*$.The exchanged instantaneous power is given by the dual pairing $P = \langle \boldsymbol{e} \mid \boldsymbol{f} \rangle$.

The generalized variables have straightforward correspondences to system variables from different physical domains. Table 2.1 summarizes the correspondences between the generalized variables and the domain-specific variables for the electric, magnetic, mechanic, hydraulic, thermal, and chemical domain.

This thesis makes use of a correspondence scheme which is known as *thermodynamic framework of domains and variables* [Breedveld, 2009, p. 24]. In this framework, the generalized state $\boldsymbol{x}$ is defined as the integral of the generalized flow over time, i.e., $\boldsymbol{x} := \int \boldsymbol{f} \mathrm{d}t$, see Table 2.1. Moreover, note that for each domain the product of the flow- and the effort-related variable has the unit of power.

Table 2.1: Correspondences between generalized variables and physical variables in the thermodynamic framework of domains and variables (cf. Breedveld [2009, p. 24])

	f	flow	e	effort	x	state
electric	I	current	V	voltage	Q	charge
magnetic	V	voltage	I	current	ψ	flux linkage
kinetic translation	F	force	v	velocity	p	momentum
potential translation	v	velocity	F	force	z	displacement
kinetic rotation	T	torque	ω	angular velocity	b	angular momentum
potential rotation	ω	angular velocity	T	torque	θ	angular displacement
elastic hydraulic	ϕ	volume flow	p	pressure	Λ	volume
kinetic hydraulic	p	pressure	ϕ	volume flow	Γ	momentum of a fluid
thermal	f_S	entropy flow	T	temperature	S	entropy
chemical	f_N	molar flow	μ	chemical potential	N	number of moles

The following example illustrates how the concept of generalized power variables contributes to a unified modeling of physical systems. The example will be successively applied in this chapter.

Example 2.14:

Consider the electromagnetic, mechanic, and hydraulic systems in Figure 2.2 (a), (b), and (c), respectively. The systems are structured as follows:

- *The electromagnetic system consists of a current source feeding a current I_0, an inductor L, a resistor R, and two capacitors C_1 and C_2.*
- *The mechanic system is composed of a Hookean spring with stiffness k, a damper with damping constant b, and two masses m_1 and m_2; an external force F_0 acts on the mass m_1.*
- *The hydraulic system consists of a pump providing a volume flow ϕ_0 and two open tanks with fluid capacitances $C_{f,1}$ and $C_{f,2}$;[1] the two tanks are connected through a pipe with fluid inertance L_f and fluid resistance R_f.*

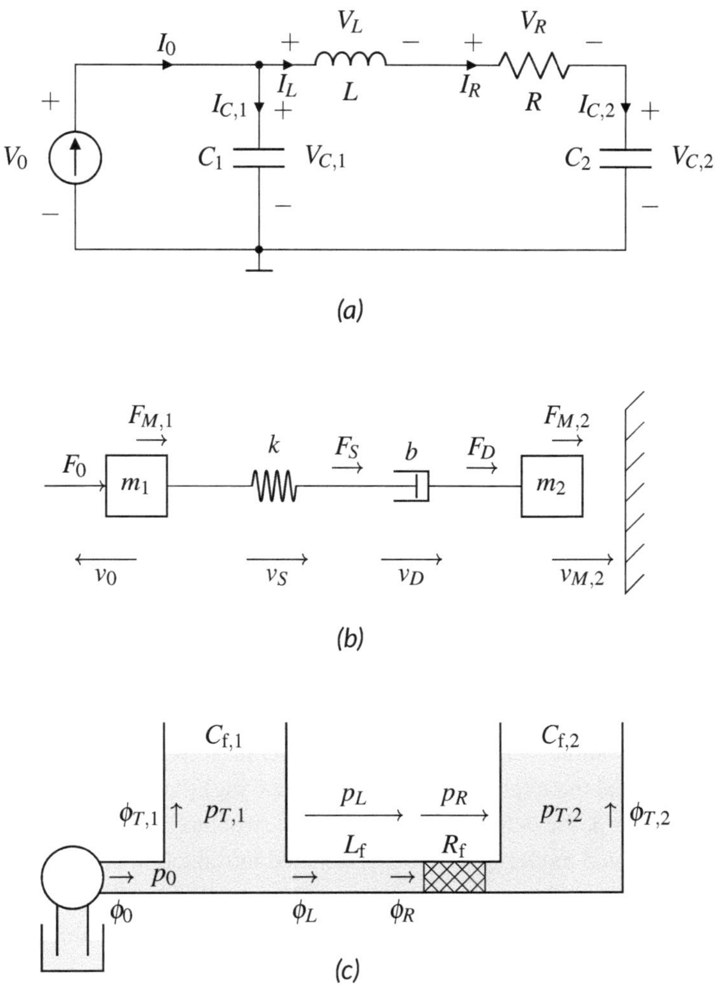

Figure 2.2: Systems in the electromagnetic (a), mechanic (b), and hydraulic domain (c)

Let us introduce the generalized power variables as follows:

	$f_{C,1}$	$f_{C,2}$	$f_{C,3}$	f_R	f_0	$e_{C,1}$	$e_{C,2}$	$e_{C,3}$	e_R	e_0
(a) electromagnetic	$I_{C,1}$	$I_{C,2}$	V_L	I_R	I_0	$V_{C,1}$	$V_{C,2}$	I_L	V_R	V_0
(b) mechanic	$F_{M,1}$	$F_{M,2}$	v_S	F_D	F_0	$v_{M,1}$	$v_{M,2}$	F_S	v_D	v_0
(c) hydraulic	$\phi_{T,1}$	$\phi_{T,2}$	p_L	ϕ_R	ϕ_0	$p_{T,1}$	$p_{T,2}$	ϕ_L	p_R	p_0

Under these correspondences the systems from Figure 2.2 (a), (b), and (c) obey the same network equations:

$$f_0 - f_{C,1} - e_{C,3} = 0, \tag{2.2a}$$
$$e_0 - f_{C,3} - e_R - e_{C,2} = 0, \tag{2.2b}$$
$$e_{C,3} - f_R = 0, \tag{2.2c}$$
$$f_R - f_{C,2} = 0, \tag{2.2d}$$
$$e_0 - e_{C,1} = 0. \tag{2.2e}$$

2.2.2 Dirac Structures

The notion of a Dirac structure is of central importance for the port-Hamiltonian framework. The Dirac structure establishes the link between the Lagrangian/Hamiltonian and the network-based point of view. It describes the network interconnections in form of a power-conserving, geometric structure. Therewith, the Dirac structure generalizes the symplectic geometry of the phase-space from the classical Hamiltonian equations of motion [van der Schaft, 2009, p. 74]. The power-conservation is ensured by relating generalized power variables $\boldsymbol{f}$ and $\boldsymbol{e}$ in such a way that the total power entering (or leaving) the Dirac structure is zero.

In Chapter 3, the determination of a Dirac structure will be the enabling step for an automated model generation. A detailed introduction into the concept of a Dirac structure can be found in Bloch and Crouch [1999] and van der Schaft and Jeltsema [2014]. The following definition is taken from the latter:

[1]The fluid capacitances $C_{f,i}$ can be calculated as $C_{f,i} = A_i/(\rho g)$, where A_i is the cross-sectional area of tank i, ρ the mass density of the (incompressible) fluid, and g the gravitational constant ($i = 1, 2$).

Definition 2.15 (Dirac structure)

Consider the flows $\boldsymbol{f} \in \mathbb{F}$ and the efforts $\boldsymbol{e} \in \mathbb{E} = \mathbb{F}^*$ where $\mathbb{F}$ is an abstract finite-dimensional vector space. A constant Dirac structure is a subspace $\mathbb{D} \subset \mathbb{F} \times \mathbb{E}$ such that

$$(i) \quad \langle \boldsymbol{e} \mid \boldsymbol{f} \rangle = 0, \quad \forall (\boldsymbol{f}, \boldsymbol{e}) \in \mathbb{D}, \tag{2.3a}$$
$$(ii) \quad \dim \mathbb{D} = \dim \mathbb{F}. \tag{2.3b}$$

Property (i) represents the power-conservation of the Dirac structure. The maximal dimension of any subspace $\mathbb{D} \subset \mathbb{F} \times \mathbb{E}$ satisfying property (i) is $\dim \mathbb{F}$ [van der Schaft, 2009, p. 56]. Hence, property (ii) requires $\mathbb{D}$ to be of maximal dimension.

The following definition extends Definition 2.15 to the situation where a Dirac structure is modulated by a state vector $\boldsymbol{x} \in \mathbb{X}$:

Definition 2.16 (Modulated Dirac structure)

A modulated Dirac structure *is a family of constant Dirac structures $\mathbb{D}(\boldsymbol{x}) \subset \mathbb{F} \times \mathbb{E}$ indexed over $\boldsymbol{x} \in \mathbb{X}$.*

Remark 2.17 (Real vector spaces). Throughout this thesis we have $\mathbb{F} = \mathbb{R}^n$. As $\mathbb{E} = (\mathbb{R}^n)^*$ is isomorphic to $\mathbb{R}^n$, we identify $\mathbb{E}$ with $\mathbb{R}^n$.

Definitions 2.15 and 2.16 are coordinate-free. For the practical work, matrix representations of Dirac structures are of paramount importance. The following definition presents two matrix representations, viz. the *kernel representation* and the *input-output representation*.

Definition 2.18 (Kernel representation)

A kernel representation of a modulated Dirac structure $\mathbb{D}(\boldsymbol{x}) \subset \mathbb{R}^n \times \mathbb{R}^n$ with $\boldsymbol{x} \in \mathbb{X}$ is

$$\mathbb{D}(\boldsymbol{x}) = \{(\boldsymbol{f}, \boldsymbol{e}) \in \mathbb{R}^n \times \mathbb{R}^n \mid \boldsymbol{F}(\boldsymbol{x})\boldsymbol{f} + \boldsymbol{E}(\boldsymbol{x})\boldsymbol{e} = \boldsymbol{0}\}, \tag{2.4}$$

where the matrices $\boldsymbol{F}(\boldsymbol{x})$ *and* $\boldsymbol{E}(\boldsymbol{x})$ *satisfy*

$$(i)\quad \boldsymbol{E}(\boldsymbol{x})\boldsymbol{F}^\top(\boldsymbol{x})+\boldsymbol{F}(\boldsymbol{x})\boldsymbol{E}^\top(\boldsymbol{x})=\boldsymbol{0}, \tag{2.5a}$$

$$(ii)\quad \operatorname{rank}(\boldsymbol{F}(\boldsymbol{x})\ \boldsymbol{E}(\boldsymbol{x}))=n, \tag{2.5b}$$

for all $\boldsymbol{x}\in\mathbb{X}$. *The power balance of (2.4) is*

$$\boldsymbol{e}^\top\boldsymbol{f}=0,\quad \forall(\boldsymbol{f},\boldsymbol{e})\in\mathbb{D}(\boldsymbol{x})\,. \tag{2.6}$$

Remark 2.19 (Uniqueness of the kernel representation). The matrices $\boldsymbol{F}(\boldsymbol{x})$ and $\boldsymbol{E}(\boldsymbol{x})$ are not uniquely determined by the kernel representation. For example, both matrices can be multiplied from the left by an arbitrary invertible matrix $\boldsymbol{T}(\boldsymbol{x})$ without changing $\mathbb{D}$.

Proposition 2.20 (Input-output representation)

Let $\mathbb{D}(\boldsymbol{x})\subset\mathbb{R}^n\times\mathbb{R}^n$ *with* $\boldsymbol{x}\in\mathbb{X}$ *be a Dirac structure in kernel representation. Let* $\operatorname{rank}(\boldsymbol{F}(\boldsymbol{x}))=m\ (\leq n)$. *We select* m *independent columns of* $\boldsymbol{F}(\boldsymbol{x})$ *and group them into a matrix* $\boldsymbol{F}_1(\boldsymbol{x})$. *The remaining* $n-m$ *columns are collected in a matrix* $\boldsymbol{F}_2(\boldsymbol{x})$. *Possibly after permutations, we write* $\boldsymbol{F}(\boldsymbol{x})=(\boldsymbol{F}_1(\boldsymbol{x})\ \boldsymbol{F}_2(\boldsymbol{x}))$. *Correspondingly, we split* $\boldsymbol{E}(\boldsymbol{x})$, $\boldsymbol{f}$, *and* $\boldsymbol{e}$ *into* $(\boldsymbol{E}_1(\boldsymbol{x})\ \boldsymbol{E}_2(\boldsymbol{x}))$, $\left(\boldsymbol{f}_1^\top\ \boldsymbol{f}_2^\top\right)^\top$, *and* $\left(\boldsymbol{e}_1^\top\ \boldsymbol{e}_2^\top\right)^\top$, *respectively. Then, the matrix* $(\boldsymbol{F}_1(\boldsymbol{x})\ \boldsymbol{E}_2(\boldsymbol{x}))$ *is invertible for all* $\boldsymbol{x}\in\mathbb{X}$ *and an input-output representation of* $\mathbb{D}(\boldsymbol{x})$ *is given by*

$$\mathbb{D}(\boldsymbol{x})=\{(\boldsymbol{f},\boldsymbol{e})\in\mathbb{R}^n\times\mathbb{R}^n\mid \boldsymbol{y}=\boldsymbol{Z}(\boldsymbol{x})\boldsymbol{u}\} \tag{2.7}$$

with

$$\boldsymbol{Z}(\boldsymbol{x})=-(\boldsymbol{F}_1(\boldsymbol{x})\ \boldsymbol{E}_2(\boldsymbol{x}))^{-1}(\boldsymbol{E}_1(\boldsymbol{x})\ \boldsymbol{F}_2(\boldsymbol{x}))\,. \tag{2.8}$$

The matrix $\boldsymbol{Z}(\boldsymbol{x})$ *satisfies* $\boldsymbol{Z}(\boldsymbol{x})=-\boldsymbol{Z}(\boldsymbol{x})^\top$ *for all* $\boldsymbol{x}\in\mathbb{X}$. *The vectors* $\boldsymbol{u}=(\boldsymbol{e}_1^\top\ \boldsymbol{f}_2^\top)^\top$ *and* $\boldsymbol{y}=(\boldsymbol{f}_1^\top\ \boldsymbol{e}_2^\top)^\top$ *are referred to as input vector and output vector, respectively. The power balance of (2.7) is*

$$\boldsymbol{u}^\top\boldsymbol{y}=\boldsymbol{e}_1^\top\boldsymbol{f}_1+\boldsymbol{f}_2^\top\boldsymbol{e}_2=0,\ \forall\left(\begin{pmatrix}\boldsymbol{f}_1\\ \boldsymbol{f}_2\end{pmatrix},\begin{pmatrix}\boldsymbol{e}_1\\ \boldsymbol{e}_2\end{pmatrix}\right)\in\mathbb{D}(\boldsymbol{x})\,. \tag{2.9}$$

Proof:
Bloch and Crouch [1999] prove this statement for a constant Dirac structure, i.e., the case where $\boldsymbol{F}(\boldsymbol{x}) = \boldsymbol{F} = \text{const.}$, $\boldsymbol{E}(\boldsymbol{x}) = \boldsymbol{E} = \text{const.}$; the pointwise application of the proof from Bloch and Crouch [1999] then proves the statement in Proposition 2.20. □

Remark 2.21 (Implicit and explicit representations). Due to the structure of the equation systems, the kernel representation (2.4) and the input-output representation (2.7) are denoted as implicit and explicit representations, respectively.

In the following, we continue with the example to illustrate the kernel and the input-output representation of a Dirac structure.

Example 2.22:

Consider the three systems from Example 2.14. By rearranging the network equations (2.2), the internal interconnection structure of each of the systems can be described by a Dirac structure in kernel representation (2.4):

$$\mathbb{D} = \{(\begin{pmatrix} f_{\mathrm{C},1} \\ f_{\mathrm{C},2} \\ f_{\mathrm{C},3} \\ f_{\mathrm{R}} \\ f_{\mathrm{O}} \end{pmatrix}, \begin{pmatrix} e_{\mathrm{C},1} \\ e_{\mathrm{C},2} \\ e_{\mathrm{C},3} \\ e_{\mathrm{R}} \\ e_{\mathrm{O}} \end{pmatrix}) \in \mathbb{R}^5 \times \mathbb{R}^5 \mid \begin{pmatrix} -1 & 0 & 0 & 0 & -1 \\ 0 & 0 & 1 & 0 & 0 \\ 1 & 0 & 0 & 1 & 1 \\ 1 & 1 & 0 & 0 & 1 \\ 0 & 0 & 0 & 0 & 0 \end{pmatrix} \begin{pmatrix} -f_{\mathrm{C},1} \\ -f_{\mathrm{C},2} \\ -f_{\mathrm{C},3} \\ -f_{\mathrm{R}} \\ f_{\mathrm{O}} \end{pmatrix} + \begin{pmatrix} 0 & 0 & 1 & 0 & 0 \\ 0 & -1 & 0 & -1 & 1 \\ 0 & 0 & 0 & 0 & 0 \\ 0 & 0 & 0 & 0 & 0 \\ -1 & 0 & 0 & 0 & 1 \end{pmatrix} \begin{pmatrix} e_{\mathrm{C},1} \\ e_{\mathrm{C},2} \\ e_{\mathrm{C},3} \\ e_{\mathrm{R}} \\ e_{\mathrm{O}} \end{pmatrix} = \begin{pmatrix} 0 \\ 0 \\ 0 \\ 0 \\ 0 \end{pmatrix} \}. \quad (2.10)$$

It can be easily verified that the matrices in (2.10) satisfy the conditions (2.5). An input-output representation of (2.10) is given by

$$\mathbb{D} = \{ (\begin{pmatrix} f_{C,1} \\ f_{C,2} \\ f_{C,3} \\ f_R \\ f_O \end{pmatrix}, \begin{pmatrix} e_{C,1} \\ e_{C,2} \\ e_{C,3} \\ e_R \\ e_O \end{pmatrix}) \in \mathbb{R}^5 \times \mathbb{R}^5 \mid \begin{pmatrix} -f_{C,1} \\ -f_{C,2} \\ -f_{C,3} \\ -f_R \\ e_O \end{pmatrix} = \left(\begin{array}{ccc:c:c} 0 & 0 & 1 & 0 & -1 \\ 0 & 0 & -1 & 0 & 0 \\ -1 & 1 & 0 & 1 & 0 \\ \hdashline 0 & 0 & -1 & 0 & 0 \\ \hdashline 1 & 0 & 0 & 0 & 0 \end{array}\right) \begin{pmatrix} e_{C,1} \\ e_{C,2} \\ e_{C,3} \\ e_R \\ f_O \end{pmatrix} \}. \quad (2.11)$$

Note that the equation systems in (2.10) and (2.11) are equivalent to (2.2). The division of the matrix in (2.11) into matrix blocks will be relevant in the following chapter.

Example 2.22 gives a first hint that the matrix representations of a Dirac structure can be derived from the network equations of the system. The structured derivation of such representations is a central aspect of Chapter 3.

2.2.3 Input-State-Output Port-Hamiltonian Systems

In this thesis, we consider the class of *explicit input-state-output PHSs*. This is a particularly important class of PHSs as it represents the starting point for the majority of controller and observer design methods from the port-Hamiltonian theory, see, e.g., Ortega et al. [2008], Venkatraman and van der Schaft [2010], Vincent et al. [2016], van der Schaft [2017], Yaghmaei and Yazdanpanah [2019b].

Explicit input-state-output PHSs have first been introduced in the 1990s; amongst others, pioneering works are from Maschke and van der Schaft [1992], Maschke et al. [1992], and van der Schaft and Maschke [1995]. Kugi [2001] was the first to use this system class to solve practical control engineering problems. Comprehensive information about the theoretical background of explicit input-state-output PHSs can be found in the textbooks from Duindam et al. [2009], van der Schaft and Jeltsema [2014], and van der Schaft [2017]. The following definition is based on the latter:

Definition 2.23 (Explicit input-state-output PHS with feedthrough)
An explicit input-state-output PHS with feedthrough is an explicit state-space model of the form

$$\dot{\boldsymbol{x}} = (\boldsymbol{J}(\boldsymbol{x}) - \boldsymbol{R}(\boldsymbol{x}))\frac{\partial H}{\partial \boldsymbol{x}}(\boldsymbol{x}) + (\boldsymbol{G}(\boldsymbol{x}) - \boldsymbol{P}(\boldsymbol{x}))\boldsymbol{u}, \tag{2.12a}$$

$$\boldsymbol{y} = (\boldsymbol{G}(\boldsymbol{x}) + \boldsymbol{P}(\boldsymbol{x}))^\top \frac{\partial H}{\partial \boldsymbol{x}}(\boldsymbol{x}) + (\boldsymbol{M}(\boldsymbol{x}) + \boldsymbol{S}(\boldsymbol{x}))\boldsymbol{u}, \tag{2.12b}$$

where $\boldsymbol{x} \in \mathbb{X} \subseteq \mathbb{R}^n$, $\boldsymbol{u} \in \mathbb{U} \subseteq \mathbb{R}^p$, and $\boldsymbol{y} \in \mathbb{Y} \subseteq \mathbb{R}^p$ are the state vector, the input vector, and the output vector, respectively. The Hamiltonian H is a continuously differentiable function with $H\colon \mathbb{X} \to \mathbb{R}$ that is bounded from below. The matrices $\boldsymbol{J}(\boldsymbol{x})$, $\boldsymbol{R}(\boldsymbol{x}) \in \mathbb{R}^{n\times n}$, $\boldsymbol{G}(\boldsymbol{x})$, $\boldsymbol{P}(\boldsymbol{x}) \in \mathbb{R}^{n\times p}$, $\boldsymbol{M}(\boldsymbol{x})$, $\boldsymbol{S}(\boldsymbol{x}) \in \mathbb{R}^{p\times p}$ satisfy $\boldsymbol{J}(\boldsymbol{x}) = -\boldsymbol{J}^\top(\boldsymbol{x})$, $\boldsymbol{M}(\boldsymbol{x}) = -\boldsymbol{M}^\top(\boldsymbol{x})$, and

$$\boldsymbol{\Theta}(\boldsymbol{x}) := \begin{pmatrix} \boldsymbol{R}(\boldsymbol{x}) & \boldsymbol{P}(\boldsymbol{x}) \\ \boldsymbol{P}^\top(\boldsymbol{x}) & \boldsymbol{S}(\boldsymbol{x}) \end{pmatrix} = \begin{pmatrix} \boldsymbol{R}(\boldsymbol{x}) & \boldsymbol{P}(\boldsymbol{x}) \\ \boldsymbol{P}^\top(\boldsymbol{x}) & \boldsymbol{S}(\boldsymbol{x}) \end{pmatrix}^\top \succeq 0, \ \forall \boldsymbol{x} \in \mathbb{X}. \tag{2.13}$$

In the remainder of this work, the class of explicit input-state-output PHSs is briefly denoted as *explicit PHSs*. The matrices, functions, and vectors in (2.12) allow for a deep physical interpretation which is an essential advantage of an explicit PHS over an ordinary state-space representation: the matrix $\boldsymbol{J}(\boldsymbol{x})$ represents the internal energy-preserving interconnection in the system; $\boldsymbol{R}(\boldsymbol{x})$ accounts for energy-dissipating effects; $\boldsymbol{G}(\boldsymbol{x})$, $\boldsymbol{P}(\boldsymbol{x})$, $\boldsymbol{M}(\boldsymbol{x})$, and $\boldsymbol{S}(\boldsymbol{x})$ specify the interaction between the system and its environment via the system ports. The Hamiltonian H is a storage function which is strongly related to the total energy contained in the system. The instantaneous power exchange between the system and its environment is given by $\boldsymbol{u}^\top \boldsymbol{y}$.

A limitation of explicit PHSs is given for systems with irreversible thermodynamic processes. For such systems, the energy function is defined implicitly as a Legendre submanifold of the thermodynamic phase-space [Arnold, 1989, Appendix 4]. This implicit definition of the energy function impedes an explicit modeling.

In the sequel, we turn our attention to a well-known special case of (2.12), viz. to explicit PHSs without feedthrough.

Definition 2.24 (Explicit PHS without feedthrough)

An explicit input-state-output PHS without feedthrough is an explicit state-space model of the form

$$\dot{x} = (\boldsymbol{J}(\boldsymbol{x}) - \boldsymbol{R}(\boldsymbol{x}))\frac{\partial H}{\partial \boldsymbol{x}}(\boldsymbol{x}) + \boldsymbol{G}(\boldsymbol{x})\boldsymbol{u}, \tag{2.14a}$$

$$\boldsymbol{y} = \boldsymbol{G}^\top(\boldsymbol{x})\frac{\partial H}{\partial \boldsymbol{x}}(\boldsymbol{x}), \tag{2.14b}$$

where $\boldsymbol{x}$, $\boldsymbol{u}$, $\boldsymbol{y}$, *and* H *as in Definition 2.23 and* $\boldsymbol{J}(\boldsymbol{x}) = -\boldsymbol{J}^\top(\boldsymbol{x})$, $\boldsymbol{R}(\boldsymbol{x}) = \boldsymbol{R}^\top(\boldsymbol{x}) \succeq 0$.

The PHS (2.14) is particularly transparent with respect to the underlying Dirac structure. The input-output representation of the Dirac structure (cf. Proposition 2.20) is specified by the skew-symmetric matrix

$$\boldsymbol{Z}(\boldsymbol{x}) = \begin{pmatrix} -\boldsymbol{J}(\boldsymbol{x}) & -\boldsymbol{G}_\mathrm{R}(\boldsymbol{x}) & -\boldsymbol{G}(\boldsymbol{x}) \\ \boldsymbol{G}_\mathrm{R}^\top(\boldsymbol{x}) & \boldsymbol{0} & \boldsymbol{0} \\ \boldsymbol{G}^\top(\boldsymbol{x}) & \boldsymbol{0} & \boldsymbol{0} \end{pmatrix}. \tag{2.15}$$

Hence, in the non-feedthrough case, the matrices $\boldsymbol{J}(\boldsymbol{x})$ and $\boldsymbol{G}(\boldsymbol{x})$ of the PHS (2.14) can directly be determined from the Dirac structure. Likewise, the matrix $\boldsymbol{R}(\boldsymbol{x})$ results from $\boldsymbol{R}(\boldsymbol{x}) = \boldsymbol{G}_\mathrm{R}(\boldsymbol{x})\tilde{\boldsymbol{R}}(\boldsymbol{x})\boldsymbol{G}_\mathrm{R}^\top(\boldsymbol{x})$ for some resistive relation $\boldsymbol{f}_\mathrm{R} = -\tilde{\boldsymbol{R}}(\boldsymbol{x})\boldsymbol{e}_\mathrm{R}$ with $\tilde{\boldsymbol{R}}(\boldsymbol{x}) = \tilde{\boldsymbol{R}}^\top(\boldsymbol{x}) \succeq 0$. Our successive example illustrates the formulation of an explicit PHS based on a Dirac structure.

Example 2.25:

Consider the three systems from Example 2.14. The energy-dissipating elements of the three systems obey a linear resistive relation of the form $f_\mathrm{R} = -De_\mathrm{R}$ *where* $D = 1/R$, $D = b$, *and* $D = 1/R_\mathrm{f}$ *for the electromagnetic, mechanic, and hydraulic system, respectively. From the resistive relation and the Dirac structure (2.11) we can calculate the following explicit PHS:*

$$\frac{\mathrm{d}}{\mathrm{d}t}\begin{pmatrix} x_1 \\ x_2 \\ x_3 \end{pmatrix} = \left(\begin{pmatrix} 0 & 0 & -1 \\ 0 & 0 & 1 \\ 1 & -1 & 0 \end{pmatrix} - \begin{pmatrix} 0 & 0 & 0 \\ 0 & 0 & 0 \\ 0 & 0 & D^{-1} \end{pmatrix}\right)\frac{\partial H}{\partial \boldsymbol{x}}(\boldsymbol{x}) + \begin{pmatrix} 1 \\ 0 \\ 0 \end{pmatrix} u, \tag{2.16a}$$

$$y = \begin{pmatrix} 1 & 0 & 0 \end{pmatrix}\frac{\partial H}{\partial \boldsymbol{x}}(\boldsymbol{x}). \tag{2.16b}$$

The states in (2.16) correspond to the generalized states of the energy-storing elements of the systems. For the electromagnetic system, the states x_1 and x_2 are given by the charges Q_1 and Q_2 on the capacitors C_1 and C_2, respectively; the state x_3 relates to the flux linkage of the inductor L. The input and output of the model are given by the current I_0 and the voltage V_0, respectively. For the mechanic and hydraulic systems, one can draw similar correspondences as summarized in the following table:

	state x_1	*state x_2*	*state x_3*	*input u*	*output y*
(a) electromag.	*charge Q_1*	*charge Q_2*	*flux linkage ψ*	*current I_0*	*voltage V_0*
(b) mechanic	*momentum p_1*	*momentum p_2*	*displacement z_0*	*force F_0*	*velocity v_0*
(c) hydraulic	*volume Λ_1*	*volume Λ_2*	*momentum Γ*	*volume flow ϕ_0*	*pressure p_0*

The Hamiltonian in (2.16) is the sum of the energies of the individual energy-storing elements and can be expressed as

$$H(\boldsymbol{x}) = \frac{1}{2}\boldsymbol{x}^\top \begin{pmatrix} q_1 & 0 & 0 \\ 0 & q_2 & 0 \\ 0 & 0 & q_3 \end{pmatrix} \boldsymbol{x}, \tag{2.17}$$

where q_1, q_2, q_3 are equal to $C_1^{-1}, C_2^{-1}, L^{-1}$ for the electromagnetic system; m_1^{-1}, m_2^{-1}, k for the mechanic system; and $C_{\mathrm{f},1}^{-1}, C_{\mathrm{f},2}^{-1}, L_{\mathrm{f}}^{-1}$ for the hydraulic system.

The previous example shows that PHSs can be used to model systems from different domains in a unifying framework. The three systems are *form equivalent* and lead to structurally identical port-Hamiltonian models. This, however, is an absolute special case. In general, different systems will lead to port-Hamiltonian models of different structures.

Example 2.25 illustrates that a PHS preserves the underlying physical structure of the system and therewith allows for a deep physical insight. Moreover, we have seen that a port-Hamiltonian modeling approach differentiates between the network model and the constitutive relations of the components. The Dirac structure is the connecting element as it links the generalized port variables to the constitutive relations of the components connected to the ports. This gives the port-Hamiltonian approach a high degree of modularity.

Another advantageous property of explicit PHSs is that they are inherently *passive*.

Proposition 2.26 (Passivity of explicit PHSs)
The explicit PHS from Definition 2.23 is passive. Moreover, if the matrix $\boldsymbol{\Theta}(\boldsymbol{x})$ *in* (2.13) *is positive-definite for all* $\boldsymbol{x} \in \mathbb{X}$*, the PHS is strictly passive.*

Proof:
The proof can be found in Appendix A.1. □

Remark 2.27 (Passivity of non-feedthrough PHSs). The claim from Proposition 2.26 applies also for the PHS from Definition 2.24. For strict passivity, the requirement of positive definiteness in (2.13) simplifies to $\boldsymbol{R}(\boldsymbol{x}) = \boldsymbol{R}^\top(\boldsymbol{x}) \succ 0$ for all $\boldsymbol{x} \in \mathbb{X}$.

The inherent passivity of PHSs facilitates the use of this system class for passivity-based control design techniques [van der Schaft, 2017]. In Chapter 4, the passivity of PHSs will be exploited for the automated model-based design of observers.

Another noteworthy property of an explicit PHS is that the system output $\boldsymbol{y}$ is determined mainly on the basis of physical considerations, cf. Example 2.25. Hence, in general, the output variables are different from the measured variables.[2] This will also be taken into account in Chapter 4.

This concludes the fundamentals chapter. Based on this groundwork, we will develop port-Hamiltonian methods for an automated model generation and observer design in the following two chapters.

[2] An exception in which the output variables coincide with the measurement variables is given for a collocated actuator/sensor configuration.

Chapter 3

Automated Model Generation

This chapter addresses the development of methods for an automated generation of explicit port-Hamiltonian models for interconnected systems. First, in Section 3.1 we review the techniques that are available in the literature for describing an interconnected systems as a PHS. We will identify a research gap in which existing methods suffer from different shortcomings impeding an automated model generation. Based on this research gap the aims of this chapter are formulated. Section 3.2 addresses these aims by deriving new methods for the modeling of interconnected systems as PHSs. These methods constitute algorithms which allow for the automated generation of explicit PHSs. The results from Section 3.2 will be discussed in Section 3.3. The chapter ends with a brief summary of the main insights in Section 3.4.

3.1 Literature Review

There have been different approaches for the derivation of port-Hamiltonian models of interconnected systems. The basic idea of these approaches is to formulate a PHS by describing the topological structure of the system as a graph. The available methods from the literature are detailed in the sequel.

A pioneering work on the graph-based derivation of PHSs stems from Kugi [2001]. The author describes electromagnetic systems as directed graphs. Based on the graph description, a method for the structured derivation of an explicit port-Hamiltonian model is proposed. Another milestone in this field stems from van der Schaft and Maschke [2013]. The authors describe various interconnected systems as open directed graphs. Based on the directed graphs, explicit port-Hamiltonian models are obtained. Fiaz et al. [2013] take up the idea from van der Schaft and Maschke [2013] and propose a systematic framework for the port-

Hamiltonian modeling of power networks. Circuit graphs are used by Falaize and Hélie [2016] to derive port-Hamiltonian models of analog circuits. The method from Falaize and Hélie [2016] has been implemented in a Python tool to enable an automated model generation of differential-algebraic equation (DAE) models, see Falaize and Hélie [2019]. Another graph approach is proposed by Scheuermann et al. [2020]. The authors use solid graphs to derive discrete-time port-Hamiltonian models for the heat transfer in metallic foams. A graph-based port-Hamiltonian model for adaptive mechanic structures has been presented by Warsewa et al. [2020]. Gernandt et al. [2021] derive a method for a graph-based modeling of nonlinear electrical circuits as implicit PHSs.

The above graph-based approaches are easy to automate and allow for a time-efficient modeling of different classes of interconnected systems. However, as pointed out in the conclusion of van der Schaft and Maschke [2013], approaches based on classical graphs are limited to conservation or balance laws within a particular physical domain. This is emphasized by the fact that each of the above-mentioned graph-based modeling techniques aims at a special class of interconnected systems. Hence, these approaches cannot be applied for a wide class of interconnected systems. In particular, they disallow for a treatment of multi-domain systems.

The bond graph methodology is a promising framework to overcome the limitations imposed by classical graph approaches [van der Schaft and Maschke, 2013]. Bond graphs are a domain-neutral graphical representation of physically networked systems. By this, bond graphs are an ideal starting point for the modeling of a large class of interconnected systems involving the electric, mechanic, hydraulic, thermal, and chemical domains [Breedveld, 2009, p. 24]. The bond graph concept was devised by Paynter [1961] and refined by Karnopp, Rosenberg, and others. For a thorough introduction into the bond graph framework refer to Borutzky [2010] or Karnopp et al. [2012]. For the derivation of a PHS, bond graphs are particularly appealing as both—bond graphs and PHSs—share the same physically unifying power- and energy-based modeling paradigm.

Rosenberg [1971] was the first to systematically derive a state-space formulation of bond graphs. The method is based on a mathematical representation of the bond-graph referred to as field representation. The formulation of a bond graph as a PHS was first investigated by Golo et al. [2003]. The authors show that each well-posed bond graph permits an implicit port-Hamiltonian formulation. Such an implicit PHS aims at a use in numerical simulations. For the design of port-Hamiltonian control methods, however, an explicit PHS is required. The transfer from an implicit to an explicit port-Hamiltonian representation is non-trivial. In particular, as will be seen later, the existence of an explicit port-Hamiltonian formulation of a bond graph is not guaranteed, even if the bond graph is well-posed. Lopes [2016] addresses the formulation of a bond graph as differential-algebraic PHS. It has been shown that such a differential-algebraic PHS can possibly be transferred into an explicit input-state-output PHS [Lopes, 2016]. Concerning this transfer, there exists a sufficient condition which, however, is restrictive as it demands some block matrices of the underlying Dirac structure to be zero. A necessary condition for the existence of an explicit port-Hamiltonian formulation of a bond graph is missing in the literature. Donaire and Junco [2009] provide a method transferring a class of causal bond graphs to an explicit input-state-output PHS. The approach is restricted to non-feedthrough systems. As with Rosenberg [1971], the starting point of Donaire and Junco [2009] is a bond graph field representation. In the field representation, the authors assume some of the block matrices to be constant or zero. Dai [2016] proposes a concept for formulating bond graphs as simulation models with port-Hamiltonian dynamics. However, the models are not formulated as input-state-output PHSs and are restricted to a use in numerical simulations.

As can be seen from the above, bond graphs are a promising starting point for the generation of port-Hamiltonian models. Nevertheless, the automated explicit port-Hamiltonian formulation of bond graphs has only been treated for various special cases in literature so far. Lopes [2016] and Donaire and Junco [2009] address this topic but are restricted to particular classes of bond graphs. Moreover, the literature lacks necessary conditions for the existence of an explicit port-Hamiltonian formulation of bond graphs. The results of Lopes [2016] suggest that an automated generation of port-Hamiltonian models from bond graphs is possible. However, a specific method which can be fully automated is missing.

The remainder of this chapter addresses the automated generation of explicit port-Hamiltonian models from bond graphs. The specific aims are:

(i) to provide methods for the automated generation of an explicit port-Hamiltonian model (2.12) based on bond graphs,

(ii) to derive necessary and sufficient conditions for the existence of the models in (i),

(iii) to implement the results from (i) and (ii) in algorithms that enable an automated generation of explicit PHSs for a large class of interconnected systems.

3.2 Main Results

The previous section stated the aims of this chapter. This section presents the the main results to reach these aims. Subsection 3.2.1 first present the solution strategy. Afterwards, subsections 3.2.2 to 3.2.6 follow a stepwise procedure which will lead to the main results to be presented in subsections 3.2.7 and 3.2.8. The results from this research have been reported by Pfeifer et al. [2019b], Pfeifer et al. [2020a], Pfeifer et al. [2020b].

3.2.1 Basic Idea and Notation

PHSs and bond graphs are related by their port- and energy-based foundations. The leading idea of the following subsections is to exploit this relationship to a high degree. To this end, we strictly distinguish between balance equations and constitutive relations. The former are described by means of a Dirac structure. Transferring this Dirac structure from an implicit to an explicit representation will be the crucial step on the way to an explicit PHS.

The structure of the approach is depicted in Figure 3.1. Subsection 3.2.2 initially presents a formal description of a bond graph. The formalized bond graph builds the basis for an algorithmic implementation of the developed methods. In Subsection 3.2.3, the power-conserving, energy-routing elements of the bond graph are described by a set of Dirac structures in implicit form. The set of Dirac structures is composed into one single Dirac structure in Subsection 3.2.4. In Subsection 3.2.5, we transfer the composite Dirac structure from an implicit to an explicit representation. Based on the constitutive relations of storages and resistors from Subsection 3.2.2 and the explicit form of the Dirac structure, an explicit port-Hamiltonian formulation of a bond graph is derived in Subsection 3.2.6. Subsection 3.2.7 summarizes the insights from the previous sections in a theorem and a corresponding algorithm. Subsection 3.2.8 refines the result from Subsection 3.2.7 for a case of particular practical interest, viz. for bond graphs containing dependent storages.

This section makes use the following notation. Let $\mathscr{G} = (\mathbb{V}, \mathbb{B})$ be a directed graph. $\mathscr{G}$ is *weakly connected* if replacing its directed edges with undirected edges yields a connected (undirected) graph. The set

$$\mathbb{B}(u) := \{(v,u),(u,v) \in \mathbb{B} \mid v \in \mathbb{V}\} \tag{3.1}$$

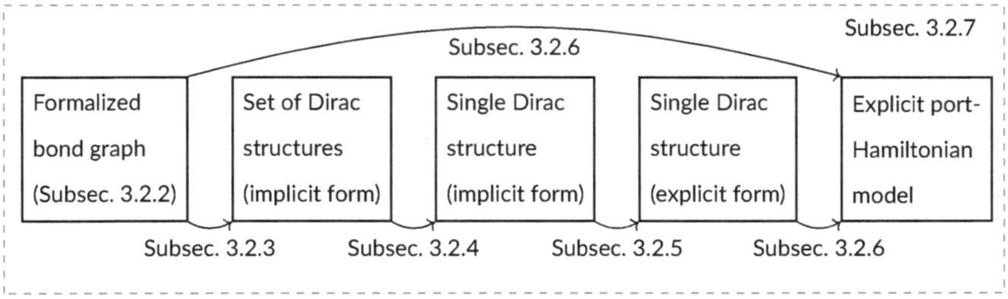

Figure 3.1: Structure of the approach for an automated generation of explicit PHSs from bond graphs

with $M(u) := |\mathbb{B}(u)|$ contains all incident edges at $u \in \mathbb{V}$. The sets

$$\overleftarrow{\mathbb{B}}(u) := \{(v,u) \in \mathbb{B} \mid v \in \mathbb{V}\}, \tag{3.2a}$$

$$\overrightarrow{\mathbb{B}}(u) := \{(u,v) \in \mathbb{B} \mid v \in \mathbb{V}\} \tag{3.2b}$$

are the ingoing and outgoing edges at $u \in \mathbb{V}$, respectively. Moreover, let $\mathbb{M}$ be a set of indices. For each $i \in \mathbb{M}$, let $\boldsymbol{A}_i \in \mathbb{R}^{n \times m_i}$ be a matrix with n rows and m_i columns. For the horizontal concatenation of all $\boldsymbol{A}_i$ we write $(\boldsymbol{A}_i)$ and append "for all $i \in \mathbb{M}$". Further, for each $i \in \mathbb{M}$, suppose a (column) vector $\boldsymbol{x}_i \in \mathbb{R}^n$. For the vertical concatenation of all $\boldsymbol{x}_i$ we write $(\boldsymbol{x}_i)$ and append "for all $i \in \mathbb{M}$".

3.2.2 Bond Graph Formalization

Let us consider K-dimensional bond graphs ($K \in \mathbb{N}_{\geq 1}$) in the generalized bond graph framework[1] with the following types of elements: storages (C), modulated resistors (R), sources of flow (Sf), sources of effort (Se), 0-junctions (0), 1-junctions (1), modulated transformers (TF) and modulated gyrators (GY). Figure 3.2 illustrates the graphical representations of the different types of elements. The set

$$\mathbb{E} := \{\mathrm{C}, \mathrm{R}, \mathrm{Sf}, \mathrm{Se}, 0, 1, \mathrm{TF}, \mathrm{GY}\} \tag{3.3}$$

collects the different types of elements. We now describe the topology of a bond graph by a directed graph. For each $\alpha \in \mathbb{E}$, let us define a set $\mathbb{V}_\alpha$ of cardinality

[1] In contrast to the standard bond graph framework, the generalized bond graph framework comprises only one type of storage elements, viz. C-type storages [Breedveld, 2009, p. 24].

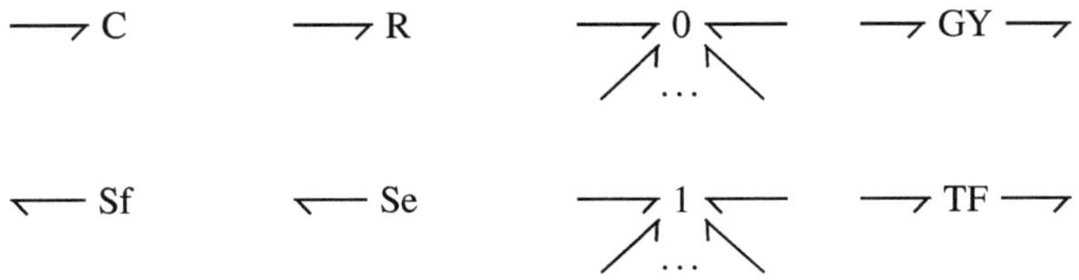

Figure 3.2: Graphical representation of different types of bond graph elements

$N_\alpha := |\mathbb{V}_\alpha|$ which contains all elements of type α. Elements of type C, R, Sf, Se are denoted as *exterior elements*; elements of type 0, 1, TF, GY are referred to as *interior elements*. The sets of exterior and interior elements are defined as

$$\mathbb{V}_\mathrm{E} := \mathbb{V}_\mathrm{C} \cup \mathbb{V}_\mathrm{R} \cup \mathbb{V}_\mathrm{Sf} \cup \mathbb{V}_\mathrm{Se}, \qquad N_\mathrm{E} := |\mathbb{V}_\mathrm{E}|, \tag{3.4a}$$

$$\mathbb{V}_\mathrm{I} := \mathbb{V}_0 \cup \mathbb{V}_1 \cup \mathbb{V}_\mathrm{TF} \cup \mathbb{V}_\mathrm{GY}, \qquad N_\mathrm{I} := |\mathbb{V}_\mathrm{I}|, \tag{3.4b}$$

respectively. The union $\mathbb{V} := \cup_{\alpha \in \mathbb{E}} \mathbb{V}_\alpha = \cup_{\alpha \in \{\mathrm{E},\mathrm{I}\}} \mathbb{V}_\alpha$ is the set of all bond graph elements ($N := |\mathbb{V}|$). The N elements of $\mathbb{V}$ are connected by a set $\mathbb{B}$ of M bonds, i.e., $M := |\mathbb{B}|$. Each bond $j \in \mathbb{B}$ carries a flow $\boldsymbol{f}_j \in \mathbb{R}^K$ and an effort $\boldsymbol{e}_j \in \left(\mathbb{R}^K\right)^*$.[2]

The directed graph $\mathscr{G} = (\mathbb{V}, \mathbb{B})$ describes the topology of the bond graph. Analogous to the naming of elements, we define sets of exterior and interior bonds:

$$\mathbb{B}_\mathrm{E} := \{(u,v),(v,u) \in \mathbb{B} \mid v \in \mathbb{V}_\mathrm{E}, u \in \mathbb{V}_\mathrm{I}\}, \qquad M_\mathrm{E} := |\mathbb{B}_\mathrm{E}|, \tag{3.5a}$$

$$\mathbb{B}_\mathrm{I} := \{(u,v) \in \mathbb{B} \mid u,v \in \mathbb{V}_\mathrm{I}\}, \qquad M_\mathrm{I} := |\mathbb{B}_\mathrm{I}|. \tag{3.5b}$$

The set $\mathbb{B}_\mathrm{E}$ contains bonds which connect an exterior element to an interior element; $\mathbb{B}_\mathrm{I}$ contains bonds which connect two interior elements with each other.[3] We consider bond graphs that are non-degenerate, i.e., bond graphs where $\mathscr{G} = (\mathbb{V}, \mathbb{B})$ is weakly connected, and where each exterior element is connected by exactly one bond to one interior element, i.e., for each $v \in \mathbb{V}_\mathrm{E}$ we have $\mathbb{V}(v) \subset \mathbb{V}_\mathrm{I}$

[2] As $\left(\mathbb{R}^K\right)^*$ is isomorphic to $\mathbb{R}^K$ it is identified with $\mathbb{R}^K$.

[3] Without loss of generality we exclude the presence of bonds interlinking two exterior elements.

with $|\mathbb{B}(v)| = 1$ and $|\mathbb{V}(v)| = 1$. Moreover, we use the bond orientation rules from standard bond graph literature [Borutzky, 2010, p. 59] in which bonds are incoming to storages and resistors and outgoing from sources of flow and effort. Without loss of generality, we assume each transformer and each gyrator to have exactly one incoming and exactly one outgoing bond in order to enable an unambiguous definition of transformer and gyrator ratios.

Next, we define and analyze an essential part of the bond graph, viz. the *junction structure*.

Definition 3.1 (Junction structure)

The junction structure of a bond graph is the subgraph $\mathcal{G}_\mathrm{I} \subset \mathcal{G}$ with $\mathcal{G}_\mathrm{I} = (\mathbb{V}_\mathrm{I}, \mathbb{B}_\mathrm{I})$.

Property 3.2 (Bond graph properties)

From the properties of a non-degenerate bond graph, it follows: $\mathcal{G}_\mathrm{I}$ is weakly connected, $\mathbb{B} = \mathbb{B}_\mathrm{E} \cup \mathbb{B}_\mathrm{I}$, and $M_\mathrm{E} = N_\mathrm{E}$.

So far, we described the topology of the bond graph by means of a directed graph. Next, we formalize the constitutive relations of the bond graph elements. To this end, let us first make the following two assumptions:

Assumption 3.3 (Modulation)

Modulation of resistors, transformers and gyrators can be expressed only in dependence on states of C-type elements and constant parameters.

Assumption 3.4 (Dissipation)

The constitutive relations of modulated resistors are linear with respect to the respective power-port variables and in Onsager form.[4]

Borutzky [2010, p. 159] has shown that bond graphs violating Assumption 3.3 cannot in general be formulated in an explicit form. Likewise, Assumption 3.4 is a well-known requirement for formulating an explicit PHS of the form (2.12) (cf. van der Schaft and Jeltsema [2014, p. 53]).

Now the constitutive relations for the different types of bond graph elements can be specified.

[4]For the notion of the Onsager form, refer to Borutzky [2010, p. 364].

- Each storage element $i \in \mathbb{V}_C$ is specified by a constitutive relation of the form
$$\boldsymbol{f}_j = \dot{\boldsymbol{x}}_i, \qquad \boldsymbol{e}_j = \frac{\partial V_i}{\partial \boldsymbol{x}_i}(\boldsymbol{x}_i), \tag{3.6}$$
with $j \in \mathbb{B}(i)$ and energy state $\boldsymbol{x}_i \in \mathbb{X}_i$, $\dim(\mathbb{X}_i) = K$ (cf. [Borutzky, 2010, p. 357]). The function $V_i\colon \mathbb{X}_i \to \mathbb{R}$, $\boldsymbol{x}_i \mapsto V_i(\boldsymbol{x}_i)$ is a differentiable storage function that is bounded from below. This storage function describes the energy contained in storage i. The energy states of all storage elements are collected in an overall energy state $\boldsymbol{x} := (\boldsymbol{x}_i)$ for all $i \in \mathbb{V}_C$ with $\boldsymbol{x} \in \mathbb{X}$, $\dim(\mathbb{X}) = KN_C$. The composite storage function $V(\boldsymbol{x}) := \sum_{i \in \mathbb{V}_C} V_i(\boldsymbol{x}_i)$ describes the energy of all storages.[5]

- Each resistive element $i \in \mathbb{V}_R$ is given by a constitutive relation in Onsager (conductance) form [Borutzky, 2010, p. 365]
$$\boldsymbol{f}_j = \boldsymbol{D}_i(\boldsymbol{x})\boldsymbol{e}_j, \tag{3.7}$$
with $j \in \mathbb{B}(i)$, $\boldsymbol{D}_i(\boldsymbol{x}) \in \mathbb{R}^{K \times K}$, and $\boldsymbol{D}_i(\boldsymbol{x}) = \boldsymbol{D}_i(\boldsymbol{x})^\top \succeq 0$. The matrices $\boldsymbol{D}_i(\boldsymbol{x})$ are collected in an overall dissipation matrix $\boldsymbol{D}(\boldsymbol{x}) := \operatorname{blkdiag}(\boldsymbol{D}_i(\boldsymbol{x}))$ for all $i \in \mathbb{V}_R$ with $\boldsymbol{D}(\boldsymbol{x}) = \boldsymbol{D}(\boldsymbol{x})^\top \succeq 0$.

- Elements i of type 0 and type 1 obey
$$i \in \mathbb{V}_0: \qquad \sum_{j \in \overleftarrow{\mathbb{B}}(i)} \boldsymbol{f}_j - \sum_{j \in \overrightarrow{\mathbb{B}}(i)} \boldsymbol{f}_j = \boldsymbol{0}, \quad \boldsymbol{e}_j = \boldsymbol{e}_k, \quad \forall j,k \in \mathbb{B}(i) \tag{3.8}$$
and
$$i \in \mathbb{V}_1: \qquad \sum_{j \in \overleftarrow{\mathbb{B}}(i)} \boldsymbol{e}_j - \sum_{j \in \overrightarrow{\mathbb{B}}(i)} \boldsymbol{e}_j = \boldsymbol{0}, \quad \boldsymbol{f}_j = \boldsymbol{f}_k, \quad \forall j,k \in \mathbb{B}(i). \tag{3.9}$$

[5]Note that from the properties of $V_i(\boldsymbol{x}_i)$ it follows that $V(\boldsymbol{x})$ is differentiable and bounded from below.

- Transformers and gyrators are determined by constitutive relations of the form

$$i \in \mathbb{V}_{\mathrm{TF}}: \qquad \boldsymbol{f}_j = \boldsymbol{U}_i(\boldsymbol{x})\boldsymbol{f}_k, \quad \boldsymbol{e}_k = \boldsymbol{U}_i^\top(\boldsymbol{x})\boldsymbol{e}_j, \tag{3.10}$$

and

$$i \in \mathbb{V}_{\mathrm{GY}}: \qquad \boldsymbol{e}_j = \boldsymbol{W}_i(\boldsymbol{x})\boldsymbol{f}_k, \quad \boldsymbol{e}_k = \boldsymbol{W}_i^\top(\boldsymbol{x})\boldsymbol{f}_j, \tag{3.11}$$

where $j \in \overleftarrow{\mathbb{B}}(i)$, $k \in \overrightarrow{\mathbb{B}}(i)$ [Borutzky, 2010, pp. 358–359]. The square matrices $\boldsymbol{U}_i(\boldsymbol{x})$ and $\boldsymbol{W}_i(\boldsymbol{x})$ have full rank K for all $\boldsymbol{x} \in \mathbb{X}$ and are assembled in a transformer matrix $\boldsymbol{U}(\boldsymbol{x}) := \mathrm{blkdiag}(\boldsymbol{U}_i(\boldsymbol{x}))$ and a gyrator matrix $\boldsymbol{W}(\boldsymbol{x}) := \mathrm{blkdiag}(\boldsymbol{W}_i(\boldsymbol{x}))$ for all $i \in \mathbb{V}_{\mathrm{TF}}$ and $\mathbb{V}_{\mathrm{GY}}$, respectively.

Source elements $i \in \mathbb{V}_{\mathrm{Sf}} \cup \mathbb{V}_{\mathrm{Se}}$ describe conditions at the system boundary. Thus, these elements are not subject to specific constitutive relations.

Now we have all prerequisites for a formal definition of a *bond graph.*

Definition 3.5 (Bond graph)

Let assumptions 3.3 and 3.4 hold. A bond graph is a tuple

$$\mathscr{BG} := (\mathbb{V}, \mathbb{B}, \mathbb{X}, V, \boldsymbol{D}, \boldsymbol{U}, \boldsymbol{W}, K), \tag{3.12}$$

with a set of elements $\mathbb{V}$, a set of bonds $\mathbb{B}$, a state-space $\mathbb{X}$, an energy storage function $V : \mathbb{X} \mapsto \mathbb{R}$, a dissipation matrix $\boldsymbol{D}(\boldsymbol{x})$, a transformer matrix $\boldsymbol{U}(\boldsymbol{x})$, a gyrator matrix $\boldsymbol{W}(\boldsymbol{x})$, and dimension K. The matrices $\boldsymbol{D}, \boldsymbol{U}, \boldsymbol{W}$ are parametrized over $\boldsymbol{x} \in \mathbb{X}$.

Remark 3.6 (Single- und multi-bond graphs). In the literature, bond graphs with $K = 1$ are referred to as single-bond graphs and bond graphs with $K > 1$ are denoted as multi-bond graphs Borutzky [2010]. Unless stated otherwise, the term "bond graph" refers to a multi-bond graph.

In the following, the formal definition of a bond graph is applied to an example system. In the course of chapters 3 and 4, this example system will continuously be used to exemplify the developed methods. This demonstrates the algorithmic nature of these methods.

Example 3.7:

Consider the K-dimensional bond graph in Figure 3.3.

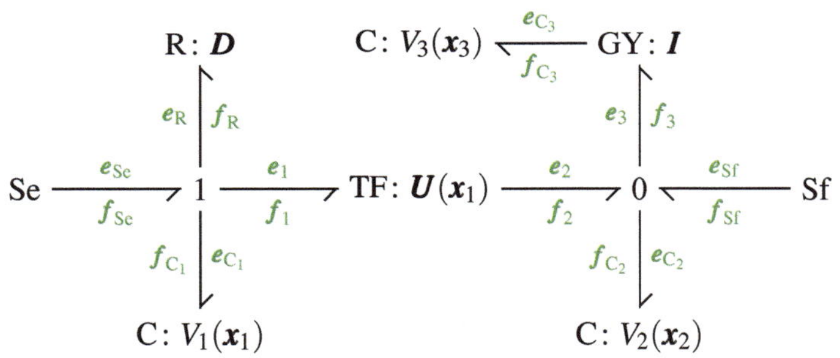

Figure 3.3: Exemplary bond graph

The elements and bonds are summarized in $\mathbb{V} = \mathbb{V}_{\mathrm{E}} \cup \mathbb{V}_{\mathrm{I}}$, $\mathbb{B} = \mathbb{B}_{\mathrm{E}} \cup \mathbb{B}_{\mathrm{I}}$, *respectively, with* $\mathbb{V}_{\mathrm{E}} = \{C_1, C_2, C_3, R, Sf, Se\}$, $\mathbb{V}_{\mathrm{I}} = \{0, 1, TF, GY\}$, $\mathbb{B}_{\mathrm{E}} = \{C_1, C_2, C_3, R, Sf, Se\}$, *and* $\mathbb{B}_{\mathrm{I}} = \{1, 2, 3\}$. *Note that in Figure 3.3 the elements of* $\mathbb{B}$ *are not explicitly highlighted but can be identified from the indices of the efforts and flows. The system state vector is* $\boldsymbol{x} = \left(\boldsymbol{x}_1^\top \, \boldsymbol{x}_2^\top \, \boldsymbol{x}_3^\top\right)^\top \in \mathbb{X}$ *where*

$$\mathbb{X} = \{ \begin{pmatrix} \boldsymbol{x}_1 \\ \boldsymbol{x}_2 \\ \boldsymbol{x}_3 \end{pmatrix} \in \mathbb{R}^{3K} \mid \|\boldsymbol{x}_1\| < \infty \}.^6 \tag{3.13}$$

Suppose an arbitrary, differentiable, non-negative storage function

$$V(\boldsymbol{x}) = V_1(\boldsymbol{x}_1) + V_2(\boldsymbol{x}_2) + V_3(\boldsymbol{x}_3). \tag{3.14}$$

The R*-type element is specified by a matrix* $\boldsymbol{D} \in \mathbb{R}^{K \times K}$ *with* $\boldsymbol{D} = \boldsymbol{D}^\top \succ 0$. *The transformer* TF *is modulated by* $\boldsymbol{x}_1$. *The transformation ratio is given by a full rank matrix* $\boldsymbol{U}(\boldsymbol{x}) = \boldsymbol{U}^\top(\boldsymbol{x}) \in \mathbb{R}^{K \times K}$ *with*

$$\boldsymbol{U}(\boldsymbol{x}_1) = \exp\left(-\kappa \operatorname{diag}(\boldsymbol{x}_1)\right). \tag{3.15}$$

where $\exp(\cdot)$ *denotes the matrix exponential. The gyrator* GY *has a constant gyration ration of* $\boldsymbol{W} = \boldsymbol{I} \in \mathbb{R}^{K \times K}$. *The formal description of the bond-graph from Figure 3.3 is then given by*

$$\mathscr{BG} = (\{C_1, C_2, C_3, R, Sf, Se, 0, 1, TF, GY\}, \{C_1, C_2, C_3, R, Sf, Se, 1, 2, 3\}, \\ \mathbb{X}, V_1(\boldsymbol{x}_1) + V_2(\boldsymbol{x}_2) + V_3(\boldsymbol{x}_3), \boldsymbol{D}, \exp(-\kappa \operatorname{diag}(\boldsymbol{x}_1)), \boldsymbol{I}, K), \tag{3.16}$$

where $\mathbb{X}$ and $\boldsymbol{D}$ as described above.

Recall that the problem addressed in this chapter is to derive an automatable modeling method that transfers a bond graph into an explicit input-state-output PHS. In this context, it is crucial that such a PHS respects the causality of the source elements of the bond graph. In other words, the input vector of the PHS has to include flows of Sf elements and efforts of Se elements, while the output vector has to include efforts of Sf elements and flows of Se elements. This is formalized in the following property:

Property 3.8 (Inputs and outputs)

Let $\mathbb{B}_\alpha = \cup_{i \in \mathbb{V}_\alpha} \mathbb{B}(i)$ for $\alpha \in \{\mathrm{Sf}, \mathrm{Se}\}$. The input vector $\boldsymbol{u}$ consists of $(\boldsymbol{f}_j)$, $(\boldsymbol{e}_k)$ while the output vector $\boldsymbol{y}$ consists of $(\boldsymbol{e}_j)$, $(\boldsymbol{f}_k)$ for all $j \in \mathbb{B}_{\mathrm{Sf}}$, $k \in \mathbb{B}_{\mathrm{Se}}$.

Based on Definition 3.5 and Property 3.8 we may now formally specify the problem under consideration.

Problem 3.9 (Automated port-Hamiltonian formulation of a bond graph)

Given a bond graph as in Definition 3.5. What is an automated modeling method that formulates the bond graph as a PHS (2.12) with Property 3.8? Under which conditions does such a port-Hamiltonian formulation of a bond graph exist?

In the following subsections 3.2.3 to 3.2.6, we elaborate a solution to Problem 3.9.

3.2.3 Description of Interior Elements as Dirac Structures

Let us first focus on the junction structure of the bond graph. The idea in this subsection is to describe each element of the junction structure by a dedicated

[6] The state-space is restricted to bounded values of $\boldsymbol{x}_1$ in order to ensure the full rank property of the matrix (3.15) for all $\boldsymbol{x} \in \mathbb{X}$.

Dirac structure. For the Dirac structure we formulate a generic pattern which makes its generation appealing for an automation. Before we consider the main lemma of this subsection, viz. Lemma 3.12, we introduce two preliminary statements. In the sequel, we will often modify flows and efforts from Dirac structures by means of permutations. If the flows and efforts of a particular Dirac structure are permuted in the same manner, this can be seen as a change of basis. Intuitively, such a change of basis should not alter the fact that the considered vector space is a Dirac structure. Proposition 3.10 and Corollary 3.11 formally analyze this intuition and provide a practical calculation law to formulate the Dirac structure in the new coordinates.

Proposition 3.10 (Orthogonal transformation)

Consider a modulated Dirac structure (2.4) and let $\boldsymbol{T}(\boldsymbol{x}) \in \mathbb{O}(n)$ be a family of orthogonal matrices parametrized over $\boldsymbol{x} \in \mathbb{X}$. Then

$$\tilde{\mathbb{D}}(\boldsymbol{x}) = \{(\tilde{\boldsymbol{f}}, \tilde{\boldsymbol{e}}) \in \mathbb{R}^n \times \mathbb{R}^n \mid \tilde{\boldsymbol{F}}(\boldsymbol{x})\tilde{\boldsymbol{f}} + \tilde{\boldsymbol{E}}(\boldsymbol{x})\tilde{\boldsymbol{e}} = \boldsymbol{0}\} \tag{3.17}$$

with $\tilde{\boldsymbol{F}}(\boldsymbol{x}) = \boldsymbol{F}(\boldsymbol{x})\boldsymbol{T}(\boldsymbol{x})^\top$, $\tilde{\boldsymbol{E}}(\boldsymbol{x}) = \boldsymbol{E}(\boldsymbol{x})\boldsymbol{T}(\boldsymbol{x})^\top$ is a modulated Dirac structure.

Proof:

Inserting $\boldsymbol{f} = \boldsymbol{T}(\boldsymbol{x})^\top \tilde{\boldsymbol{f}}$ and $\boldsymbol{e} = \boldsymbol{T}(\boldsymbol{x})^\top \tilde{\boldsymbol{e}}$ into (2.4) gives (3.17). Equation (3.17) is a Dirac structure as it fulfills (2.5):

$$(i)\ \tilde{\boldsymbol{F}}(\boldsymbol{x})\tilde{\boldsymbol{E}}^\top(\boldsymbol{x}) + \tilde{\boldsymbol{E}}(\boldsymbol{x})\tilde{\boldsymbol{F}}^\top(\boldsymbol{x}) = \boldsymbol{F}(\boldsymbol{x})\boldsymbol{E}^\top(\boldsymbol{x}) + \boldsymbol{E}(\boldsymbol{x})\boldsymbol{F}^\top(\boldsymbol{x}) = \boldsymbol{0}, \tag{3.18a}$$

$$\begin{aligned}(ii)\ \operatorname{rank}\left(\tilde{\boldsymbol{F}}(\boldsymbol{x}) \quad \tilde{\boldsymbol{E}}(\boldsymbol{x})\right) &= \operatorname{rank}\left(\left(\boldsymbol{F}(\boldsymbol{x}) \quad \boldsymbol{E}(\boldsymbol{x})\right)\boldsymbol{T}^\top(\boldsymbol{x})\right) \\ &= \operatorname{rank}\left(\boldsymbol{F}(\boldsymbol{x}) \quad \boldsymbol{E}(\boldsymbol{x})\right) = n. \end{aligned} \tag{3.18b}$$

□

Corollary 3.11 (Equivalent Dirac structures)

Given two vector spaces $\mathbb{D}_i(\boldsymbol{x}) = \{(\boldsymbol{f}_i, \boldsymbol{e}_i) \in \mathbb{R}^n \times \mathbb{R}^n \mid \boldsymbol{F}_i(\boldsymbol{x})\boldsymbol{f}_i + \boldsymbol{E}_i(\boldsymbol{x})\boldsymbol{e}_i = \boldsymbol{0}\}$ with $\boldsymbol{x} \in \mathbb{X}$, $i \in \{1, 2\}$. If for every $\boldsymbol{x} \in \mathbb{X}$ there exists a $\boldsymbol{T}(\boldsymbol{x}) \in \mathbb{O}(n)$ such that

$(\boldsymbol{f}_1, \boldsymbol{e}_1) \mapsto (\boldsymbol{T}(\boldsymbol{x})\boldsymbol{f}_1, \boldsymbol{T}(\boldsymbol{x})\boldsymbol{e}_1)$ is a bijection between $\mathbb{D}_1(\boldsymbol{x})$ and $\mathbb{D}_2(\boldsymbol{x})$, then "$\mathbb{D}_1(\boldsymbol{x})$ is a Dirac structure" is equivalent to "$\mathbb{D}_2(\boldsymbol{x})$ is a Dirac structure".

Proof:
The proof follows directly from a twofold application of Lemma 3.10. □

Now follows the main lemma of this subsection. This lemma makes use of the following notation: $\boldsymbol{0}_K$ is the $K \times K$-dimensional zero matrix; $\boldsymbol{0}_K^{a,b}$ is an $(a \times b)$ block matrix of zero matrices $\boldsymbol{0}_K$. Analogously, $\boldsymbol{I}_K$ is the $K \times K$-dimensional identity matrix and $\boldsymbol{I}_K^{a \times b}$ is an $(a \times b)$ block matrix of identity matrices $\boldsymbol{I}_K$.

Lemma 3.12 (Set of Dirac structures)

Given a bond graph (3.12) with the set of interior elements $\mathbb{V}_\mathrm{I} = \mathbb{V}_0 \cup \mathbb{V}_1 \cup \mathbb{V}_\mathrm{TF} \cup \mathbb{V}_\mathrm{GY}$, $N_\mathrm{I} = |\mathbb{V}_\mathrm{I}|$. The constitutive relations of all elements of $\mathbb{V}_\mathrm{I}$ can be described by a set of Dirac structures $\mathbb{DS}$ with $|\mathbb{DS}| = N_\mathrm{I}$. For each element $i \in \mathbb{V}_\mathrm{I}$ there exists a corresponding element $\mathbb{D}_i(\boldsymbol{x}) \in \mathbb{DS}$ with

$$\begin{aligned} \mathbb{D}_i(\boldsymbol{x}) = \{ (\begin{pmatrix} (\boldsymbol{f}_j) \\ (\boldsymbol{f}_k) \end{pmatrix}, \begin{pmatrix} (\boldsymbol{e}_j) \\ (\boldsymbol{e}_k) \end{pmatrix}) \in \mathbb{R}^{K \cdot M(i)} \times \mathbb{R}^{K \cdot M(i)} \mid \\ \boldsymbol{F}_i(\boldsymbol{x}) \begin{pmatrix} (\boldsymbol{f}_j) \\ -(\boldsymbol{f}_k) \end{pmatrix} + \boldsymbol{E}_i(\boldsymbol{x}) \begin{pmatrix} (\boldsymbol{e}_j) \\ (\boldsymbol{e}_k) \end{pmatrix} = \boldsymbol{0} \}, \end{aligned} \tag{3.19}$$

where $j \in \overleftarrow{\mathbb{B}}(i), k \in \overrightarrow{\mathbb{B}}(i)$, and $\mathbb{B}(i) = \overleftarrow{\mathbb{B}}(i) \cup \overrightarrow{\mathbb{B}}(i)$ with $M(i) := |\mathbb{B}(i)|$. Depending on the type of i, the matrices $\boldsymbol{F}_i(\boldsymbol{x})$ and $\boldsymbol{E}_i(\boldsymbol{x})$ in (3.19) are of one of the following forms:

$$i \in \mathbb{V}_0 : \boldsymbol{F}_i(\boldsymbol{x}) = \begin{pmatrix} \boldsymbol{I}_K^{1\times M(i)} \\ \boldsymbol{0}_K^{(M(i)-1)\times M(i)} \end{pmatrix}, \tag{3.20a}$$

$$\boldsymbol{E}_i(\boldsymbol{x}) = \begin{pmatrix} \boldsymbol{0}_K & \boldsymbol{0}_K^{1\times(M(i)-1)} \\ \boldsymbol{I}_K^{(M(i)-1)\times 1} & -\boldsymbol{I}_{K(M(i)-1)}, \end{pmatrix}, \tag{3.20b}$$

$$i \in \mathbb{V}_1 : \boldsymbol{F}_i(\boldsymbol{x}) = \begin{pmatrix} \boldsymbol{0}_K & \boldsymbol{0}_K^{1\times(M(i)-1)} \\ \boldsymbol{I}_K^{(M(i)-1)\times 1} & -\boldsymbol{I}_{K(M(i)-1)} \end{pmatrix} \begin{pmatrix} \boldsymbol{I}_{K\cdot|\overleftarrow{\mathbb{B}}(i)|} & \boldsymbol{0} \\ \boldsymbol{0} & -\boldsymbol{I}_{K\cdot|\overrightarrow{\mathbb{B}}(i)|} \end{pmatrix}, \tag{3.20c}$$

$$\boldsymbol{E}_i(\boldsymbol{x}) = \begin{pmatrix} \boldsymbol{I}_K^{1\times M(i)} \\ \boldsymbol{0}_K^{(M(i)-1)\times M(i)} \end{pmatrix} \begin{pmatrix} \boldsymbol{I}_{K\cdot|\overleftarrow{\mathbb{B}}(i)|} & \boldsymbol{0} \\ \boldsymbol{0} & -\boldsymbol{I}_{K\cdot|\overrightarrow{\mathbb{B}}(i)|} \end{pmatrix}, \tag{3.20d}$$

$$i \in \mathbb{V}_{\mathrm{TF}} : \boldsymbol{F}_i(\boldsymbol{x}) = \begin{pmatrix} \boldsymbol{I}_K & \boldsymbol{U}_i(\boldsymbol{x}) \\ \boldsymbol{0}_K & \boldsymbol{0}_K \end{pmatrix}, \tag{3.20e}$$

$$\boldsymbol{E}_i(\boldsymbol{x}) = \begin{pmatrix} \boldsymbol{0}_K & \boldsymbol{0}_K \\ -\boldsymbol{U}_i^\top(\boldsymbol{x}) & \boldsymbol{I}_K \end{pmatrix}, \tag{3.20f}$$

$$i \in \mathbb{V}_{\mathrm{GY}} : \boldsymbol{F}_i(\boldsymbol{x}) = \begin{pmatrix} \boldsymbol{0}_K & \boldsymbol{W}_i(\boldsymbol{x}) \\ -\boldsymbol{W}_i^\top(\boldsymbol{x}) & \boldsymbol{0}_K \end{pmatrix}, \tag{3.20g}$$

$$\boldsymbol{E}_i(\boldsymbol{x}) = \begin{pmatrix} \boldsymbol{I}_K & \boldsymbol{0}_K \\ \boldsymbol{0}_K & \boldsymbol{I}_K \end{pmatrix}. \tag{3.20h}$$

Proof:

First, we prove that each element $\mathbb{D}_i(\boldsymbol{x}) \in \mathbb{DS}$ describes the constitutive relations of the corresponding interior element $i \in \mathbb{V}_\mathrm{I}$. Second, we show that the elements $\mathbb{D}_i(\boldsymbol{x}) \in \mathbb{DS}$ define Dirac structures.

For $i \in \mathbb{V}_0 \cup \mathbb{V}_{\mathrm{TF}}, \cup \mathbb{V}_{\mathrm{GY}}$, we insert (3.20a), (3.20b); (3.20e), (3.20f); (3.20g), (3.20h) into the equation system of (3.19) and obtain the constitutive relations (3.8), (3.10), (3.11), respectively. Analogously, for $i \in \mathbb{V}_1$ we may write (3.9) as $\tilde{\boldsymbol{F}}_i\tilde{\boldsymbol{f}}_i + \tilde{\boldsymbol{E}}_i\tilde{\boldsymbol{e}}_i = \boldsymbol{0}$ with

$$\begin{pmatrix} \boldsymbol{0}_K & \boldsymbol{0}_K^{1\times(M(i)-1)} \\ \boldsymbol{I}_K^{(M(i)-1)\times 1} & -\boldsymbol{I}_{K(M(i)-1)} \end{pmatrix} \begin{pmatrix} (\tilde{\boldsymbol{f}}_j) \\ (\tilde{\boldsymbol{f}}_k) \end{pmatrix} + \begin{pmatrix} \boldsymbol{I}_K^{1\times M(i)} \\ \boldsymbol{0}_K^{(M(i)-1)\times M(i)} \end{pmatrix} \begin{pmatrix} (\tilde{\boldsymbol{e}}_j) \\ -(\tilde{\boldsymbol{e}}_k) \end{pmatrix} = \boldsymbol{0}, \tag{3.21}$$

for all $j \in \overleftarrow{\mathbb{B}}(i), k \in \overrightarrow{\mathbb{B}}(i)$. To bring (3.21) to the form of the equation system in (3.19), we perform a change of coordinates $\boldsymbol{f}_i = \boldsymbol{T}_i^\top \tilde{\boldsymbol{f}}_i$, $\boldsymbol{e}_i = \boldsymbol{T}_i^\top \tilde{\boldsymbol{e}}_i$, with $\boldsymbol{T}_i = \text{blkdiag}(\boldsymbol{I}_{K\cdot|\overleftarrow{\mathbb{B}}(i)|}, -\boldsymbol{I}_{K\cdot|\overrightarrow{\mathbb{B}}(i)|})$ which yields ((3.20c),(3.20d)).

Inserting the matrices $\boldsymbol{F}_i(\boldsymbol{x})$ and $\boldsymbol{E}_i(\boldsymbol{x})$ from (3.20a), (3.20b); (3.20e), (3.20f); (3.20g), (3.20h) into (2.5) shows that these matrices indeed define Dirac structures. Analogously, the matrices $\tilde{\boldsymbol{F}}_i$ and $\tilde{\boldsymbol{E}}_i$ from (3.21) define a Dirac structure. As $\boldsymbol{T}_i \in \mathbb{O}(M(i))$, by Corollary 3.11 the matrices (3.20c), (3.20d) then also define a Dirac structure. □

Remark 3.13 (Dirac structures for 0- and 1-type elements). In Lemma 3.12, for $i \in \mathbb{V}_0 \cup \mathbb{V}_1$ we have $\boldsymbol{F} = \text{const.}$ and $\boldsymbol{E} = \text{const.}$, i.e., the matrices $\boldsymbol{F}(\boldsymbol{x})$ and $\boldsymbol{E}(\boldsymbol{x})$ are independent of $\boldsymbol{x}$. The corresponding Dirac structures are thus non-modulated Dirac structures.

With the following example we continue Example 3.7 and illustrate the generation of Dirac structures according to Lemma 3.12:

Example 3.14:

Consider the K-dimensional bond graph from Example 3.7 with its formal description (3.16). The set of interior elements is given by $\mathbb{V}_\mathrm{I} = \{0, 1, TF, GY\}$ with $N_\mathrm{I} = 4$. Define a corresponding set of Dirac structures $\mathbb{DS} = \{\mathbb{D}_0, \mathbb{D}_1, \mathbb{D}_{TF}, \mathbb{D}_{GY}\}$ where the elements $\mathbb{D}_i$ are of the form (3.19) ($i \in \mathbb{V}_\mathrm{I}$). By applying the calculation rules from the above Lemma 3.12, the equation systems of the Dirac structures read: [7]

$$\mathbb{D}_0: \begin{pmatrix} I & I & I & I \\ 0 & 0 & 0 & 0 \\ 0 & 0 & 0 & 0 \\ 0 & 0 & 0 & 0 \end{pmatrix} \begin{pmatrix} f_{Sf} \\ f_2 \\ -f_{C_2} \\ -f_3 \end{pmatrix} + \begin{pmatrix} 0 & 0 & 0 & 0 \\ I & -I & 0 & 0 \\ I & 0 & -I & 0 \\ I & 0 & 0 & -I \end{pmatrix} \begin{pmatrix} e_{Sf} \\ e_2 \\ e_{C_2} \\ e_3 \end{pmatrix} = 0, \tag{3.22a}$$

$$\mathbb{D}_1: \begin{pmatrix} 0 & 0 & 0 & 0 \\ I & I & 0 & 0 \\ I & 0 & I & 0 \\ I & 0 & 0 & I \end{pmatrix} \begin{pmatrix} f_{Se} \\ -f_{C_1} \\ -f_R \\ -f_1 \end{pmatrix} + \begin{pmatrix} I & -I & -I & -I \\ 0 & 0 & 0 & 0 \\ 0 & 0 & 0 & 0 \\ 0 & 0 & 0 & 0 \end{pmatrix} \begin{pmatrix} e_{Se} \\ e_{C_1} \\ e_R \\ e_1 \end{pmatrix} = 0, \tag{3.22b}$$

$$\mathbb{D}_{TF}: \begin{pmatrix} I & U(x_1) \\ 0 & 0 \end{pmatrix} \begin{pmatrix} f_1 \\ -f_2 \end{pmatrix} + \begin{pmatrix} 0 & 0 \\ -U(x_1) & I \end{pmatrix} \begin{pmatrix} e_1 \\ e_2 \end{pmatrix} = 0, \tag{3.22c}$$

$$\mathbb{D}_{GY}: \begin{pmatrix} 0 & I \\ -I & 0 \end{pmatrix} \begin{pmatrix} f_3 \\ -f_{C_3} \end{pmatrix} + \begin{pmatrix} I & 0 \\ 0 & I \end{pmatrix} \begin{pmatrix} e_3 \\ e_{C_3} \end{pmatrix} = 0, \tag{3.22d}$$

where $U(x_1) = \exp(-\kappa \operatorname{diag}(x_1))$ *(cf. (3.15)).*

3.2.4 Composition into a Single Dirac Structure

From Lemma 3.12 we obtain N_I Dirac structures which describe the constitutive equations of the N_I interior elements of the bond graph. Next, we compose these N_I Dirac structures obtained into one single Dirac structure. To this end, we follow the idea from Batlle et al. [2011] who proposed to describe the interconnection of some Dirac structures again by a Dirac structure—a so-called *interconnection Dirac structure* [Batlle et al., 2011].

The N_I Dirac structures from Lemma 3.12 are formulated such that the vectors and matrices are ordered according to ingoing and outgoing bonds. For the determination of an interconnection Dirac structure, we first rewrite these Dirac structures such that the interconnection variables (i.e., the flows and efforts of interior bonds) become visible.

[7]Throughout this example, square zero matrices and identity matrices are of dimension K.

Proposition 3.15 (Rewriting the Dirac structures)
Consider the situation from Lemma 3.12 with the set of Dirac structures $\mathbb{DS}$. Let $\mathbb{B}_{\mathrm{I}}(i)$ and $\mathbb{B}_{\mathrm{E}}(i)$ denote the sets of interior and exterior bonds at $i \in \mathbb{V}_{\mathrm{I}}$, respectively. For all $i \in \mathbb{V}_{\mathrm{I}}$, the elements $\mathbb{D}_i \in \mathbb{DS}$ can then be written such that the vectors and matrices of $\mathbb{D}_i$ are sorted by exterior and interior bonds:

$$
\begin{aligned}
\mathbb{D}_i(\boldsymbol{x}) = \{ & (\begin{pmatrix} (\boldsymbol{f}_j) \\ (\boldsymbol{f}_k) \end{pmatrix}, \begin{pmatrix} (\boldsymbol{e}_j) \\ (\boldsymbol{e}_k) \end{pmatrix}) \in \mathbb{R}^{K \cdot M(i)} \times \mathbb{R}^{K \cdot M(i)} \mid \\
& ((\boldsymbol{F}_j(\boldsymbol{x}))\ (\boldsymbol{F}_k(\boldsymbol{x}))) \begin{pmatrix} (\varepsilon(j)\boldsymbol{f}_j) \\ (\varepsilon(k)\boldsymbol{f}_k) \end{pmatrix} + ((\boldsymbol{E}_j(\boldsymbol{x}))\ (\boldsymbol{E}_k(\boldsymbol{x}))) \begin{pmatrix} (\boldsymbol{e}_j) \\ (\boldsymbol{e}_k) \end{pmatrix} = \boldsymbol{0} \},
\end{aligned} \tag{3.23}
$$

for all $j \in \mathbb{B}_{\mathrm{E}}(i), k \in \mathbb{B}_{\mathrm{I}}(i)$ where $\varepsilon \colon \mathbb{B}(i) \to \{-1, 1\}$ is a sign function with

$$
\varepsilon(b) = \begin{cases} 1, & b \in \overleftarrow{\mathbb{B}}(i), \\ -1, & b \in \overrightarrow{\mathbb{B}}(i). \end{cases} \tag{3.24}
$$

Proof:
From Property 3.2 we know that $\mathbb{B} = \mathbb{B}_{\mathrm{E}} \cup \mathbb{B}_{\mathrm{I}}$. It follows that for each $i \in \mathbb{V}_{\mathrm{I}}$ we can permute the vectors and matrices of $\mathbb{D}_i \in \mathbb{DS}$ in (3.19) such that they are sorted by exterior and interior bonds. The rest follows from Corollary 3.11. □

Based on Proposition 3.15, we may now formulate an interconnection Dirac structure by relating the flows and efforts of the interior bonds of (3.23).

Proposition 3.16 (Interconnection Dirac stucture)
Consider the Dirac structures (3.23). For each $i \in \mathbb{V}_{\mathrm{I}}$, let us define $\boldsymbol{f}_i^{\mathrm{IC}} := (\varepsilon(k)\boldsymbol{f}_k)$ and $\boldsymbol{e}_i^{\mathrm{IC}} := (\boldsymbol{e}_k)$ for all $k \in \mathbb{B}_{\mathrm{I}}(i)$. For the Dirac structures (3.23), we can obtain an interconnection Dirac structure of the following form:

$$
\mathbb{D}_{\mathrm{IC}} = \{ (\boldsymbol{f}^{\mathrm{IC}}, \boldsymbol{e}^{\mathrm{IC}}) \in \mathbb{R}^{2KM_{\mathrm{I}}} \times \mathbb{R}^{2KM_{\mathrm{I}}} \mid (\boldsymbol{F}_i^{\mathrm{IC}})\ (\boldsymbol{f}_i^{\mathrm{IC}}) + (\boldsymbol{E}_i^{\mathrm{IC}})\ (\boldsymbol{e}_i^{\mathrm{IC}}) = \boldsymbol{0}, \forall i \in \mathbb{V}_{\mathrm{I}} \}, \tag{3.25}
$$

where $\boldsymbol{f}^{\mathrm{IC}} := (\boldsymbol{f}_i^{\mathrm{IC}})$, $\boldsymbol{e}^{\mathrm{IC}} := (\boldsymbol{e}_i^{\mathrm{IC}})$, and $\boldsymbol{F}_i^{\mathrm{IC}}, \boldsymbol{E}_i^{\mathrm{IC}} \in \mathbb{R}^{2KM_{\mathrm{I}} \times KM_{\mathrm{I}}(i)}$ for all $i \in \mathbb{V}_{\mathrm{I}}$.

Proof:
We have to show that (i) the equation system in (3.25) indeed describes the interconnections between the Dirac structures (3.23) and (ii) that (3.25) is a Dirac structure.
The junction structure consists of N_{I} interior elements which are connected by M_{I} interior bonds (cf. Definition 3.1). Each interior bond is incident to two interior elements. Thus, for each $k \in \mathbb{B}_{\mathrm{I}}$ the flow $\boldsymbol{f}_k$ appears exactly twice in $\boldsymbol{f}^{\mathrm{IC}}$: once with a positive sign and once with a negative sign. Analogously, for each $k \in \mathbb{B}_{\mathrm{I}}$ the effort $\boldsymbol{e}_k$ appears exactly twice in $\boldsymbol{e}^{\mathrm{IC}}$, both times with a positive sign. Let us equate these variables appearing twice by setting

$$\begin{pmatrix} \boldsymbol{I}_{KM_{\mathrm{I}}} & \boldsymbol{I}_{KM_{\mathrm{I}}} \\ \boldsymbol{0}_{KM_{\mathrm{I}}} & \boldsymbol{0}_{KM_{\mathrm{I}}} \end{pmatrix} \begin{pmatrix} (\boldsymbol{f}_k) \\ -(\boldsymbol{f}_k) \end{pmatrix} + \begin{pmatrix} \boldsymbol{0}_{KM_{\mathrm{I}}} & \boldsymbol{0}_{KM_{\mathrm{I}}} \\ \boldsymbol{I}_{KM_{\mathrm{I}}} & -\boldsymbol{I}_{KM_{\mathrm{I}}} \end{pmatrix} \begin{pmatrix} (\boldsymbol{e}_k) \\ (\boldsymbol{e}_k) \end{pmatrix} = \boldsymbol{0}, \tag{3.26}$$

for all $k \in \mathbb{B}_{\mathrm{I}}$. The matrices in (3.26) satisfy (2.5) and can thus be related to a Dirac structure. By permutations, we rearrange the entries of the vectors in (3.26) such that they are in the same order as in $\boldsymbol{f}^{\mathrm{IC}}$ and $\boldsymbol{e}^{\mathrm{IC}}$ (cf. Corollary 3.11). [8] Finally, we rename the columns of the resulting matrices according to their affiliation to elements of $\mathbb{V}_{\mathrm{I}}$ and obtain the equation system in (3.25). □

Proposition 3.16 provides a constructive[9] way for determining an interconnection Dirac structure for the N_{I} Dirac structures describing the interior elements. The interconnection Dirac structure exclusively contains flows and efforts of the interior bonds. The overall number of Dirac structures is then $N_{\mathrm{I}}+1$, i.e., N_{I} Dirac structures from the interior elements plus one interconnection Dirac structure. In the next lemma, we compose the $N_{\mathrm{I}}+1$ Dirac structures into one single Dirac structure.

Lemma 3.17 (Composition of Dirac structures)

Consider N_{I} Dirac structures of the form (3.23) with a corresponding interconnection Dirac structure (3.25). Then, the $N_{\mathrm{I}}+1$ Dirac structures can be merged into a single Dirac structure:

[8] Note that by $2M_{\mathrm{I}} = \sum_{i\in\mathbb{V}_{\mathrm{I}}} M_{\mathrm{I}}(i)$, the sizes of the matrices in (3.26) and (3.25) are equal.
[9] As a matter of fact, the construction rules are mainly in the proof of Proposition 3.16

$$\mathbb{D}(\boldsymbol{x}) = \{(\begin{pmatrix} \boldsymbol{f}_{\mathrm{C}} \\ \boldsymbol{f}_{\mathrm{R}} \\ \boldsymbol{f}_{\mathrm{Sf}} \\ \boldsymbol{f}_{\mathrm{Se}} \end{pmatrix}, \begin{pmatrix} \boldsymbol{e}_{\mathrm{C}} \\ \boldsymbol{e}_{\mathrm{R}} \\ \boldsymbol{e}_{\mathrm{Sf}} \\ \boldsymbol{e}_{\mathrm{Se}} \end{pmatrix}) \in \mathbb{R}^{KN_{\mathrm{E}}} \times \mathbb{R}^{KN_{\mathrm{E}}} \mid \underbrace{\left(\boldsymbol{F}_{\mathrm{C}}(\boldsymbol{x})\ \boldsymbol{F}_{\mathrm{R}}(\boldsymbol{x})\ \boldsymbol{F}_{\mathrm{Sf}}(\boldsymbol{x})\ \boldsymbol{F}_{\mathrm{Se}}(\boldsymbol{x})\right)}_{=\boldsymbol{F}(\boldsymbol{x})} \begin{pmatrix} -\boldsymbol{f}_{\mathrm{C}} \\ -\boldsymbol{f}_{\mathrm{R}} \\ \boldsymbol{f}_{\mathrm{Sf}} \\ \boldsymbol{f}_{\mathrm{Se}} \end{pmatrix} + \underbrace{\left(\boldsymbol{E}_{\mathrm{C}}(\boldsymbol{x})\ \boldsymbol{E}_{\mathrm{R}}(\boldsymbol{x})\ \boldsymbol{E}_{\mathrm{Sf}}(\boldsymbol{x})\ \boldsymbol{E}_{\mathrm{Se}}(\boldsymbol{x})\right)}_{=\boldsymbol{E}(\boldsymbol{x})} \begin{pmatrix} \boldsymbol{e}_{\mathrm{C}} \\ \boldsymbol{e}_{\mathrm{R}} \\ \boldsymbol{e}_{\mathrm{Sf}} \\ \boldsymbol{e}_{\mathrm{Se}} \end{pmatrix} \}, \quad (3.27)$$

where $\boldsymbol{f}_{\alpha}, \boldsymbol{e}_{\alpha} \in \mathbb{R}^{KN_{\alpha}}$ and $\boldsymbol{F}_{\alpha}(\boldsymbol{x}), \boldsymbol{E}_{\alpha}(\boldsymbol{x}) \in \mathbb{R}^{KN_{\mathrm{E}} \times KN_{\alpha}}$ for $\alpha \in \{\mathrm{C}, \mathrm{R}, \mathrm{Sf}, \mathrm{Se}\}$.[10]
The composite Dirac structure (3.27) can be calculated as follows:
Define a full-rank matrix $\boldsymbol{\Gamma}^{\top}(\boldsymbol{x}) \in \mathbb{R}^{2KM_{\mathrm{I}} \times K(2M_{\mathrm{I}}+M_{\mathrm{E}})}$ as a $(1 \times N_{\mathrm{I}})$ block matrix $\boldsymbol{\Gamma}^{\top}(\boldsymbol{x}) = (\boldsymbol{\Gamma}_i^{\top}(\boldsymbol{x}))$ of matrices $\boldsymbol{\Gamma}_i^{\top}(\boldsymbol{x}) \in \mathbb{R}^{2KM_{\mathrm{I}} \times KM(i)}$ for all $i \in \mathbb{V}_{\mathrm{I}}$ with

$$\boldsymbol{\Gamma}_i^{\top}(\boldsymbol{x}) = \boldsymbol{F}_i^{\mathrm{IC}}(\boldsymbol{E}_k(\boldsymbol{x}))^{\top} + \boldsymbol{E}_i^{\mathrm{IC}}(\boldsymbol{F}_k(\boldsymbol{x}))^{\top}, \ \forall k \in \mathbb{B}_{\mathrm{I}}(i). \quad (3.28)$$

Choose a matrix $\boldsymbol{\Lambda}(\boldsymbol{x}) \in \mathbb{R}^{KM_{\mathrm{E}} \times K(2M_{\mathrm{I}}+M_{\mathrm{E}})}$ such that $\mathrm{im}(\boldsymbol{\Lambda}^{\top}(\boldsymbol{x})) = \ker(\boldsymbol{\Gamma}^{\top}(\boldsymbol{x}))$ for all $\boldsymbol{x} \in \mathbb{X}$. Since $\mathrm{rank}(\boldsymbol{\Gamma}^{\top}(\boldsymbol{x})) = 2KM_{\mathrm{I}}$, we have $\dim(\ker(\boldsymbol{\Gamma}^{\top}(\boldsymbol{x}))) = KM_{\mathrm{E}}$ and such a matrix $\boldsymbol{\Lambda}(\boldsymbol{x})$ always exists. Matrix $\boldsymbol{\Lambda}(\boldsymbol{x})$ can be written as a $(1 \times N_{\mathrm{I}})$ block matrix $(\boldsymbol{\Lambda}_i(\boldsymbol{x}))$ of matrices $\boldsymbol{\Lambda}_i(\boldsymbol{x}) \in \mathbb{R}^{KM_{\mathrm{E}} \times KM(i)}$ for all $i \in \mathbb{V}_{\mathrm{I}}$. Then, a composite Dirac structure is given by:

$$\mathbb{D}(\boldsymbol{x}) = \{((\boldsymbol{f}_j), (\boldsymbol{e}_j)) \in \mathbb{R}^{KM_{\mathrm{E}}} \times \mathbb{R}^{KM_{\mathrm{E}}} \mid \underbrace{(\boldsymbol{\Lambda}_i(\boldsymbol{x})(\boldsymbol{F}_j(\boldsymbol{x})))}_{=\tilde{\boldsymbol{F}}(\boldsymbol{x})}(\boldsymbol{f}_j) + \underbrace{(\boldsymbol{\Lambda}_i(\boldsymbol{x})(\boldsymbol{E}_j(\boldsymbol{x})))}_{=\tilde{\boldsymbol{E}}(\boldsymbol{x})}(\boldsymbol{e}_j) = \boldsymbol{0}\}, \quad (3.29)$$

for all $j \in \mathbb{B}_{\mathrm{E}}(i)$, $i \in \mathbb{V}_{\mathrm{I}}$, where $\tilde{\boldsymbol{F}}(\boldsymbol{x})$, $\tilde{\boldsymbol{E}}(\boldsymbol{x})$ are square matrices of size KM_{E}. The Dirac structure (3.29) relates the flows $\boldsymbol{f}_j$ and efforts $\boldsymbol{e}_j$ of only the exterior bonds $j \in \mathbb{B}_{\mathrm{E}}$. Hence, by permutations we can easily obtain (3.27).

Proof:

Batlle et al. [2011] consider the general composition of an arbitrary number of Dirac structures coupled by any interconnection Dirac structure. The authors show that (3.29) is a calculation rule for a composite Dirac structure in which the interconnection variables have been eliminated [Batlle et al., 2011, Eq. (44),(45)]. In the case of Lemma 3.17, the interconnection variables are the flows and efforts of the interior bonds. These variables have been eliminated from the composite Dirac structure (3.29). Hence, (3.29) contains only flows and efforts corresponding to exterior bonds. By permutations, we can then rewrite (3.29) as (3.27) (cf. Corollary 3.11). □

Remark 3.18 (Signs of $\boldsymbol{f}_{\mathrm{C}}$ and $\boldsymbol{f}_{\mathrm{R}}$). The negative sign of $\boldsymbol{f}_{\mathrm{C}}$ and $\boldsymbol{f}_{\mathrm{R}}$ in the equation system of (3.27) stems from the bond orientation rules in which bonds are incoming to storages and resistors (cf. Subection 3.2.2).

Remark 3.19 (Automated composition). The calculations in this subsection can be easily automated in standard computer algebra systems. In particular, this includes the determination of a matrix $\boldsymbol{\Lambda}$ satisfying $\mathrm{im}(\boldsymbol{\Lambda}^\top) = \ker(\boldsymbol{\Gamma}^\top)$ in Lemma 3.17 [Batlle et al., 2011].

In the following, we proceed with our example and illustrate the composition of the Dirac structures.

Example 3.20:

Consider the four Dirac structures specified by (3.22) *from Example 3.14. By permutations, we may rewrite the equation systems from* (3.22) *in the form* (3.23)*:*

[10]Note that $M_{\mathrm{E}} = N_{\mathrm{E}}$ (cf. Property 3.2).

$$\mathbb{D}_0: \begin{pmatrix} \boldsymbol{I} & \boldsymbol{I} & \boldsymbol{I} & \boldsymbol{I} \\ \boldsymbol{0} & \boldsymbol{0} & \boldsymbol{0} & \boldsymbol{0} \\ \boldsymbol{0} & \boldsymbol{0} & \boldsymbol{0} & \boldsymbol{0} \\ \boldsymbol{0} & \boldsymbol{0} & \boldsymbol{0} & \boldsymbol{0} \end{pmatrix} \begin{pmatrix} -\boldsymbol{f}_{C_2} \\ \boldsymbol{f}_{Sf} \\ -\boldsymbol{f}_3 \\ \boldsymbol{f}_2 \end{pmatrix} + \begin{pmatrix} \boldsymbol{0} & \boldsymbol{0} & \boldsymbol{0} & \boldsymbol{0} \\ \boldsymbol{0} & \boldsymbol{I} & \boldsymbol{0} & -\boldsymbol{I} \\ -\boldsymbol{I} & \boldsymbol{I} & \boldsymbol{0} & \boldsymbol{0} \\ \boldsymbol{0} & \boldsymbol{I} & -\boldsymbol{I} & \boldsymbol{0} \end{pmatrix} \begin{pmatrix} \boldsymbol{e}_{C_2} \\ \boldsymbol{e}_{Sf} \\ \boldsymbol{e}_3 \\ \boldsymbol{e}_2 \end{pmatrix} = \boldsymbol{0}, \tag{3.30a}$$

$$\mathbb{D}_1: \begin{pmatrix} \boldsymbol{0} & \boldsymbol{0} & \boldsymbol{0} & \boldsymbol{0} \\ \boldsymbol{I} & \boldsymbol{0} & \boldsymbol{I} & \boldsymbol{0} \\ \boldsymbol{0} & \boldsymbol{I} & \boldsymbol{I} & \boldsymbol{0} \\ \boldsymbol{0} & \boldsymbol{0} & \boldsymbol{I} & \boldsymbol{I} \end{pmatrix} \begin{pmatrix} -\boldsymbol{f}_{C_1} \\ -\boldsymbol{f}_{R} \\ \boldsymbol{f}_{Se} \\ -\boldsymbol{f}_1 \end{pmatrix} + \begin{pmatrix} -\boldsymbol{I} & -\boldsymbol{I} & \boldsymbol{I} & -\boldsymbol{I} \\ \boldsymbol{0} & \boldsymbol{0} & \boldsymbol{0} & \boldsymbol{0} \\ \boldsymbol{0} & \boldsymbol{0} & \boldsymbol{0} & \boldsymbol{0} \\ \boldsymbol{0} & \boldsymbol{0} & \boldsymbol{0} & \boldsymbol{0} \end{pmatrix} \begin{pmatrix} \boldsymbol{e}_{C_1} \\ \boldsymbol{e}_{R} \\ \boldsymbol{e}_{Se} \\ \boldsymbol{e}_1 \end{pmatrix} = \boldsymbol{0}, \tag{3.30b}$$

$$\mathbb{D}_{TF}: \begin{pmatrix} \boldsymbol{I} & \boldsymbol{U}(\boldsymbol{x}_1) \\ \boldsymbol{0} & \boldsymbol{0} \end{pmatrix} \begin{pmatrix} \boldsymbol{f}_1 \\ -\boldsymbol{f}_2 \end{pmatrix} + \begin{pmatrix} \boldsymbol{0} & \boldsymbol{0} \\ -\boldsymbol{U}(\boldsymbol{x}_1) & \boldsymbol{I} \end{pmatrix} \begin{pmatrix} \boldsymbol{e}_1 \\ \boldsymbol{e}_2 \end{pmatrix} = \boldsymbol{0}, \tag{3.30c}$$

$$\mathbb{D}_{GY}: \begin{pmatrix} \boldsymbol{I} & \boldsymbol{0} \\ \boldsymbol{0} & -\boldsymbol{I} \end{pmatrix} \begin{pmatrix} -\boldsymbol{f}_{C_3} \\ \boldsymbol{f}_3 \end{pmatrix} + \begin{pmatrix} \boldsymbol{0} & \boldsymbol{I} \\ \boldsymbol{I} & \boldsymbol{0} \end{pmatrix} \begin{pmatrix} \boldsymbol{e}_{C_3} \\ \boldsymbol{e}_3 \end{pmatrix} = \boldsymbol{0}, \tag{3.30d}$$

where $\boldsymbol{U}(\boldsymbol{x}_1) = \exp(-\kappa\,\mathrm{diag}(\boldsymbol{x}_1))$. *The dashed lines indicate the partitioning of the flow and effort vectors with respect to the affiliation of the variables to exterior and interior bonds. Next, we determine an interconnection Dirac structure* $\mathbb{D}_{\mathrm{IC}}$ *(3.25). From Proposition 3.16, one obtains the following equation system for* $\mathbb{D}_{\mathrm{IC}}$*:*

$$
\underbrace{\begin{pmatrix} I & 0 & 0 & 0 & 0 & I \\ 0 & 0 & I & I & 0 & 0 \\ 0 & I & 0 & 0 & I & 0 \\ 0 & 0 & 0 & 0 & 0 & 0 \\ 0 & 0 & 0 & 0 & 0 & 0 \\ 0 & 0 & 0 & 0 & 0 & 0 \end{pmatrix}}_{F_0^{\mathrm{IC}}\ F_1^{\mathrm{IC}}\ F_{TF}^{\mathrm{IC}}\ F_{GY}^{\mathrm{IC}}}
\begin{pmatrix} -f_3 \\ f_2 \\ \hdashline -f_1 \\ \hdashline f_1 \\ -f_2 \\ \hdashline f_3 \end{pmatrix}
+ \underbrace{\begin{pmatrix} 0 & 0 & 0 & 0 & 0 & 0 \\ 0 & 0 & 0 & 0 & 0 & 0 \\ 0 & 0 & 0 & 0 & 0 & 0 \\ -I & 0 & 0 & 0 & 0 & I \\ 0 & 0 & -I & I & 0 & 0 \\ 0 & I & 0 & 0 & -I & 0 \end{pmatrix}}_{E_0^{\mathrm{IC}}\ E_1^{\mathrm{IC}}\ E_{TF}^{\mathrm{IC}}\ E_{GY}^{\mathrm{IC}}}
\begin{pmatrix} e_3 \\ e_2 \\ \hdashline e_1 \\ \hdashline e_1 \\ e_2 \\ \hdashline e_3 \end{pmatrix} = \mathbf{0}. \quad (3.31)
$$

Equation (3.31) *relates flow and efforts only of interior bonds. In this case, the dashed lines separate the vector entries with respect to their origin from the different interior elements. As can be seen, the matrices in* (3.31) *are simply column-wise permutations of the matrices in* (3.26). *Now we can compose the Dirac structures according to Lemma 3.17. By using* (3.30) *and* (3.31) *we calculate* (3.28) *and obtain the following expression for* $\boldsymbol{\Gamma}^{\top}(\boldsymbol{x})$*:*

$$
\begin{pmatrix}
0 & 0 & 0 & -I & 0 & 0 & 0 & 0 & 0 & 0 & I & 0 \\
0 & 0 & 0 & 0 & -I & 0 & 0 & 0 & 0 & -U(x_1) & 0 & 0 \\
0 & -I & 0 & 0 & 0 & 0 & 0 & 0 & 0 & I & 0 & 0 \\
-I & 0 & 0 & 0 & 0 & 0 & 0 & 0 & 0 & 0 & 0 & -I \\
0 & 0 & 0 & 0 & 0 & 0 & 0 & -I & I & 0 & 0 & 0 \\
I & 0 & 0 & 0 & 0 & 0 & 0 & 0 & -U(x_1) & 0 & 0 & 0
\end{pmatrix}. \quad (3.32)
$$

with the column blocks $\underbrace{\quad}_{\Gamma_0}$ (columns 1–4), $\underbrace{\quad}_{\Gamma_1}$ (columns 5–8), $\underbrace{\quad}_{\Gamma_{TF}(x_1)}$ (columns 9–10), $\underbrace{\quad}_{\Gamma_{GY}}$ (columns 11–12).

With a computer algebra system, we evaluate $\mathrm{im}(\boldsymbol{\Lambda}^\top(\boldsymbol{x}_1)) = \ker(\boldsymbol{\Gamma}^\top(\boldsymbol{x}_1))$ *and obtain the following expression for* $\boldsymbol{\Lambda}^\top(\boldsymbol{x}_1)$*:*

$$\underbrace{\begin{pmatrix} -\boldsymbol{I} & \boldsymbol{0} & \boldsymbol{0} & \boldsymbol{0} \\ \boldsymbol{0} & \boldsymbol{0} & \boldsymbol{0} & \boldsymbol{I} \\ \boldsymbol{0} & \boldsymbol{I} & \boldsymbol{0} & \boldsymbol{0} \\ \boldsymbol{0} & \boldsymbol{0} & \boldsymbol{0} & \boldsymbol{0} \\ \boldsymbol{0} & \boldsymbol{0} & \boldsymbol{0} & \boldsymbol{0} \\ \boldsymbol{0} & \boldsymbol{0} & \boldsymbol{I} & \boldsymbol{0} \end{pmatrix}}_{\boldsymbol{\Lambda}_0} \quad \underbrace{\begin{matrix} \boldsymbol{0} & \boldsymbol{0} & \boldsymbol{0} & -\boldsymbol{U}^{-1}(\boldsymbol{x}_1) \\ \boldsymbol{0} & \boldsymbol{0} & \boldsymbol{0} & \boldsymbol{0} \\ -\boldsymbol{U}(\boldsymbol{x}_1) & \boldsymbol{0} & \boldsymbol{0} & \boldsymbol{0} \\ \boldsymbol{0} & \boldsymbol{0} & \boldsymbol{I} & \boldsymbol{0} \\ \boldsymbol{0} & \boldsymbol{I} & \boldsymbol{0} & \boldsymbol{0} \\ \boldsymbol{0} & \boldsymbol{0} & \boldsymbol{0} & \boldsymbol{0} \end{matrix}}_{\boldsymbol{\Lambda}_1(\boldsymbol{x}_1)} \quad \underbrace{\begin{matrix} -\boldsymbol{U}^{-1}(\boldsymbol{x}_1) & \boldsymbol{0} \\ \boldsymbol{0} & \boldsymbol{0} \\ \boldsymbol{0} & \boldsymbol{I} \\ \boldsymbol{0} & \boldsymbol{0} \\ \boldsymbol{0} & \boldsymbol{0} \\ \boldsymbol{0} & \boldsymbol{0} \end{matrix}}_{\boldsymbol{\Lambda}_{TF}(\boldsymbol{x}_1)} \quad \underbrace{\begin{matrix} \boldsymbol{0} & \boldsymbol{I} \\ \boldsymbol{I} & \boldsymbol{0} \\ \boldsymbol{0} & \boldsymbol{0} \\ \boldsymbol{0} & \boldsymbol{0} \\ \boldsymbol{0} & \boldsymbol{0} \\ \boldsymbol{0} & \boldsymbol{0} \end{matrix}}_{\boldsymbol{\Lambda}_{GY}} \tag{3.33}$$

Finally, with (3.33) and (3.30) we calculate (3.29) and obtain a single Dirac structure describing the equations of the junction structure. The equation system of the composed Dirac structure reads

$$\Bigg(\underbrace{\begin{matrix} \boldsymbol{0} & -\boldsymbol{I} & \boldsymbol{0} \\ \boldsymbol{0} & \boldsymbol{0} & \boldsymbol{I} \\ \boldsymbol{0} & \boldsymbol{0} & \boldsymbol{0} \\ \boldsymbol{0} & \boldsymbol{0} & \boldsymbol{0} \\ \boldsymbol{I} & \boldsymbol{0} & \boldsymbol{0} \\ \boldsymbol{0} & \boldsymbol{0} & \boldsymbol{0} \end{matrix}}_{\boldsymbol{F}_{\mathrm{C}}} \; \underbrace{\begin{matrix} \boldsymbol{0} \\ \boldsymbol{0} \\ \boldsymbol{0} \\ \boldsymbol{I} \\ \boldsymbol{0} \\ \boldsymbol{0} \end{matrix}}_{\boldsymbol{F}_{\mathrm{R}}} \; \underbrace{\begin{matrix} -\boldsymbol{I} \\ \boldsymbol{0} \\ \boldsymbol{0} \\ \boldsymbol{0} \\ \boldsymbol{0} \\ \boldsymbol{0} \end{matrix}}_{\boldsymbol{F}_{\mathrm{Sf}}} \; \underbrace{\begin{matrix} -\boldsymbol{U}^{-1}(\boldsymbol{x}_1) \\ \boldsymbol{0} \\ \boldsymbol{0} \\ \boldsymbol{I} \\ \boldsymbol{I} \\ \boldsymbol{0} \end{matrix}}_{\boldsymbol{F}_{\mathrm{Se}}(\boldsymbol{x}_1)}\Bigg) \begin{pmatrix} -\boldsymbol{f}_{C_1} \\ -\boldsymbol{f}_{C_2} \\ -\boldsymbol{f}_{C_3} \\ -\boldsymbol{f}_{R} \\ \boldsymbol{f}_{Sf} \\ \boldsymbol{f}_{Se} \end{pmatrix} + \Bigg(\underbrace{\begin{matrix} \boldsymbol{0} & \boldsymbol{0} & \boldsymbol{I} \\ \boldsymbol{0} & \boldsymbol{0} & \boldsymbol{0} \\ \boldsymbol{U}(\boldsymbol{x}_1) & \boldsymbol{0} & \boldsymbol{0} \\ \boldsymbol{0} & \boldsymbol{0} & \boldsymbol{0} \\ \boldsymbol{0} & \boldsymbol{0} & \boldsymbol{0} \\ \boldsymbol{0} & -\boldsymbol{I} & \boldsymbol{0} \end{matrix}}_{\boldsymbol{E}_{\mathrm{C}}(\boldsymbol{x}_1)} \; \underbrace{\begin{matrix} \boldsymbol{0} \\ \boldsymbol{0} \\ \boldsymbol{U}(\boldsymbol{x}_1) \\ \boldsymbol{0} \\ \boldsymbol{0} \\ \boldsymbol{0} \end{matrix}}_{\boldsymbol{E}_{\mathrm{R}}(\boldsymbol{x}_1)} \; \underbrace{\begin{matrix} \boldsymbol{0} \\ \boldsymbol{I} \\ \boldsymbol{I} \\ \boldsymbol{0} \\ \boldsymbol{0} \\ \boldsymbol{I} \end{matrix}}_{\boldsymbol{E}_{\mathrm{Sf}}} \; \underbrace{\begin{matrix} \boldsymbol{0} \\ \boldsymbol{0} \\ -\boldsymbol{U}(\boldsymbol{x}_1) \\ \boldsymbol{0} \\ \boldsymbol{0} \\ \boldsymbol{0} \end{matrix}}_{\boldsymbol{E}_{\mathrm{Se}}(\boldsymbol{x}_1)}\Bigg) \begin{pmatrix} \boldsymbol{e}_{C_1} \\ \boldsymbol{e}_{C_2} \\ \boldsymbol{e}_{C_3} \\ \boldsymbol{e}_{R} \\ \boldsymbol{e}_{Sf} \\ \boldsymbol{e}_{Se} \end{pmatrix} = \boldsymbol{0}. \tag{3.34}$$

It can be seen that (3.34) relates flows and efforts of only exterior bonds. Moreover, (3.34) is in the form of the equation system from (3.27).

3.2.5 Explicit Representation of the Dirac Structure

The Dirac structure obtained from Lemma 3.17 is in kernel representation which is an implicit form. For the derivation of an explicit PHS, the natural next step is to transfer the Dirac structure from an implicit form into an explicit form. In the following, we consider methods for transferring a Dirac structure from an (implicit) kernel representation into an (explicit) input-output representation (2.7). The basic idea to this end is simple: we rearrange the implicit equations in (3.27) such that a suitable explicit representation is obtained.

Bloch and Crouch [1999] show that for each Dirac structure in kernel representation there exists an input-output representation and vice versa. However, as with the kernel representation (cf. Remark 2.19), the input-output representation of a Dirac structure is not unique. In particular, not all input-output representations of a Dirac structure allow for a subsequent derivation of an explicit PHS with a suitable choice of inputs and outputs, viz. an explicit PHS with Property 3.8. The inputs and outputs of an explicit PHS are determined by the inputs and outputs of the underlying explicit form of the Dirac structure. Thus, based on Property 3.8, we deduce that the input vector of the input-output representation has to contain all flows corresponding to Sf elements and all efforts corresponding to Se elements; the output vector has to contain the corresponding conjugated variables. This is formalized in the following property:

Property 3.21 (Inputs and outputs of the Dirac structure)
Let $\mathbb{B}_\alpha = \cup_{i \in \mathbb{V}_\alpha} \mathbb{B}(i)$ *for* $\alpha \in \{\mathrm{Sf}, \mathrm{Se}\}$. *The input vector of the explicit Dirac structure includes* $(\boldsymbol{f}_j)$, $(\boldsymbol{e}_k)$ *while the output vector includes* $(\boldsymbol{e}_j)$, $(\boldsymbol{f}_k)$ *for all* $j \in \mathbb{B}_{\mathrm{Sf}}$, $k \in \mathbb{B}_{\mathrm{Se}}$.

We seek to formulate the Dirac structure (3.27) in an explicit representation that has Property 3.21. To this end, it is essential that the bond graph does not contain dependent sources. Dependent sources occur if the values imposed by two sources are conflicting. For example, think of an electric circuit with two voltage sources in parallel or two current sources in series. Intuitively, for such systems, it is impossible to find a model which has both conflicting variables as inputs. Hence, we assume the source elements of the bond graph to be independent—which is ensured by the following assumption:

Assumption 3.22 (Independent sources)
The matrices in (3.27) fulfill

$$\operatorname{rank}\left(\boldsymbol{E}_{\mathrm{Sf}}(\boldsymbol{x})\ \boldsymbol{F}_{\mathrm{Se}}(\boldsymbol{x})\right) = K\left(N_{\mathrm{Sf}}+N_{\mathrm{Se}}\right), \quad \forall \boldsymbol{x} \in \mathbb{X}. \tag{3.35}$$

In Appendix B.1, is is proven that the source elements from the bond graph are independent if and only if Assumption 3.22 is met. Conversely, if Assumption 3.22 is violated, the junction structure implies a dependency between source elements. Dependent sources are serious modeling inconsistencies which may occur from physically implausible structures in the bond graph. They must be resolved by modifications of the bond graph [Karnopp et al., 2012, p. 169]. Based on the exclusion of dependent sources in Assumption 3.22, one can now state the following lemma:

Lemma 3.23 (Input-output representation)
Consider a Dirac structure (3.27) which fulfills Assumption 3.22. The Dirac structure can be formulated in an input-output representation with Property 3.21:

$$\mathbb{D}(\boldsymbol{x}) = \left\{\left(\begin{pmatrix}\boldsymbol{f}_{\mathrm{C}}\\ \boldsymbol{f}_{\mathrm{R}}\\ \boldsymbol{f}_{\mathrm{Sf}}\\ \boldsymbol{f}_{\mathrm{Se}}\end{pmatrix}, \begin{pmatrix}\boldsymbol{e}_{\mathrm{C}}\\ \boldsymbol{e}_{\mathrm{R}}\\ \boldsymbol{e}_{\mathrm{Sf}}\\ \boldsymbol{e}_{\mathrm{Se}}\end{pmatrix}\right) \in \mathbb{R}^{KN_{\mathrm{E}}} \times \mathbb{R}^{KN_{\mathrm{E}}} \mid \begin{pmatrix}\boldsymbol{y}_{\mathrm{C}}\\ \boldsymbol{y}_{\mathrm{R}}\\ \boldsymbol{y}_{\mathrm{P}}\end{pmatrix} = \boldsymbol{Z}(\boldsymbol{x}) \begin{pmatrix}\boldsymbol{u}_{\mathrm{C}}\\ \boldsymbol{u}_{\mathrm{R}}\\ \boldsymbol{u}_{\mathrm{P}}\end{pmatrix}\right\}, \tag{3.36}$$

where $\boldsymbol{Z}(\boldsymbol{x})$ is skew-symmetric for all $\boldsymbol{x} \in \mathbb{X}$ and

$$\boldsymbol{u}_{\mathrm{C}} = \begin{pmatrix}\boldsymbol{e}_{\mathrm{C},1}\\ -\boldsymbol{f}_{\mathrm{C},2}\end{pmatrix}, \quad \boldsymbol{u}_{\mathrm{R}} = \begin{pmatrix}\boldsymbol{e}_{\mathrm{R},1}\\ -\boldsymbol{f}_{\mathrm{R},2}\end{pmatrix}, \quad \boldsymbol{u}_{\mathrm{P}} = \begin{pmatrix}\boldsymbol{f}_{\mathrm{Sf}}\\ \boldsymbol{e}_{\mathrm{Se}}\end{pmatrix}, \tag{3.37a}$$

$$\boldsymbol{y}_{\mathrm{C}} = \begin{pmatrix}-\boldsymbol{f}_{\mathrm{C},1}\\ \boldsymbol{e}_{\mathrm{C},2}\end{pmatrix}, \quad \boldsymbol{y}_{\mathrm{R}} = \begin{pmatrix}-\boldsymbol{f}_{\mathrm{R},1}\\ \boldsymbol{e}_{\mathrm{R},2}\end{pmatrix}, \quad \boldsymbol{y}_{\mathrm{P}} = \begin{pmatrix}\boldsymbol{e}_{\mathrm{Sf}}\\ \boldsymbol{f}_{\mathrm{Se}}\end{pmatrix}. \tag{3.37b}$$

The matrix $\boldsymbol{Z}(\boldsymbol{x})$ exists for all $\boldsymbol{x} \in \mathbb{X}$ and is given by:

$$\begin{aligned}\boldsymbol{Z}(\boldsymbol{x}) = &\left(\boldsymbol{F}_{\mathrm{C},1}(\boldsymbol{x}) \quad \boldsymbol{E}_{\mathrm{C},2}(\boldsymbol{x}) \quad \boldsymbol{F}_{\mathrm{R},1}(\boldsymbol{x}) \quad \boldsymbol{E}_{\mathrm{R},2}(\boldsymbol{x}) \quad \boldsymbol{E}_{\mathrm{Sf}}(\boldsymbol{x}) \quad \boldsymbol{F}_{\mathrm{Se}}(\boldsymbol{x})\right)^{-1} \\ &\cdot \left(\boldsymbol{E}_{\mathrm{C},1}(\boldsymbol{x}) \quad \boldsymbol{F}_{\mathrm{C},2}(\boldsymbol{x}) \quad \boldsymbol{E}_{\mathrm{R},1}(\boldsymbol{x}) \quad \boldsymbol{F}_{\mathrm{R},2}(\boldsymbol{x}) \quad \boldsymbol{F}_{\mathrm{Sf}}(\boldsymbol{x}) \quad \boldsymbol{E}_{\mathrm{Se}}(\boldsymbol{x})\right).\end{aligned} \tag{3.38}$$

The matrices in (3.38) can be obtained from splitting (possibly after some permutations) $\boldsymbol{F}_{\mathrm{C}}(\boldsymbol{x})$ and $\boldsymbol{F}_{\mathrm{R}}(\boldsymbol{x})$ into $(\boldsymbol{F}_{\mathrm{C},1}(\boldsymbol{x})\ \boldsymbol{F}_{\mathrm{C},2}(\boldsymbol{x}))$ and $(\boldsymbol{F}_{\mathrm{R},1}(\boldsymbol{x})\ \boldsymbol{F}_{\mathrm{R},2}(\boldsymbol{x}))$, respectively, such that

$$\begin{aligned}&(i)\ \left(\boldsymbol{F}_{\mathrm{C},1}(\boldsymbol{x}) \quad \boldsymbol{F}_{\mathrm{R},1}(\boldsymbol{x}) \quad \boldsymbol{E}_{\mathrm{Sf}}(\boldsymbol{x}) \quad \boldsymbol{F}_{\mathrm{Se}}(\boldsymbol{x})\right) \textit{ has full column rank} && \text{(3.39a)}\\ &(ii)\ \operatorname{rank}\left(\boldsymbol{F}_{\mathrm{C},1}(\boldsymbol{x}) \quad \boldsymbol{F}_{\mathrm{R},1}(\boldsymbol{x}) \quad \boldsymbol{E}_{\mathrm{Sf}}(\boldsymbol{x}) \quad \boldsymbol{F}_{\mathrm{Se}}(\boldsymbol{x})\right) = \\ &\qquad \operatorname{rank}\left(\boldsymbol{F}_{\mathrm{C}}(\boldsymbol{x}) \quad \boldsymbol{F}_{\mathrm{R}}(\boldsymbol{x}) \quad \boldsymbol{E}_{\mathrm{Sf}}(\boldsymbol{x}) \quad \boldsymbol{F}_{\mathrm{Se}}(\boldsymbol{x})\right), && \text{(3.39b)}\end{aligned}$$

for all $\boldsymbol{x} \in \mathbb{X}$. According to the manner in which $\boldsymbol{F}_{\mathrm{C}}(\boldsymbol{x})$ and $\boldsymbol{F}_{\mathrm{R}}(\boldsymbol{x})$ are split, we partition $\boldsymbol{E}_{\mathrm{C}}(\boldsymbol{x})$ and $\boldsymbol{E}_{\mathrm{R}}(\boldsymbol{x})$ into $(\boldsymbol{E}_{\mathrm{C},1}(\boldsymbol{x})\ \boldsymbol{E}_{\mathrm{C},2}(\boldsymbol{x}))$ and $(\boldsymbol{E}_{\mathrm{R},1}(\boldsymbol{x})\ \boldsymbol{E}_{\mathrm{R},2}(\boldsymbol{x}))$, respectively. In the same way, we split $\boldsymbol{f}_{\mathrm{C}}$, $\boldsymbol{e}_{\mathrm{C}}$ and $\boldsymbol{f}_{\mathrm{R}}$, $\boldsymbol{e}_{\mathrm{R}}$.

Proof:
Let Assumption 3.22 hold. For the sake of notation, we neglect the argument $\boldsymbol{x}$ and the supplement "for all $\boldsymbol{x} \in \mathbb{X}$" in this proof. We apply the ideas from Bloch and Crouch [1999, Theorem 4] to show that we always find decompositions $(\boldsymbol{F}_{\mathrm{R},1}, \boldsymbol{F}_{\mathrm{R},2})$, $(\boldsymbol{F}_{\mathrm{C},1}, \boldsymbol{F}_{\mathrm{C},2})$, $(\boldsymbol{E}_{\mathrm{C},1}, \boldsymbol{E}_{\mathrm{C},2})$, $(\boldsymbol{E}_{\mathrm{R},1}, \boldsymbol{E}_{\mathrm{R},2})$ for $\boldsymbol{F}_{\mathrm{C}}$, $\boldsymbol{F}_{\mathrm{R}}$, $\boldsymbol{E}_{\mathrm{C}}$, $\boldsymbol{E}_{\mathrm{R}}$, respectively, such that

$$\operatorname{rank}\left(\boldsymbol{F}_{\mathrm{C},1} \quad \boldsymbol{E}_{\mathrm{C},2} \quad \boldsymbol{F}_{\mathrm{R},1} \quad \boldsymbol{E}_{\mathrm{R},2} \quad \boldsymbol{E}_{\mathrm{Sf}} \quad \boldsymbol{F}_{\mathrm{Se}}\right) = K N_{\mathrm{E}} \tag{3.40}$$

holds. Choose decompositions of $\boldsymbol{F}_{\mathrm{C}}$ and $\boldsymbol{F}_{\mathrm{R}}$ (possibly after some permutations) such that the conditions in (3.39) are fulfilled. Let $\boldsymbol{F}_{\mathrm{C},2}$, $\boldsymbol{F}_{\mathrm{R},2}$ denote the rest of $\boldsymbol{F}_{\mathrm{C}}$, $\boldsymbol{F}_{\mathrm{R}}$, respectively, i.e., $\boldsymbol{F}_{\mathrm{C}} = (\boldsymbol{F}_{\mathrm{C},1}\ \boldsymbol{F}_{\mathrm{C},2})$, $\boldsymbol{F}_{\mathrm{R}} = (\boldsymbol{F}_{\mathrm{R},1}\ \boldsymbol{F}_{\mathrm{R},2})$. Next, split $\boldsymbol{E}_{\mathrm{C}}$ and $\boldsymbol{E}_{\mathrm{R}}$ according to the decomposition chosen for $\boldsymbol{F}_{\mathrm{C}}$ and $\boldsymbol{F}_{\mathrm{R}}$, respectively, into $\boldsymbol{E}_{\mathrm{C}} = (\boldsymbol{E}_{\mathrm{C},1}\ \boldsymbol{E}_{\mathrm{C},2})$ and $\boldsymbol{E}_{\mathrm{R}} = (\boldsymbol{E}_{\mathrm{R},1}\ \boldsymbol{E}_{\mathrm{R},2})$. By (3.39a) the matrix $(\boldsymbol{F}_{\mathrm{C},1}\ \boldsymbol{F}_{\mathrm{R},1}\ \boldsymbol{E}_{\mathrm{Sf}}\ \boldsymbol{F}_{\mathrm{Se}})$ has full column rank. Thus, its adjoint $(\boldsymbol{F}_{\mathrm{C},1}\ \boldsymbol{F}_{\mathrm{R},1}\ \boldsymbol{E}_{\mathrm{Sf}}\ \boldsymbol{F}_{\mathrm{Se}})^{\top}$

is surjective. In particular we have

$$\begin{aligned}
&\operatorname{im}\begin{pmatrix}\boldsymbol{E}_{\mathrm{C},1} & \boldsymbol{E}_{\mathrm{R},1} & \boldsymbol{F}_{\mathrm{Sf}} & \boldsymbol{E}_{\mathrm{Se}}\end{pmatrix} \\
&\quad= \operatorname{im}\left(\begin{pmatrix}\boldsymbol{E}_{\mathrm{C},1} & \boldsymbol{E}_{\mathrm{R},1} & \boldsymbol{F}_{\mathrm{Sf}} & \boldsymbol{E}_{\mathrm{Se}}\end{pmatrix}\cdot\begin{pmatrix}\boldsymbol{F}_{\mathrm{C},1} & \boldsymbol{F}_{\mathrm{R},1} & \boldsymbol{E}_{\mathrm{Sf}} & \boldsymbol{F}_{\mathrm{Se}}\end{pmatrix}^\top\right) \\
&\qquad\qquad= \operatorname{im}\left(\boldsymbol{E}_{\mathrm{C},1}\boldsymbol{F}_{\mathrm{C},1}^\top + \boldsymbol{E}_{\mathrm{R},1}\boldsymbol{F}_{\mathrm{R},1}^\top + \boldsymbol{F}_{\mathrm{Sf}}\boldsymbol{E}_{\mathrm{Sf}}^\top + \boldsymbol{E}_{\mathrm{Se}}\boldsymbol{F}_{\mathrm{Se}}^\top\right). \qquad (3.41)
\end{aligned}$$

For (3.27), equation (2.5a) reads

$$\begin{aligned}
\mathbf{0} &= \boldsymbol{E}\boldsymbol{F}^\top + \boldsymbol{F}\boldsymbol{E}^\top \\
&= \boldsymbol{E}_{\mathrm{C},1}\boldsymbol{F}_{\mathrm{C},1}^\top + \boldsymbol{E}_{\mathrm{C},2}\boldsymbol{F}_{\mathrm{C},2}^\top + \boldsymbol{E}_{\mathrm{R},1}\boldsymbol{F}_{\mathrm{R},1}^\top + \boldsymbol{E}_{\mathrm{R},2}\boldsymbol{F}_{\mathrm{R},2}^\top + \boldsymbol{E}_{\mathrm{Sf}}\boldsymbol{F}_{\mathrm{Sf}}^\top + \boldsymbol{E}_{\mathrm{Se}}\boldsymbol{F}_{\mathrm{Se}}^\top + \\
&\quad\ \boldsymbol{F}_{\mathrm{C},1}\boldsymbol{E}_{\mathrm{C},1}^\top + \boldsymbol{F}_{\mathrm{C},2}\boldsymbol{E}_{\mathrm{C},2}^\top + \boldsymbol{F}_{\mathrm{R},1}\boldsymbol{E}_{\mathrm{R},1}^\top + \boldsymbol{F}_{\mathrm{R},2}\boldsymbol{E}_{\mathrm{R},2}^\top + \boldsymbol{F}_{\mathrm{Sf}}\boldsymbol{E}_{\mathrm{Sf}}^\top + \boldsymbol{F}_{\mathrm{Se}}\boldsymbol{E}_{\mathrm{Se}}^\top,
\end{aligned} \qquad (3.42)$$

from which follows

$$\begin{aligned}
&\operatorname{im}\left(\boldsymbol{E}_{\mathrm{C},1}\boldsymbol{F}_{\mathrm{C},1}^\top + \boldsymbol{E}_{\mathrm{R},1}\boldsymbol{F}_{\mathrm{R},1}^\top + \boldsymbol{F}_{\mathrm{Sf}}\boldsymbol{E}_{\mathrm{Sf}}^\top + \boldsymbol{E}_{\mathrm{Se}}\boldsymbol{F}_{\mathrm{Se}}^\top\right) \\
&\quad= \operatorname{im}\Big(\boldsymbol{F}_{\mathrm{C},1}\boldsymbol{E}_{\mathrm{C},1}^\top + \boldsymbol{F}_{\mathrm{C},2}\boldsymbol{E}_{\mathrm{C},2}^\top + \boldsymbol{E}_{\mathrm{C},2}\boldsymbol{F}_{\mathrm{C},2}^\top + \boldsymbol{F}_{\mathrm{R},1}\boldsymbol{E}_{\mathrm{R},1}^\top \\
&\qquad\qquad + \boldsymbol{F}_{\mathrm{R},2}\boldsymbol{E}_{\mathrm{R},2}^\top + \boldsymbol{E}_{\mathrm{R},2}\boldsymbol{F}_{\mathrm{R},2}^\top + \boldsymbol{E}_{\mathrm{Sf}}\boldsymbol{F}_{\mathrm{Sf}}^\top + \boldsymbol{F}_{\mathrm{Se}}\boldsymbol{E}_{\mathrm{Se}}^\top\Big) \\
&\quad\subseteq \operatorname{im}\Big(\boldsymbol{F}_{\mathrm{C},1}\boldsymbol{E}_{\mathrm{C},1}^\top \;\; \boldsymbol{F}_{\mathrm{C},2}\boldsymbol{E}_{\mathrm{C},2}^\top \;\; \boldsymbol{E}_{\mathrm{C},2}\boldsymbol{F}_{\mathrm{C},2}^\top \;\; \boldsymbol{F}_{\mathrm{R},1}\boldsymbol{E}_{\mathrm{R},1}^\top \\
&\qquad\qquad \boldsymbol{F}_{\mathrm{R},2}\boldsymbol{E}_{\mathrm{R},2}^\top \;\; \boldsymbol{E}_{\mathrm{R},2}\boldsymbol{F}_{\mathrm{R},2}^\top \;\; \boldsymbol{E}_{\mathrm{Sf}}\boldsymbol{F}_{\mathrm{Sf}}^\top \;\; \boldsymbol{F}_{\mathrm{Se}}\boldsymbol{E}_{\mathrm{Se}}^\top\Big) \qquad (3.43) \\
&\quad\subseteq \operatorname{im}\begin{pmatrix}\boldsymbol{F}_{\mathrm{C},1} & \boldsymbol{F}_{\mathrm{C},2} & \boldsymbol{E}_{\mathrm{C},2} & \boldsymbol{F}_{\mathrm{R},1} & \boldsymbol{F}_{\mathrm{R},2} & \boldsymbol{E}_{\mathrm{R},2} & \boldsymbol{E}_{\mathrm{Sf}} & \boldsymbol{F}_{\mathrm{Se}}\end{pmatrix} \\
&\quad\overset{(3.39)}{=} \operatorname{im}\begin{pmatrix}\boldsymbol{F}_{\mathrm{C},1} & \boldsymbol{E}_{\mathrm{C},2} & \boldsymbol{F}_{\mathrm{R},1} & \boldsymbol{E}_{\mathrm{R},2} & \boldsymbol{E}_{\mathrm{Sf}} & \boldsymbol{F}_{\mathrm{Se}}\end{pmatrix}.
\end{aligned}$$

Combining (3.41) and (3.43) we can derive

$$\begin{aligned}
\operatorname{im}\begin{pmatrix}\boldsymbol{F} & \boldsymbol{E}\end{pmatrix} &= \operatorname{im}\begin{pmatrix}\boldsymbol{E}_{\mathrm{C},1} & \boldsymbol{E}_{\mathrm{R},1} & \boldsymbol{F}_{\mathrm{Sf}} & \boldsymbol{E}_{\mathrm{Se}}\end{pmatrix} + \\
&\qquad \operatorname{im}\begin{pmatrix}\boldsymbol{F}_{\mathrm{C},1} & \boldsymbol{F}_{\mathrm{C},2} & \boldsymbol{E}_{\mathrm{C},2} & \boldsymbol{F}_{\mathrm{R},1} & \boldsymbol{F}_{\mathrm{R},2} & \boldsymbol{E}_{\mathrm{R},2} & \boldsymbol{E}_{\mathrm{Sf}} & \boldsymbol{F}_{\mathrm{Se}}\end{pmatrix} \qquad (3.44) \\
&\subseteq \operatorname{im}\begin{pmatrix}\boldsymbol{F}_{\mathrm{C},1} & \boldsymbol{E}_{\mathrm{C},2} & \boldsymbol{F}_{\mathrm{R},1} & \boldsymbol{E}_{\mathrm{R},2} & \boldsymbol{E}_{\mathrm{Sf}} & \boldsymbol{F}_{\mathrm{Se}}\end{pmatrix} \subseteq \operatorname{im}\begin{pmatrix}\boldsymbol{F} & \boldsymbol{E}\end{pmatrix}.
\end{aligned}$$

Thus, equality holds in the above formula and we have

$$\operatorname{rank}\begin{pmatrix}\boldsymbol{F}_{\mathrm{C},1} & \boldsymbol{E}_{\mathrm{C},2} & \boldsymbol{F}_{\mathrm{R},1} & \boldsymbol{E}_{\mathrm{R},2} & \boldsymbol{E}_{\mathrm{Sf}} & \boldsymbol{F}_{\mathrm{Se}}\end{pmatrix} = \operatorname{rank}\begin{pmatrix}\boldsymbol{F} & \boldsymbol{E}\end{pmatrix} \overset{(2.5b)}{=} KN_{\mathrm{E}}. \tag{3.45}$$

Hence, the square matrix $(\boldsymbol{F}_{\mathrm{C},1}\ \boldsymbol{E}_{\mathrm{C},2}\ \boldsymbol{F}_{\mathrm{R},1}\ \boldsymbol{E}_{\mathrm{R},2}\ \boldsymbol{E}_{\mathrm{Sf}}\ \boldsymbol{F}_{\mathrm{Se}})$ has full rank and is invertible. From Bloch and Crouch [1999] it follows that under the above rank condition (3.45) the kernel representation (3.27) can be formulated as the input-output representation (3.36). The matrix $\boldsymbol{Z}$ can be determined by the following calculation law which is formally derived in Appendix B.2:

$$\begin{aligned}\boldsymbol{Z} = -\boldsymbol{Z}^{\top} = &-\begin{pmatrix}\boldsymbol{F}_{\mathrm{C},1} & \boldsymbol{E}_{\mathrm{C},2} & \boldsymbol{F}_{\mathrm{R},1} & \boldsymbol{E}_{\mathrm{R},2} & \boldsymbol{E}_{\mathrm{Sf}} & \boldsymbol{F}_{\mathrm{Se}}\end{pmatrix}^{-1} \\ &\cdot \begin{pmatrix}\boldsymbol{E}_{\mathrm{C},1} & \boldsymbol{F}_{\mathrm{C},2} & \boldsymbol{E}_{\mathrm{R},1} & \boldsymbol{F}_{\mathrm{R},2} & \boldsymbol{F}_{\mathrm{Sf}} & \boldsymbol{E}_{\mathrm{Se}}\end{pmatrix}.\end{aligned} \tag{3.46}$$

□

We continue with our example to illustrate the reformulation of a Dirac structure in an explicit input-output form.

Example 3.24:

Consider the Dirac structure from Example 3.20 given in an implicit representation:

$$\mathbb{D}(\boldsymbol{x}_1) = \{(\begin{pmatrix} \boldsymbol{f}_{C_1} \\ \boldsymbol{f}_{C_2} \\ \boldsymbol{f}_{C_3} \\ \boldsymbol{f}_{R} \\ \boldsymbol{f}_{Sf} \\ \boldsymbol{f}_{Se} \end{pmatrix}, \begin{pmatrix} \boldsymbol{e}_{C_1} \\ \boldsymbol{e}_{C_2} \\ \boldsymbol{e}_{C_3} \\ \boldsymbol{e}_{R} \\ \boldsymbol{e}_{Sf} \\ \boldsymbol{e}_{Se} \end{pmatrix}) \in \mathbb{R}^{6K} \times \mathbb{R}^{6K} \mid$$

$$\underbrace{\begin{pmatrix} \boldsymbol{0} & -\boldsymbol{I} & \boldsymbol{0} \\ \boldsymbol{0} & \boldsymbol{0} & \boldsymbol{I} \\ \boldsymbol{0} & \boldsymbol{0} & \boldsymbol{0} \\ \boldsymbol{0} & \boldsymbol{0} & \boldsymbol{0} \\ \boldsymbol{I} & \boldsymbol{0} & \boldsymbol{0} \\ \boldsymbol{0} & \boldsymbol{0} & \boldsymbol{0} \end{pmatrix}}_{\boldsymbol{F}_{\mathrm{C}}}\!\!\!\!\!\!\underbrace{\begin{matrix} \boldsymbol{0} \\ \boldsymbol{0} \\ \boldsymbol{0} \\ \boldsymbol{I} \\ \boldsymbol{0} \\ \boldsymbol{0} \end{matrix}}_{\boldsymbol{F}_{\mathrm{R}}}\ \underbrace{\begin{matrix} -\boldsymbol{I} \\ \boldsymbol{0} \\ \boldsymbol{0} \\ \boldsymbol{0} \\ \boldsymbol{0} \\ \boldsymbol{0} \end{matrix}}_{\boldsymbol{F}_{\mathrm{Sf}}}\ \underbrace{\begin{matrix} -\boldsymbol{U}^{-1}(\boldsymbol{x}_1) \\ \boldsymbol{0} \\ \boldsymbol{0} \\ \boldsymbol{I} \\ \boldsymbol{I} \\ \boldsymbol{0} \end{matrix}}_{\boldsymbol{F}_{\mathrm{Se}}(\boldsymbol{x}_1)} \begin{pmatrix} -\boldsymbol{f}_{C_1} \\ -\boldsymbol{f}_{C_2} \\ -\boldsymbol{f}_{C_3} \\ -\boldsymbol{f}_{R} \\ \boldsymbol{f}_{Sf} \\ \boldsymbol{f}_{Se} \end{pmatrix}$$

$$+ \underbrace{\begin{matrix} \boldsymbol{0} & \boldsymbol{0} & \boldsymbol{I} \\ \boldsymbol{0} & \boldsymbol{0} & \boldsymbol{0} \\ \boldsymbol{U}(\boldsymbol{x}_1) & \boldsymbol{0} & \boldsymbol{0} \\ \boldsymbol{0} & \boldsymbol{0} & \boldsymbol{0} \\ \boldsymbol{0} & \boldsymbol{0} & \boldsymbol{0} \\ \boldsymbol{0} & -\boldsymbol{I} & \boldsymbol{0} \end{matrix}}_{\boldsymbol{E}_{\mathrm{C}}(\boldsymbol{x}_1)}\ \underbrace{\begin{matrix} \boldsymbol{0} \\ \boldsymbol{0} \\ \boldsymbol{U}(\boldsymbol{x}_1) \\ \boldsymbol{0} \\ \boldsymbol{0} \\ \boldsymbol{0} \end{matrix}}_{\boldsymbol{E}_{\mathrm{R}}(\boldsymbol{x}_1)}\ \underbrace{\begin{matrix} \boldsymbol{0} \\ \boldsymbol{I} \\ \boldsymbol{I} \\ \boldsymbol{0} \\ \boldsymbol{0} \\ \boldsymbol{I} \end{matrix}}_{\boldsymbol{E}_{\mathrm{Sf}}}\ \underbrace{\begin{matrix} \boldsymbol{0} \\ \boldsymbol{0} \\ -\boldsymbol{U}(\boldsymbol{x}_1) \\ \boldsymbol{0} \\ \boldsymbol{0} \\ \boldsymbol{0} \end{matrix}}_{\boldsymbol{E}_{\mathrm{Se}}(\boldsymbol{x}_1)} \begin{pmatrix} \boldsymbol{e}_{C_1} \\ \boldsymbol{e}_{C_2} \\ \boldsymbol{e}_{C_3} \\ \boldsymbol{e}_{R} \\ \boldsymbol{e}_{Sf} \\ \boldsymbol{e}_{Se} \end{pmatrix} = \boldsymbol{0}\}. \quad (3.47)$$

Recall that $N_{\mathrm{C}} = 3$, $N_{\mathrm{R}} = 1 = N_{\mathrm{Sf}} = 1 = N_{\mathrm{Se}} = 1$ *and that* $\boldsymbol{U}(\boldsymbol{x}_1)$ *has full rank (see Example 3.7). With the matrices in (3.47), Assumption 3.22 is fulfilled as* $\operatorname{rank}(\boldsymbol{E}_{\mathrm{Sf}}\, \boldsymbol{F}_{\mathrm{Se}}(\boldsymbol{x}_1)) = 2K$. *Hence, we can apply Lemma 3.23 to determine an explicit representation of (3.47). First, we search splittings of* $\boldsymbol{F}_{\mathrm{C}}(\boldsymbol{x})$ *and*

$\boldsymbol{F}_\mathrm{R}(\boldsymbol{x})$ such that (3.39a) and (3.39b) are fulfilled. Take

$$\boldsymbol{F}_{\mathrm{C},1} = \begin{pmatrix} -\boldsymbol{I} & \boldsymbol{0} \\ \boldsymbol{0} & \boldsymbol{I} \\ \boldsymbol{0} & \boldsymbol{0} \\ \boldsymbol{0} & \boldsymbol{0} \\ \boldsymbol{0} & \boldsymbol{0} \\ \boldsymbol{0} & \boldsymbol{0} \end{pmatrix}, \quad \boldsymbol{F}_{\mathrm{C},2} = \begin{pmatrix} \boldsymbol{0} \\ \boldsymbol{0} \\ \boldsymbol{0} \\ \boldsymbol{0} \\ \boldsymbol{I} \\ \boldsymbol{0} \end{pmatrix}, \tag{3.48}$$

and $\boldsymbol{F}_{\mathrm{R},1} = \boldsymbol{F}_\mathrm{R}$, i.e., $\boldsymbol{F}_{\mathrm{R},2}$ is an empty matrix. It can be easily verified that this choice satisfies the conditions in (3.39). Thus, we may evaluate (3.38) with this splitting choice to determine the matrix $\boldsymbol{Z}(\boldsymbol{x})$ and obtain the following explicit representation of (3.47):

$$\mathbb{D}(\boldsymbol{x}_1) = \{(\begin{pmatrix} \boldsymbol{f}_{C_1} \\ \boldsymbol{f}_{C_2} \\ \boldsymbol{f}_{C_3} \\ \boldsymbol{f}_R \\ \boldsymbol{f}_{Sf} \\ \boldsymbol{f}_{Se} \end{pmatrix}, \begin{pmatrix} \boldsymbol{e}_{C_1} \\ \boldsymbol{e}_{C_2} \\ \boldsymbol{e}_{C_3} \\ \boldsymbol{e}_R \\ \boldsymbol{e}_{Sf} \\ \boldsymbol{e}_{Se} \end{pmatrix}) \in \mathbb{R}^{6K} \times \mathbb{R}^{6K} \mid$$
$$\begin{pmatrix} -\boldsymbol{f}_{C_2} \\ -\boldsymbol{f}_{C_3} \\ \boldsymbol{e}_{C_1} \\ -\boldsymbol{f}_R \\ \boldsymbol{e}_{Sf} \\ \boldsymbol{f}_{Se} \end{pmatrix} = \begin{pmatrix} \boldsymbol{0} & -\boldsymbol{I} & -\boldsymbol{U}^{-1}(\boldsymbol{x}_1) & \boldsymbol{0} & \boldsymbol{I} & \boldsymbol{0} \\ \boldsymbol{I} & \boldsymbol{0} & \boldsymbol{0} & \boldsymbol{0} & \boldsymbol{0} & \boldsymbol{0} \\ \boldsymbol{U}^{-1}(\boldsymbol{x}_1) & \boldsymbol{0} & \boldsymbol{0} & \boldsymbol{I} & \boldsymbol{0} & -\boldsymbol{I} \\ \boldsymbol{0} & \boldsymbol{0} & -\boldsymbol{I} & \boldsymbol{0} & \boldsymbol{0} & \boldsymbol{0} \\ -\boldsymbol{I} & \boldsymbol{0} & \boldsymbol{0} & \boldsymbol{0} & \boldsymbol{0} & \boldsymbol{0} \\ \boldsymbol{0} & \boldsymbol{0} & \boldsymbol{I} & \boldsymbol{0} & \boldsymbol{0} & \boldsymbol{0} \end{pmatrix} \begin{pmatrix} \boldsymbol{e}_{C_2} \\ \boldsymbol{e}_{C_3} \\ -\boldsymbol{f}_{C_1} \\ \boldsymbol{e}_R \\ \boldsymbol{f}_{Sf} \\ \boldsymbol{e}_{Se} \end{pmatrix} \}. \tag{3.49}$$

As can be seen from the above example, Lemma 3.23 assigns the variables of the source elements appropriately to inputs and outputs of the Dirac structure (cf. (3.37)). By this, the resulting Dirac structure has Property 3.21. It follows from Lemma 3.23, that Assumption 3.22 is a sufficient condition for the existence of a Dirac structure representation with Property 3.21. Of course, it would be desirable to show that Assumption 3.22 is also necessary. However, for a proof of necessity, the non-uniqueness of an input-output representation of a Dirac structure is a stumbling block. As a remedy, the next proposition shows that there is some kind of uniqueness, viz. that the input-output representation is unique for a certain arrangement of variables.

Proposition 3.25 (Uniqueness of $\boldsymbol{Z}(\boldsymbol{x})$)

For any given order of the variables in (3.37), the matrix $\boldsymbol{Z}(\boldsymbol{x})$ in (3.36) is unique. This statement is independent of Assumption 3.22.

Proof:

For the sake of releasing notational burden we will suppress the argument $\boldsymbol{x}$ to the matrices during the proof. Let $\boldsymbol{Z}$ and $\boldsymbol{Z}' \in \mathbb{R}^{KN_\mathrm{E} \times KN_\mathrm{E}}$ be two matrices fulfilling the equations

$$\begin{pmatrix} \boldsymbol{y}_\mathrm{C} \\ \boldsymbol{y}_\mathrm{R} \\ \boldsymbol{y}_\mathrm{P} \end{pmatrix} = \boldsymbol{Z} \begin{pmatrix} \boldsymbol{u}_\mathrm{C} \\ \boldsymbol{u}_\mathrm{R} \\ \boldsymbol{u}_\mathrm{P} \end{pmatrix} \quad \text{and} \quad \begin{pmatrix} \boldsymbol{y}_\mathrm{C} \\ \boldsymbol{y}_\mathrm{R} \\ \boldsymbol{y}_\mathrm{P} \end{pmatrix} = \boldsymbol{Z}' \begin{pmatrix} \boldsymbol{u}_\mathrm{C} \\ \boldsymbol{u}_\mathrm{R} \\ \boldsymbol{u}_\mathrm{P} \end{pmatrix}. \tag{3.50}$$

Recall (3.37) and that $\dim \mathbb{D} = KN_\mathrm{E}$. As $(\boldsymbol{y}_\mathrm{C}^\top \ \boldsymbol{y}_\mathrm{R}^\top \ \boldsymbol{y}_\mathrm{P}^\top)^\top$ linearly depends on $(\boldsymbol{u}_\mathrm{C}^\top \ \boldsymbol{u}_\mathrm{R}^\top \ \boldsymbol{u}_\mathrm{P}^\top)^\top$, we have that $\mathbb{D}$ is isomorphic to $\mathbb{R}^{KN_\mathrm{E}}$ via

$$\mathbb{R}^{KN_\mathrm{E}} \to \mathbb{D}, \quad \begin{pmatrix} \boldsymbol{u}_\mathrm{C} \\ \boldsymbol{u}_\mathrm{R} \\ \boldsymbol{u}_\mathrm{P} \end{pmatrix} \mapsto \left(\begin{pmatrix} \boldsymbol{f}_{\mathrm{C},1} \\ \boldsymbol{f}_{\mathrm{C},2} \\ \boldsymbol{f}_{\mathrm{R},1} \\ \boldsymbol{f}_{\mathrm{R},2} \\ \boldsymbol{f}_\mathrm{Sf} \\ \boldsymbol{f}_\mathrm{Se} \end{pmatrix}, \begin{pmatrix} \boldsymbol{e}_{\mathrm{C},1} \\ \boldsymbol{e}_{\mathrm{C},2} \\ \boldsymbol{e}_{\mathrm{R},1} \\ \boldsymbol{e}_{\mathrm{R},2} \\ \boldsymbol{e}_\mathrm{Sf} \\ \boldsymbol{e}_\mathrm{Se} \end{pmatrix} \right) \quad \text{where} \begin{pmatrix} \boldsymbol{y}_\mathrm{C} \\ \boldsymbol{y}_\mathrm{R} \\ \boldsymbol{y}_\mathrm{P} \end{pmatrix} = \boldsymbol{Z} \begin{pmatrix} \boldsymbol{u}_\mathrm{C} \\ \boldsymbol{u}_\mathrm{R} \\ \boldsymbol{u}_\mathrm{P} \end{pmatrix}, \tag{3.51}$$

$$\mathbb{D} \to \mathbb{R}^{KN_\mathrm{E}}, \quad \left(\begin{pmatrix} \boldsymbol{e}_{\mathrm{C},1} \\ \boldsymbol{e}_{\mathrm{C},2} \\ \boldsymbol{e}_{\mathrm{R},1} \\ \boldsymbol{e}_{\mathrm{R},2} \\ \boldsymbol{e}_\mathrm{Sf} \\ \boldsymbol{e}_\mathrm{Se} \end{pmatrix}, \begin{pmatrix} \boldsymbol{f}_{\mathrm{C},1} \\ \boldsymbol{f}_{\mathrm{C},2} \\ \boldsymbol{f}_{\mathrm{R},1} \\ \boldsymbol{f}_{\mathrm{R},2} \\ \boldsymbol{f}_\mathrm{Sf} \\ \boldsymbol{f}_\mathrm{Se} \end{pmatrix} \right) \mapsto \begin{pmatrix} \boldsymbol{u}_\mathrm{C} \\ \boldsymbol{u}_\mathrm{R} \\ \boldsymbol{u}_\mathrm{P} \end{pmatrix}. \tag{3.52}$$

From (3.50) it follows that

$$\boldsymbol{Z}\begin{pmatrix}\boldsymbol{u}_{\mathrm{C}}\\ \boldsymbol{u}_{\mathrm{R}}\\ \boldsymbol{u}_{\mathrm{P}}\end{pmatrix}=\boldsymbol{Z}'\begin{pmatrix}\boldsymbol{u}_{\mathrm{C}}\\ \boldsymbol{u}_{\mathrm{R}}\\ \boldsymbol{u}_{\mathrm{P}}\end{pmatrix} \tag{3.53}$$

and thus $\boldsymbol{Z}=\boldsymbol{Z}'$ as $\left(\boldsymbol{u}_{\mathrm{C}}^{\top}\ \boldsymbol{u}_{\mathrm{P}}^{\top}\ \boldsymbol{u}_{\mathrm{R}}^{\top}\right)^{\top}$ ranges over all of $\mathbb{R}^{KN_{\mathrm{E}}}$. □

Based on Proposition 3.25, we can now show that Assumption 3.22 is also a necessary condition for the existence of an explicit Dirac structure representation with Property 3.21.

Proposition 3.26 (Existence condition)

Assumption 3.22 is a necessary and sufficient condition for the existence of an input-output representation of (3.27) which has Property 3.21. This statement is true independent of the specific realization of $\boldsymbol{F}(\boldsymbol{x})$ and $\boldsymbol{E}(\boldsymbol{x})$ in (3.27) (cf. Remark 2.19).

Proof:

From the proof of Lemma 3.23 it follows that Assumption 3.22 is a sufficient condition for transferring (3.27) into an input-output representation with Property 3.21. So it is left to show that it is also necessary. To this end, we use the uniqueness of $\boldsymbol{Z}(\boldsymbol{x})$ from Proposition 3.25. For the sake of brevity, we neglect the argument $\boldsymbol{x}$ and the supplement "for all $\boldsymbol{x}\in\mathbb{X}$" in this proof. Moreover, we give a shorthand to two matrices:

$$\boldsymbol{X}=\left(\boldsymbol{F}_{\mathrm{C},1}\quad \boldsymbol{E}_{\mathrm{C},2}\quad \boldsymbol{F}_{\mathrm{R},1}\quad \boldsymbol{E}_{\mathrm{R},2}\quad \boldsymbol{E}_{\mathrm{Sf}}\quad \boldsymbol{F}_{\mathrm{Se}}\right)\in\mathbb{R}^{KN_{\mathrm{E}}\times KN_{\mathrm{E}}}, \tag{3.54a}$$
$$\boldsymbol{Y}=\left(\boldsymbol{E}_{\mathrm{C},1}\quad \boldsymbol{F}_{\mathrm{C},2}\quad \boldsymbol{E}_{\mathrm{R},1}\quad \boldsymbol{F}_{\mathrm{R},2}\quad \boldsymbol{F}_{\mathrm{Sf}}\quad \boldsymbol{E}_{\mathrm{Se}}\right)\in\mathbb{R}^{KN_{\mathrm{E}}\times KN_{\mathrm{E}}}. \tag{3.54b}$$

Assume we can write $\mathbb{D}$ in both forms (3.27) and (3.36). Moreover, Assumption 3.22 is fulfilled if $\boldsymbol{X}$ has full rank. Note that in the situation of Lemma 3.23 we have $\boldsymbol{Z}=-\boldsymbol{X}^{-1}\boldsymbol{Y}$ which gives a hint that we should prove and use $\boldsymbol{X}\boldsymbol{Z}=-\boldsymbol{Y}$

along the way. As an element $(\boldsymbol{f}, \boldsymbol{e})$ of $\mathbb{D}$ fulfills the equations in (3.27), we have

$$\boldsymbol{F}\begin{pmatrix} -\boldsymbol{f}_{\mathrm{C},1} \\ -\boldsymbol{f}_{\mathrm{C},2} \\ -\boldsymbol{f}_{\mathrm{R},1} \\ -\boldsymbol{f}_{\mathrm{R},2} \\ \boldsymbol{f}_{\mathrm{Sf}} \\ \boldsymbol{f}_{\mathrm{Se}} \end{pmatrix} + \boldsymbol{E}\begin{pmatrix} \boldsymbol{e}_{\mathrm{C},1} \\ \boldsymbol{e}_{\mathrm{C},2} \\ \boldsymbol{e}_{\mathrm{R},1} \\ \boldsymbol{e}_{\mathrm{R},2} \\ \boldsymbol{e}_{\mathrm{Sf}} \\ \boldsymbol{e}_{\mathrm{Se}} \end{pmatrix} = \boldsymbol{0} \tag{3.55}$$

or equivalently—after reordering[11]:

$$\boldsymbol{X}\begin{pmatrix} \boldsymbol{y}_{\mathrm{C}} \\ \boldsymbol{y}_{\mathrm{R}} \\ \boldsymbol{y}_{\mathrm{P}} \end{pmatrix} = -\boldsymbol{Y}\begin{pmatrix} \boldsymbol{u}_{\mathrm{C}} \\ \boldsymbol{u}_{\mathrm{R}} \\ \boldsymbol{u}_{\mathrm{P}} \end{pmatrix}. \tag{3.56}$$

The same element $(\boldsymbol{f}, \boldsymbol{e})$ also fulfills (3.36), i.e., we have

$$\begin{pmatrix} \boldsymbol{y}_{\mathrm{C}} \\ \boldsymbol{y}_{\mathrm{R}} \\ \boldsymbol{y}_{\mathrm{P}} \end{pmatrix} = \boldsymbol{Z}\begin{pmatrix} \boldsymbol{u}_{\mathrm{C}} \\ \boldsymbol{u}_{\mathrm{R}} \\ \boldsymbol{u}_{\mathrm{P}} \end{pmatrix}, \tag{3.57}$$

where $\boldsymbol{Z}$ is unique according to Proposition 3.25. By multiplying with $\boldsymbol{X}$ from right we obtain

$$\boldsymbol{X}\begin{pmatrix} \boldsymbol{y}_{\mathrm{C}} \\ \boldsymbol{y}_{\mathrm{R}} \\ \boldsymbol{y}_{\mathrm{P}} \end{pmatrix} = \boldsymbol{X}\boldsymbol{Z}\begin{pmatrix} \boldsymbol{u}_{\mathrm{C}} \\ \boldsymbol{u}_{\mathrm{R}} \\ \boldsymbol{u}_{\mathrm{P}} \end{pmatrix}. \tag{3.58}$$

Combining (3.56) and (3.58) yields

$$\boldsymbol{X}\boldsymbol{Z}\begin{pmatrix} \boldsymbol{u}_{\mathrm{C}} \\ \boldsymbol{u}_{\mathrm{R}} \\ \boldsymbol{u}_{\mathrm{P}} \end{pmatrix} = -\boldsymbol{Y}\begin{pmatrix} \boldsymbol{u}_{\mathrm{C}} \\ \boldsymbol{u}_{\mathrm{R}} \\ \boldsymbol{u}_{\mathrm{P}} \end{pmatrix}. \tag{3.59}$$

[11] cf. (B.5) and (B.6)

Whence $\boldsymbol{X}\boldsymbol{Z} = -\boldsymbol{Y}$ since $(\boldsymbol{u}_{\mathrm{C}}^\top \ \boldsymbol{u}_{\mathrm{R}}^\top \ \boldsymbol{u}_{\mathrm{P}}^\top)^\top$ ranges over all of $\mathbb{R}^{KN_{\mathrm{E}}}$. Let us now investigate the rank of $\boldsymbol{X}$. First, note that $\mathrm{im}\,\boldsymbol{X} = \mathrm{im}(\boldsymbol{X}\ \boldsymbol{X}\boldsymbol{Z})$ as $\mathrm{im}\,\boldsymbol{X}\boldsymbol{Z} \subseteq \mathrm{im}\,\boldsymbol{X}$. From this the statement that $\boldsymbol{X}$ has full rank follows:

$$\mathrm{rank}\,\boldsymbol{X} = \mathrm{rank}\begin{pmatrix}\boldsymbol{X} & \boldsymbol{X}\boldsymbol{Z}\end{pmatrix} \overset{(3.59)}{=} \mathrm{rank}\begin{pmatrix}\boldsymbol{X} & -\boldsymbol{Y}\end{pmatrix} \overset{(3.54)}{=} \mathrm{rank}\begin{pmatrix}\boldsymbol{F} & \boldsymbol{E}\end{pmatrix} \overset{(2.3b)}{=} KN_{\mathrm{E}}. \quad (3.60)$$

Note that (3.60) holds for any realization of $\boldsymbol{F}$ and $\boldsymbol{E}$. Moreover, every submatrix in $\boldsymbol{X}$ must have full column rank. In particular, Assumption 3.22 holds. □

Lemma 3.23 presents a method which allows to convert the Dirac structure (3.27) to an explicit form (3.36). The independence of source elements in Assumption 3.22 was proven to be necessary and sufficient for the existence of (3.36). Note that Lemma 3.23 does not make any assumptions on the presence of dependent storages. For the aim of deriving an explicit PHS, dependent storages are particularly unpleasant: they contribute to the energy in a system without having a dedicated representation in the system's state vector. Systems with dependent storages lead in general to mathematical models in forms of DAEs [Borutzky, 2010, p. 142]. Therefore, we now consider an important special case of (3.36) which—in addition to independent sources—also requires independent storages. Later, we will relax the latter restriction and allow for dependent storages.

Assumption 3.27 (Independent storages and sources)
The matrices in (3.27) fulfill

$$\mathrm{rank}\left(\boldsymbol{F}_{\mathrm{C}}(\boldsymbol{x})\ \boldsymbol{E}_{\mathrm{Sf}}(\boldsymbol{x})\ \boldsymbol{F}_{\mathrm{Se}}(\boldsymbol{x})\right) = K\left(N_{\mathrm{C}} + N_{\mathrm{Sf}} + N_{\mathrm{Se}}\right), \quad \forall \boldsymbol{x} \in \mathbb{X}. \quad (3.61)$$

Note that Assumption 3.27 implies Assumption 3.22. The subsequent corollary makes use of Assumption 3.27 and addresses an important special case of Lemma 3.23.

Corollary 3.28 (Input-output representation under Assumption 3.27)
Consider a Dirac structure (3.27) which fulfills Assumption 3.27. The Dirac structure (3.27) can then be formulated in an input-output representation

$$\mathbb{D}(\boldsymbol{x}) = \{(\begin{pmatrix} \boldsymbol{f}_{\mathrm{C}} \\ \boldsymbol{f}_{\mathrm{R}} \\ \boldsymbol{f}_{\mathrm{Sf}} \\ \boldsymbol{f}_{\mathrm{Se}} \end{pmatrix}, \begin{pmatrix} \boldsymbol{e}_{\mathrm{C}} \\ \boldsymbol{e}_{\mathrm{R}} \\ \boldsymbol{e}_{\mathrm{Sf}} \\ \boldsymbol{e}_{\mathrm{Se}} \end{pmatrix}) \in \mathbb{R}^{KN_{\mathrm{E}}} \times \mathbb{R}^{KN_{\mathrm{E}}} \mid \begin{pmatrix} \boldsymbol{y}_{\mathrm{C}} \\ \boldsymbol{y}_{\mathrm{R}} \\ \boldsymbol{y}_{\mathrm{P}} \end{pmatrix} = \underbrace{\begin{pmatrix} \boldsymbol{Z}_{\mathrm{CC}}(\boldsymbol{x}) & -\boldsymbol{Z}_{\mathrm{CR}}(\boldsymbol{x}) & -\boldsymbol{Z}_{\mathrm{CP}}(\boldsymbol{x}) \\ \boldsymbol{Z}_{\mathrm{CR}}^{\top}(\boldsymbol{x}) & \boldsymbol{Z}_{\mathrm{RR}}(\boldsymbol{x}) & -\boldsymbol{Z}_{\mathrm{RP}}(\boldsymbol{x}) \\ \boldsymbol{Z}_{\mathrm{CP}}^{\top}(\boldsymbol{x}) & \boldsymbol{Z}_{\mathrm{RP}}^{\top}(\boldsymbol{x}) & \boldsymbol{Z}_{\mathrm{PP}}(\boldsymbol{x}) \end{pmatrix}}_{\boldsymbol{Z}(\boldsymbol{x})} \begin{pmatrix} \boldsymbol{u}_{\mathrm{C}} \\ \boldsymbol{u}_{\mathrm{R}} \\ \boldsymbol{u}_{\mathrm{P}} \end{pmatrix} \}, \quad (3.62)$$

where $\boldsymbol{Z}(\boldsymbol{x}) = -\boldsymbol{Z}^{\top}(\boldsymbol{x})$ *for all* $\boldsymbol{x} \in \mathbb{X}$ *and*

$$\boldsymbol{u}_{\mathrm{C}} = \boldsymbol{e}_{\mathrm{C}}, \qquad \boldsymbol{u}_{\mathrm{R}} = \begin{pmatrix} \boldsymbol{e}_{\mathrm{R},1} \\ -\boldsymbol{f}_{\mathrm{R},2} \end{pmatrix}, \qquad \boldsymbol{u}_{\mathrm{P}} = \begin{pmatrix} \boldsymbol{f}_{\mathrm{Sf}} \\ \boldsymbol{e}_{\mathrm{Se}} \end{pmatrix}, \quad (3.63a)$$

$$\boldsymbol{y}_{\mathrm{C}} = -\boldsymbol{f}_{\mathrm{C}}, \qquad \boldsymbol{y}_{\mathrm{R}} = \begin{pmatrix} -\boldsymbol{f}_{\mathrm{R},1} \\ \boldsymbol{e}_{\mathrm{R},2} \end{pmatrix}, \qquad \boldsymbol{y}_{\mathrm{P}} = \begin{pmatrix} \boldsymbol{e}_{\mathrm{Sf}} \\ \boldsymbol{f}_{\mathrm{Se}} \end{pmatrix}. \quad (3.63b)$$

The matrix $\boldsymbol{Z}(\boldsymbol{x})$ *is given by (3.38) with* $\boldsymbol{F}_{\mathrm{C},1}(\boldsymbol{x}) = \boldsymbol{F}_{\mathrm{C}}(\boldsymbol{x})$, $\boldsymbol{E}_{\mathrm{C},1}(\boldsymbol{x}) = \boldsymbol{E}_{\mathrm{C}}(\boldsymbol{x})$ *and* $(\boldsymbol{F}_{\mathrm{R},1}(\boldsymbol{x})\ \boldsymbol{F}_{\mathrm{R},2}(\boldsymbol{x}))$ *a splitting of* $\boldsymbol{F}_{\mathrm{R}}(\boldsymbol{x})$ *such that (3.39) is fulfilled. Moreover, Assumption 3.27 is necessary and sufficient for the existence of the representation (3.62) with vectors as in (3.63).*

Proof:
The proof of Corollary 3.28 follows directly from Lemma 3.23 under Assumption 3.27, which also shows that Assumption 3.27 is a sufficient condition. The proof for the necessity is the same as the one given for Proposition 3.26. □

Note that Corollary 3.28 requires only a splitting of R-type power variables; the splitting of C-type variables disappears.[12] The corollary is illustrated in the following example which continues the previous ones:

Example 3.29:

Recall the Dirac structure from Example 3.24 in implicit representation (3.47). Assumption 3.27 is satisfied as $\operatorname{rank}(\boldsymbol{F}_{\mathrm{C}}\ \boldsymbol{E}_{\mathrm{Sf}}\ \boldsymbol{F}_{\mathrm{Se}}(\boldsymbol{x}_1)) = 5K$ *for all* $\boldsymbol{x}_1$*. Hence, we can apply Corollary 3.28 to determine an explicit representation of (3.47). The splitting* $\boldsymbol{F}_{\mathrm{R},2} = \boldsymbol{F}_{\mathrm{R}}$ *satisfies the conditions in (3.39). By using this splitting we obtain the following explicit representation of (3.47):*

$$\mathbb{D}(\boldsymbol{x}_1) = \{(\begin{pmatrix} \boldsymbol{f}_{C_1} \\ \boldsymbol{f}_{C_2} \\ \boldsymbol{f}_{C_3} \\ \boldsymbol{f}_R \\ \boldsymbol{f}_{Sf} \\ \boldsymbol{f}_{Se} \end{pmatrix}, \begin{pmatrix} \boldsymbol{e}_{C_1} \\ \boldsymbol{e}_{C_2} \\ \boldsymbol{e}_{C_3} \\ \boldsymbol{e}_R \\ \boldsymbol{e}_{Sf} \\ \boldsymbol{e}_{Se} \end{pmatrix}) \in \mathbb{R}^{6K} \times \mathbb{R}^{6K} \mid$$
$$\begin{pmatrix} -\boldsymbol{f}_{C_1} \\ -\boldsymbol{f}_{C_2} \\ -\boldsymbol{f}_{C_3} \\ \boldsymbol{e}_R \\ \boldsymbol{e}_{Sf} \\ \boldsymbol{f}_{Se} \end{pmatrix} = \left(\begin{array}{ccc:c:c:c} \mathbf{0} & \mathbf{0} & \mathbf{0} & \boldsymbol{I} & \mathbf{0} & \mathbf{0} \\ \mathbf{0} & \mathbf{0} & \boldsymbol{I} & \boldsymbol{U}^{-1}(\boldsymbol{x}_1) & -\boldsymbol{I} & \mathbf{0} \\ \mathbf{0} & -\boldsymbol{I} & \mathbf{0} & \mathbf{0} & \mathbf{0} & \mathbf{0} \\ \hdashline -\boldsymbol{I} & -\boldsymbol{U}^{-1}(\boldsymbol{x}_1) & \mathbf{0} & \mathbf{0} & \mathbf{0} & \boldsymbol{I} \\ \hdashline \mathbf{0} & \boldsymbol{I} & \mathbf{0} & \mathbf{0} & \mathbf{0} & \mathbf{0} \\ \hdashline \mathbf{0} & \mathbf{0} & \mathbf{0} & -\boldsymbol{I} & \mathbf{0} & \mathbf{0} \end{array}\right) \begin{pmatrix} \boldsymbol{e}_{C_1} \\ \boldsymbol{e}_{C_2} \\ \boldsymbol{e}_{C_3} \\ -\boldsymbol{f}_R \\ \boldsymbol{f}_{Sf} \\ \boldsymbol{e}_{Se} \end{pmatrix} \}. \tag{3.64}$$

The dashed lines indicate the matrix blocks according to (3.62). Compared to (3.49) in Example 3.29, (3.64) is another explicit representation of (3.47). In particular, in (3.64) all storage flows $\boldsymbol{f}_{\mathrm{C}}$ *are on the left side of the equation system.*

The previous example illustrates that Corollary 3.28 yields an explicit representation in which the flows $\boldsymbol{f}_{\mathrm{C}}$ are on the left side of the equation system. Recalling the constitutive relations of C-type elements (3.6), in particular $\boldsymbol{f}_{\mathrm{C}} = \dot{\boldsymbol{x}}$, gives us a hint why this choice is useful: it paves the way for a dynamics equation of form $\dot{\boldsymbol{x}} = \boldsymbol{f}(\boldsymbol{x}, \boldsymbol{u})$. The next subsection is devoted to deriving such a model.

[12] In fact, this is the property with which Corollary 3.28 particularizes Lemma 3.23.

3.2.6 Formulation of an Explicit Port-Hamiltonian System

Based on the Dirac structure from Corollary 3.28 we now derive an explicit PHS. To this end, we merge the explicit form of the Dirac structure with the constitutive relations of storages and resistors.

Consider a Dirac structure of the form (3.62) with vectors as in (3.63). Let us first elaborate the constitutive relations of the storage elements and the resistive elements of the bond graph. For C-type elements, the constitutive relations read

$$\boldsymbol{y}_{\mathrm{C}} \overset{(3.63)}{=} -\boldsymbol{f}_{\mathrm{C}} \overset{(3.6)}{=} -\dot{\boldsymbol{x}}, \qquad \boldsymbol{u}_{\mathrm{C}} \overset{(3.63)}{=} \boldsymbol{e}_{\mathrm{C}} \overset{(3.6)}{=} \frac{\partial V}{\partial \boldsymbol{x}}(\boldsymbol{x}). \tag{3.65}$$

For R-type elements, the constitutive relation (3.7) from Subsection 3.2.2 read $\boldsymbol{f}_{\mathrm{R}} = \boldsymbol{D}(\boldsymbol{x})\,\boldsymbol{e}_{\mathrm{R}}$ with $\boldsymbol{D}(\boldsymbol{x}) = \boldsymbol{D}(\boldsymbol{x})^{\top} \succeq 0$. In the Dirac structure (3.62), the vectors $\boldsymbol{f}_{\mathrm{R}}$ and $\boldsymbol{e}_{\mathrm{R}}$ are split and their components are assigned to inputs $\boldsymbol{u}_{\mathrm{R}}$ and outputs $\boldsymbol{y}_{\mathrm{R}}$. In the following, make the assumption that this splitting can also be applied to $\boldsymbol{f}_{\mathrm{R}} = \boldsymbol{D}(\boldsymbol{x})\,\boldsymbol{e}_{\mathrm{R}}$ without loosing the positive semi-definiteness of $\boldsymbol{D}(\boldsymbol{x})$:

Assumption 3.30 (Resistive relation in input-output form)
The resistive relations can be reorganized as

$$\boldsymbol{u}_{\mathrm{R}} = -\tilde{\boldsymbol{R}}(\boldsymbol{x})\,\boldsymbol{y}_{\mathrm{R}}, \tag{3.66}$$

with $\tilde{\boldsymbol{R}}(\boldsymbol{x}) = \tilde{\boldsymbol{R}}(\boldsymbol{x})^{\top} \succeq 0$ *and* $\boldsymbol{u}_{\mathrm{R}}$, $\boldsymbol{y}_{\mathrm{R}}$ *as in (3.63).*

In Appendix B.3 it is argued that Assumption 3.30 is not restrictive. The negative sign in (3.66) accounts for the opposite orientations of the vectors $(\boldsymbol{f}_{\mathrm{R}}, \boldsymbol{e}_{\mathrm{R}})$ and $(\boldsymbol{u}_{\mathrm{R}}, \boldsymbol{y}_{\mathrm{R}})$ (see (3.37)). Now we can merge the Dirac structure and the constitutive relations of storages and resistors into an explicit PHS.

Lemma 3.31 (Port-Hamiltonian formulation)

Given an explicit Dirac structure (3.62) and constitutive relations of storages as in (3.65). Let Assumption 3.30 hold, that is, let the constitutive relations of resistive elements be given as in (3.66). Then, (3.62), (3.65), and (3.66) can be written as explicit input-state-output PHS of the form (2.12). The state vector and the Hamiltonian of the PHS are given by $\boldsymbol{x}$ *and* $H(\boldsymbol{x}) = V(\boldsymbol{x})$ *from (3.65), respectively. The input vector of the PHS is* $\boldsymbol{u} = \boldsymbol{u}_{\mathrm{P}}$*; the output vector is* $\boldsymbol{y} = \boldsymbol{y}_{\mathrm{P}}$*. The matrices of the PHS are calculated as:*

$$\boldsymbol{J}(\boldsymbol{x}) = -\tfrac{1}{2}\boldsymbol{Z}_{\mathrm{CR}}(\boldsymbol{x})\left(\tilde{\boldsymbol{K}}(\boldsymbol{x})\tilde{\boldsymbol{R}}(\boldsymbol{x}) - \tilde{\boldsymbol{R}}(\boldsymbol{x})\tilde{\boldsymbol{K}}^{\top}(\boldsymbol{x})\right)\boldsymbol{Z}_{\mathrm{CR}}^{\top}(\boldsymbol{x}) - \boldsymbol{Z}_{\mathrm{CC}}(\boldsymbol{x}), \tag{3.67a}$$

$$\boldsymbol{R}(\boldsymbol{x}) = \tfrac{1}{2}\boldsymbol{Z}_{\mathrm{CR}}(\boldsymbol{x})\left(\tilde{\boldsymbol{K}}(\boldsymbol{x})\tilde{\boldsymbol{R}}(\boldsymbol{x}) + \tilde{\boldsymbol{R}}(\boldsymbol{x})\tilde{\boldsymbol{K}}^{\top}(\boldsymbol{x})\right)\boldsymbol{Z}_{\mathrm{CR}}^{\top}(\boldsymbol{x}), \tag{3.67b}$$

$$\boldsymbol{G}(\boldsymbol{x}) = \tfrac{1}{2}\boldsymbol{Z}_{\mathrm{CR}}(\boldsymbol{x})\left(\tilde{\boldsymbol{K}}(\boldsymbol{x})\tilde{\boldsymbol{R}}(\boldsymbol{x}) - \tilde{\boldsymbol{R}}(\boldsymbol{x})\tilde{\boldsymbol{K}}^{\top}(\boldsymbol{x})\right)\boldsymbol{Z}_{\mathrm{RP}}(\boldsymbol{x}) + \boldsymbol{Z}_{\mathrm{CP}}(\boldsymbol{x}), \tag{3.67c}$$

$$\boldsymbol{P}(\boldsymbol{x}) = -\tfrac{1}{2}\boldsymbol{Z}_{\mathrm{CR}}(\boldsymbol{x})\left(\tilde{\boldsymbol{K}}(\boldsymbol{x})\tilde{\boldsymbol{R}}(\boldsymbol{x}) + \tilde{\boldsymbol{R}}(\boldsymbol{x})\tilde{\boldsymbol{K}}^{\top}(\boldsymbol{x})\right)\boldsymbol{Z}_{\mathrm{RP}}(\boldsymbol{x}), \tag{3.67d}$$

$$\boldsymbol{M}(\boldsymbol{x}) = \tfrac{1}{2}\boldsymbol{Z}_{\mathrm{RP}}^{\top}(\boldsymbol{x})\left(\tilde{\boldsymbol{K}}(\boldsymbol{x})\tilde{\boldsymbol{R}}(\boldsymbol{x}) - \tilde{\boldsymbol{R}}(\boldsymbol{x})\tilde{\boldsymbol{K}}^{\top}(\boldsymbol{x})\right)\boldsymbol{Z}_{\mathrm{RP}}(\boldsymbol{x}) + \boldsymbol{Z}_{\mathrm{PP}}(\boldsymbol{x}), \tag{3.67e}$$

$$\boldsymbol{S}(\boldsymbol{x}) = \tfrac{1}{2}\boldsymbol{Z}_{\mathrm{RP}}^{\top}(\boldsymbol{x})\left(\tilde{\boldsymbol{K}}(\boldsymbol{x})\tilde{\boldsymbol{R}}(\boldsymbol{x}) + \tilde{\boldsymbol{R}}(\boldsymbol{x})\tilde{\boldsymbol{K}}^{\top}(\boldsymbol{x})\right)\boldsymbol{Z}_{\mathrm{RP}}(\boldsymbol{x}), \tag{3.67f}$$

where

$$\tilde{\boldsymbol{K}}(\boldsymbol{x}) = \left(\boldsymbol{I} + \tilde{\boldsymbol{R}}(\boldsymbol{x})\boldsymbol{Z}_{\mathrm{RR}}(\boldsymbol{x})\right)^{-1}. \tag{3.67g}$$

Remark 3.32 (Existence of $\tilde{\boldsymbol{K}}$). The existence of matrix $\tilde{\boldsymbol{K}}(\boldsymbol{x})$ in (3.67g) will be discussed later in Lemma 3.33.

Proof:

The proof follows four steps: (i) we eliminate the resistive variables in (3.62); (ii) we decompose the structure obtained from (i) into symmetric and skew-symmetric parts; (iii) we substitute storage variables with (3.65); (iv) we show that the definiteness condition (2.13) holds. Again, we omit the argument $\boldsymbol{x}$ and the supplement "for all $\boldsymbol{x} \in \mathbb{X}$" for all matrices in this proof.

Substituting the second row from the equation system in (3.62) into (3.66) yields

$$\begin{aligned} \boldsymbol{u}_{\mathrm{R}} &= -\tilde{\boldsymbol{R}}\boldsymbol{Z}_{\mathrm{CR}}^{\top}\boldsymbol{u}_{\mathrm{C}} + \tilde{\boldsymbol{R}}\boldsymbol{Z}_{\mathrm{RP}}\boldsymbol{u}_{\mathrm{P}} - \tilde{\boldsymbol{R}}\boldsymbol{Z}_{\mathrm{RR}}\boldsymbol{u}_{\mathrm{R}} \\ \Leftrightarrow \left(\boldsymbol{I} + \tilde{\boldsymbol{R}}\boldsymbol{Z}_{\mathrm{RR}}\right)\boldsymbol{u}_{\mathrm{R}} &= -\tilde{\boldsymbol{R}}\boldsymbol{Z}_{\mathrm{CR}}^{\top}\boldsymbol{u}_{\mathrm{C}} + \tilde{\boldsymbol{R}}\boldsymbol{Z}_{\mathrm{RP}}\boldsymbol{u}_{\mathrm{P}} \\ \Leftrightarrow \boldsymbol{u}_{\mathrm{R}} &= -\tilde{\boldsymbol{K}}\tilde{\boldsymbol{R}}\boldsymbol{Z}_{\mathrm{CR}}^{\top}\boldsymbol{u}_{\mathrm{C}} + \tilde{\boldsymbol{K}}\tilde{\boldsymbol{R}}\boldsymbol{Z}_{\mathrm{RP}}\boldsymbol{u}_{\mathrm{P}}, \end{aligned} \tag{3.68}$$

with $\tilde{\boldsymbol{K}}$ as in (3.67g). Inserting (3.68) into the first and third row from the linear equation system in (3.62) yields

$$\begin{pmatrix} \boldsymbol{y}_{\mathrm{C}} \\ \boldsymbol{y}_{\mathrm{P}} \end{pmatrix} = \left[\begin{pmatrix} \boldsymbol{Z}_{\mathrm{CC}} & -\boldsymbol{Z}_{\mathrm{CP}} \\ \boldsymbol{Z}_{\mathrm{CP}}^{\top} & \boldsymbol{Z}_{\mathrm{PP}} \end{pmatrix} + \begin{pmatrix} \boldsymbol{Z}_{\mathrm{CR}} \\ -\boldsymbol{Z}_{\mathrm{RP}}^{\top} \end{pmatrix} \tilde{\boldsymbol{K}}\tilde{\boldsymbol{R}} \begin{pmatrix} \boldsymbol{Z}_{\mathrm{CR}}^{\top} & -\boldsymbol{Z}_{\mathrm{RP}} \end{pmatrix}\right] \begin{pmatrix} \boldsymbol{u}_{\mathrm{C}} \\ \boldsymbol{u}_{\mathrm{P}} \end{pmatrix}. \tag{3.69}$$

The first addend in the square bracket is a skew-symmetric matrix. The second addend is decomposed into a skew-symmetric and a symmetric matrix. Using this decomposition and $\tilde{\boldsymbol{R}} = \tilde{\boldsymbol{R}}^{\top}$, (3.69) reads

$$\begin{aligned} \begin{pmatrix} \boldsymbol{y}_{\mathrm{C}} \\ \boldsymbol{y}_{\mathrm{P}} \end{pmatrix} = \Bigg[&\begin{pmatrix} \boldsymbol{Z}_{\mathrm{CC}} & -\boldsymbol{Z}_{\mathrm{CP}} \\ \boldsymbol{Z}_{\mathrm{CP}}^{\top} & \boldsymbol{Z}_{\mathrm{PP}} \end{pmatrix} + \frac{1}{2}\begin{pmatrix} \boldsymbol{Z}_{\mathrm{CR}} \\ -\boldsymbol{Z}_{\mathrm{RP}}^{\top} \end{pmatrix}\left(\tilde{\boldsymbol{K}}\tilde{\boldsymbol{R}} - \tilde{\boldsymbol{R}}\tilde{\boldsymbol{K}}^{\top}\right)\begin{pmatrix} \boldsymbol{Z}_{\mathrm{CR}}^{\top} & -\boldsymbol{Z}_{\mathrm{RP}} \end{pmatrix} \\ &+ \frac{1}{2}\begin{pmatrix} \boldsymbol{Z}_{\mathrm{CR}} \\ -\boldsymbol{Z}_{\mathrm{RP}}^{\top} \end{pmatrix}\left(\tilde{\boldsymbol{K}}\tilde{\boldsymbol{R}} + \tilde{\boldsymbol{R}}\tilde{\boldsymbol{K}}^{\top}\right)\begin{pmatrix} \boldsymbol{Z}_{\mathrm{CR}}^{\top} & -\boldsymbol{Z}_{\mathrm{RP}} \end{pmatrix}\Bigg] \begin{pmatrix} \boldsymbol{u}_{\mathrm{C}} \\ \boldsymbol{u}_{\mathrm{P}} \end{pmatrix}. \end{aligned} \tag{3.70}$$

Equation (3.70) can be written as

$$\begin{pmatrix} \boldsymbol{y}_{\mathrm{C}} \\ \boldsymbol{y}_{\mathrm{P}} \end{pmatrix} = \Bigg[\underbrace{\begin{pmatrix} -\boldsymbol{J} & -\boldsymbol{G} \\ \boldsymbol{G}^{\top} & \boldsymbol{M} \end{pmatrix}}_{=\boldsymbol{\Psi}} + \underbrace{\begin{pmatrix} \boldsymbol{R} & \boldsymbol{P} \\ \boldsymbol{P}^{\top} & \boldsymbol{S} \end{pmatrix}}_{=\boldsymbol{\Theta}}\Bigg] \begin{pmatrix} \boldsymbol{u}_{\mathrm{C}} \\ \boldsymbol{u}_{\mathrm{P}} \end{pmatrix}, \tag{3.71}$$

with $\boldsymbol{J}$, $\boldsymbol{G}$, $\boldsymbol{M}$, $\boldsymbol{R}$, $\boldsymbol{P}$, $\boldsymbol{S}$ as in (3.67) and $\boldsymbol{\Psi} = -\boldsymbol{\Psi}^{\top}$, $\boldsymbol{\Theta} = \boldsymbol{\Theta}^{\top}$. Inserting (3.65) into (3.71) then yields (2.12). Using the idea of van der Schaft and Jeltsema [2014,

p. 56], we prove the positive semi-definiteness of $\boldsymbol{\Theta}$ (cf. (2.13)):

$$\begin{pmatrix}\boldsymbol{u}_\mathrm{C}^\top & \boldsymbol{u}_\mathrm{P}^\top\end{pmatrix}\boldsymbol{\Theta}\begin{pmatrix}\boldsymbol{u}_\mathrm{C}\\ \boldsymbol{u}_\mathrm{P}\end{pmatrix} = \begin{pmatrix}\boldsymbol{u}_\mathrm{C}^\top & \boldsymbol{u}_\mathrm{P}^\top\end{pmatrix}\boldsymbol{\Theta}\begin{pmatrix}\boldsymbol{u}_\mathrm{C}\\ \boldsymbol{u}_\mathrm{P}\end{pmatrix} + \underbrace{\begin{pmatrix}\boldsymbol{u}_\mathrm{C}^\top & \boldsymbol{u}_\mathrm{P}^\top\end{pmatrix}\boldsymbol{\Psi}\begin{pmatrix}\boldsymbol{u}_\mathrm{C}\\ \boldsymbol{u}_\mathrm{P}\end{pmatrix}}_{=0} \tag{3.72}$$

$$= \begin{pmatrix}\boldsymbol{u}_\mathrm{C}^\top & \boldsymbol{u}_\mathrm{P}^\top\end{pmatrix}(\boldsymbol{\Theta}+\boldsymbol{\Psi})\begin{pmatrix}\boldsymbol{u}_\mathrm{C}\\ \boldsymbol{u}_\mathrm{P}\end{pmatrix} \tag{3.73}$$

$$= \begin{pmatrix}\boldsymbol{u}_\mathrm{C}^\top & \boldsymbol{u}_\mathrm{P}^\top\end{pmatrix}\begin{pmatrix}\boldsymbol{y}_\mathrm{C}\\ \boldsymbol{y}_\mathrm{P}\end{pmatrix} \overset{(2.3a),(3.66)}{=} \boldsymbol{y}_\mathrm{R}^\top \tilde{\boldsymbol{R}} \boldsymbol{y}_\mathrm{R} \geq 0. \tag{3.74}$$

□

In the previous lemma, the calculation of the matrices of the PHS requires the existence of a matrix $\tilde{\boldsymbol{K}}(\boldsymbol{x}) = (\boldsymbol{I}+\tilde{\boldsymbol{R}}(\boldsymbol{x})\boldsymbol{Z}_\mathrm{RR}(\boldsymbol{x}))^{-1}$, where $\tilde{\boldsymbol{R}}(\boldsymbol{x}) = \tilde{\boldsymbol{R}}^\top(\boldsymbol{x}) \succeq 0$ and $\boldsymbol{Z}_\mathrm{RR}(\boldsymbol{x}) = -\boldsymbol{Z}_\mathrm{RR}^\top(\boldsymbol{x})$ for all $\boldsymbol{x} \in \mathbb{X}$. The matrix $\tilde{\boldsymbol{K}}(\boldsymbol{x})$ (or related expressions) has appeared in previous publications addressing the derivation of state-space formulations of bond graphs, e.g., Rosenberg [1971, eq. (7)], Wellstead [1979, p. 199], Donaire and Junco [2009, eq. (14)], and Lopes [2016, Remark 2]. However, to the best of our knowledge, the existence of $\tilde{\boldsymbol{K}}$ has not been discussed so far. The following lemma shows that $\tilde{\boldsymbol{K}}(\boldsymbol{x})$ always exists:

Lemma 3.33 (Existence of $\tilde{\boldsymbol{K}}$)

Let $\boldsymbol{X},\boldsymbol{Y} \in \mathbb{R}^{p\times p}$ with $\boldsymbol{X} = \boldsymbol{X}^\top \succeq 0$ and $\boldsymbol{Y} = -\boldsymbol{Y}^\top$. Then, the matrix $\boldsymbol{K} := (\boldsymbol{I}+\boldsymbol{X}\boldsymbol{Y})$ is regular. In particular $\tilde{\boldsymbol{K}} := \boldsymbol{K}^{-1}$ always exists.

Proof:

The idea of the proof is to show that (i) we can (without loss of generality) regard $\boldsymbol{X}$ to be diagonal; [13] (ii) the matrix $\boldsymbol{K}$ is invertible. For (ii) we investigate first the case of $\boldsymbol{X}$ being positive-definite. Afterwards, we generalize to the case of $\boldsymbol{X}$ being positive semi-definite.

Recall that $\boldsymbol{X}$ is symmetric and positive semi-definite and that $\boldsymbol{Y}$ is skew-symmetric. Without loss of generality we may assume $\boldsymbol{X}$ to be diagonal. Indeed, since $\boldsymbol{X}$ is a symmetric and real matrix, there exists (by the Spectral Theorem) an or-

[13] Note that this does not immediately follow from $\boldsymbol{X} = \boldsymbol{X}^\top \succeq 0$ but must be analyzed in relation to $\boldsymbol{I}+\boldsymbol{X}\boldsymbol{Y}$.

thogonal matrix $\boldsymbol{T} \in \mathbb{O}(p)$ such that $\boldsymbol{T}\boldsymbol{X}\boldsymbol{T}^\top$ is diagonal. Moreover, $\boldsymbol{I}+\boldsymbol{X}\boldsymbol{Y}$ is invertible if and only if $\boldsymbol{T}(\boldsymbol{I}+\boldsymbol{X}\boldsymbol{Y})\boldsymbol{T}^\top = \boldsymbol{I}+(\boldsymbol{T}\boldsymbol{X}\boldsymbol{T}^\top)(\boldsymbol{T}\boldsymbol{Y}\boldsymbol{T}^\top) = \boldsymbol{I}+\tilde{\boldsymbol{X}}\tilde{\boldsymbol{Y}}$ is invertible, where $\tilde{\boldsymbol{X}} = \boldsymbol{T}\boldsymbol{X}\boldsymbol{T}^\top$ is diagonal and positive semi-definite and $\tilde{\boldsymbol{Y}} = \boldsymbol{T}\boldsymbol{Y}\boldsymbol{T}^\top$ is skew-symmetric. Thus, we can assume $\boldsymbol{X}$ to be diagonal in the remainder of the proof.
The matrix $\boldsymbol{I}+\boldsymbol{X}\boldsymbol{Y}$ is regular if and only if 0 is not an eigenvalue of it, that is if -1 is not an eigenvalue of $\boldsymbol{X}\boldsymbol{Y}$. We will show that the only possible real-valued eigenvalue of $\boldsymbol{X}\boldsymbol{Y}$ is 0. Throughout this proof we use $\operatorname{Spec}(\boldsymbol{X})$ to denote the (real) spectrum of $\boldsymbol{X}$, i.e., the set of real eigenvalues of $\boldsymbol{X}$.

- Case 1: $\boldsymbol{X}$ is positive-definite. Let $\sqrt{\boldsymbol{X}}$ be a diagonal matrix which is a square root of $\boldsymbol{X}$, i.e., $\sqrt{\boldsymbol{X}}\sqrt{\boldsymbol{X}} = \boldsymbol{X}$. Such a matrix exists and is invertible since $\boldsymbol{X}$ is diagonal and positive-definite. Because the spectrum of a matrix is invariant under conjugation, we have

$$\begin{aligned}\operatorname{Spec}(\boldsymbol{X}\boldsymbol{Y}) &= \operatorname{Spec}\left(\sqrt{\boldsymbol{X}}^{-1}\boldsymbol{X}\boldsymbol{Y}\sqrt{\boldsymbol{X}}\right) \\ &= \operatorname{Spec}\left(\sqrt{\boldsymbol{X}}\boldsymbol{Y}\sqrt{\boldsymbol{X}}\right) = \operatorname{Spec}\left(\sqrt{\boldsymbol{X}}\boldsymbol{Y}\sqrt{\boldsymbol{X}}^\top\right) \subseteq \{0\}, \qquad (3.75)\end{aligned}$$

  where the last inclusion holds since $\sqrt{\boldsymbol{X}}\boldsymbol{Y}\sqrt{\boldsymbol{X}}^\top$ is real and skew-symmetric. Thus, -1 is not an eigenvalue of $\boldsymbol{X}\boldsymbol{Y}$ and $\boldsymbol{I}+\boldsymbol{X}\boldsymbol{Y}$ is invertible.

- Case 2: $\boldsymbol{X}$ is (genuine) positive semi-definite. By the same conjugation argument as at the beginning of the proof (this time with a permutation matrix) we may assume without loss of generality that $\boldsymbol{X}$ is of the form

$$\boldsymbol{X} = \begin{pmatrix} \boldsymbol{X}' & \mathbf{0} \\ \mathbf{0} & \mathbf{0} \end{pmatrix}, \qquad (3.76)$$

  where $\boldsymbol{X}' \in \mathbb{R}^{\ell\times\ell}$ is a positive-definite diagonal matrix. With the same block decomposition we write $\boldsymbol{Y}$ as

$$\boldsymbol{Y} = \begin{pmatrix} \boldsymbol{Y}' & \boldsymbol{Y}'' \\ * & * \end{pmatrix}, \qquad \text{where } \boldsymbol{Y}' \in \mathbb{R}^{\ell\times\ell}. \qquad (3.77)$$

We have

$$\boldsymbol{X}\boldsymbol{Y} = \begin{pmatrix} \boldsymbol{X}'\boldsymbol{Y}' & \boldsymbol{X}'\boldsymbol{Y}'' \\ \boldsymbol{0} & \boldsymbol{0} \end{pmatrix}. \tag{3.78}$$

Thus, $\operatorname{Spec}(\boldsymbol{X}\boldsymbol{Y}) = \operatorname{Spec}(\boldsymbol{X}'\boldsymbol{Y}') \cup \operatorname{Spec}(\boldsymbol{0}) \subseteq \{0\}$, where the last inclusion uses case 1 applied to $\boldsymbol{X}'\boldsymbol{Y}'$. Hence, $\boldsymbol{I} + \boldsymbol{X}\boldsymbol{Y}$ is invertible.

□

Let us illustrate the insights from this subsection by deriving a PHS for the example system.

Example 3.34:

Consider the Dirac structure from Example 3.29 in explicit representation (3.64). From Example 3.7 recall the constitutive relations of storages and resistors: the three storage elements are described by an arbitrary differentiable, non-negative storage function $V(\boldsymbol{x})$; the resistive element has been specified by $\boldsymbol{f}_{\mathrm{R}} = \boldsymbol{D}\boldsymbol{e}_{\mathrm{R}}$ where $\boldsymbol{D} = \boldsymbol{D}^{\top} \succ 0$ arbitrary. For an input-output splitting of R*-type variables as in* (3.64), *we can write*

$$\boldsymbol{f}_{\mathrm{R}} = \boldsymbol{D}\boldsymbol{e}_{\mathrm{R}} \overset{(3.64)}{\Leftrightarrow} \boldsymbol{u}_{\mathrm{R}} = -\underbrace{\boldsymbol{D}}_{=\tilde{\boldsymbol{R}}} \boldsymbol{y}_{\mathrm{R}}. \tag{3.79}$$

Thus, Assumption 3.30 is fulfilled. With the calculation rules from Lemma 3.31, we then obtain an explicit PHS of the form (2.12):

$$\begin{pmatrix}\dot{\boldsymbol{x}}_1\\ \dot{\boldsymbol{x}}_2\\ \dot{\boldsymbol{x}}_3\end{pmatrix} = \left(\begin{pmatrix}\mathbf{0} & \mathbf{0} & \mathbf{0}\\ \mathbf{0} & \mathbf{0} & -\boldsymbol{I}\\ \mathbf{0} & \boldsymbol{I} & \mathbf{0}\end{pmatrix} - \begin{pmatrix}\boldsymbol{D} & \boldsymbol{D}\boldsymbol{U}^{-1}(\boldsymbol{x}_1) & \mathbf{0}\\ \boldsymbol{U}^{-1}(\boldsymbol{x}_1)\boldsymbol{D} & \boldsymbol{U}^{-1}(\boldsymbol{x}_1)\boldsymbol{D}\boldsymbol{U}^{-1}(\boldsymbol{x}_1) & \mathbf{0}\\ \mathbf{0} & \mathbf{0} & \mathbf{0}\end{pmatrix}\right)\frac{\partial H}{\partial \boldsymbol{x}}(\boldsymbol{x})$$
$$+\left(\begin{pmatrix}\mathbf{0} & \mathbf{0}\\ \boldsymbol{I} & \mathbf{0}\\ \mathbf{0} & \mathbf{0}\end{pmatrix} - \begin{pmatrix}\mathbf{0} & -\boldsymbol{D}\\ \mathbf{0} & -\boldsymbol{U}^{-1}(\boldsymbol{x}_1)\boldsymbol{D}\\ \mathbf{0} & \mathbf{0}\end{pmatrix}\right)\begin{pmatrix}\boldsymbol{f}_{Sf}\\ \boldsymbol{e}_{Se}\end{pmatrix} \tag{3.80a}$$

$$\begin{pmatrix}\boldsymbol{e}_{Sf}\\ \boldsymbol{f}_{Se}\end{pmatrix} = \left(\begin{pmatrix}\mathbf{0} & \mathbf{0}\\ \boldsymbol{I} & \mathbf{0}\\ \mathbf{0} & \mathbf{0}\end{pmatrix} + \begin{pmatrix}\mathbf{0} & -\boldsymbol{D}\\ \mathbf{0} & -\boldsymbol{U}^{-1}(\boldsymbol{x}_1)\boldsymbol{D}\\ \mathbf{0} & \mathbf{0}\end{pmatrix}\right)^{\top}\frac{\partial H}{\partial \boldsymbol{x}}(\boldsymbol{x}) + \begin{pmatrix}\mathbf{0} & \mathbf{0}\\ \mathbf{0} & \boldsymbol{D}\end{pmatrix}\begin{pmatrix}\boldsymbol{f}_{Sf}\\ \boldsymbol{e}_{Se}\end{pmatrix}. \tag{3.80b}$$

In (3.80) we have $H(\boldsymbol{x}) = V(\boldsymbol{x})$, $\boldsymbol{M}(\boldsymbol{x}) = \mathbf{0}$, $\boldsymbol{U}(\boldsymbol{x}_1) = \exp(-\kappa \operatorname{diag}(\boldsymbol{x}_1))$, and $\boldsymbol{x} \in \mathbb{X}$. The variables from the source elements are correctly allocated to inputs and outputs; hence, the (3.80) has Property 3.8. By the symmetry of $\boldsymbol{D}$ and $\boldsymbol{U}$ it can be seen that the matrices in (3.80) indeed have the desired symmetry and skew-symmetry properties from Definition 2.23.

Investigating $\boldsymbol{R}(\boldsymbol{x}_1)$ for the case $\boldsymbol{D} = \boldsymbol{I}$ gives:

$$\begin{pmatrix}\boldsymbol{x}_1^\top & \boldsymbol{x}_2^\top & \boldsymbol{x}_3^\top\end{pmatrix}\boldsymbol{R}(\boldsymbol{x}_1)\begin{pmatrix}\boldsymbol{x}_1\\ \boldsymbol{x}_2\\ \boldsymbol{x}_3\end{pmatrix} = \|\boldsymbol{x}_1 + \boldsymbol{U}^{-1}(\boldsymbol{x}_1)\boldsymbol{x}_2\|^2 \geq 0, \quad \forall \boldsymbol{x} \in \mathbb{X}. \tag{3.81}$$

Hence, for $\boldsymbol{D} = \boldsymbol{I}$ the matrix $\boldsymbol{R}(\boldsymbol{x}_1)$ is positive semi-definite which verifies the definiteness condition 2.13.

The above example illustrates how the methods from this subsection can be used to derive an explicit PHS based on a Dirac structure and the constitutive relations of storages and resistors. In the next subsection, these methods are embedded into a main theorem which addresses the derivation of an explicit PHS from a bond graph.

3.2.7 Theorem and Algorithm

This subsection first provides a theorem which summarizes the insights from subsections 3.2.3 to 3.2.6. The theorem addresses the existence of an explicit port-Hamiltonian formulation of a bond graph. Afterwards, we assemble the corresponding methods in an algorithm which calculates an explicit PHS based on a given bond graph. The theorem and the algorithm are major results of this chapter.

Theorem 3.35 (Explicit Port-Hamiltonian Formulation of Bond Graphs)
Given a K-dimensional bond graph as in Definition 3.5. The junction structure of the bond graph can be described by a Dirac structure in implicit form:

$$\mathbb{D}(\boldsymbol{x}) = \{(\begin{pmatrix} \boldsymbol{f}_{\mathrm{C}} \\ \boldsymbol{f}_{\mathrm{R}} \\ \boldsymbol{f}_{\mathrm{Sf}} \\ \boldsymbol{f}_{\mathrm{Se}} \end{pmatrix}, \begin{pmatrix} \boldsymbol{e}_{\mathrm{C}} \\ \boldsymbol{e}_{\mathrm{R}} \\ \boldsymbol{e}_{\mathrm{Sf}} \\ \boldsymbol{e}_{\mathrm{Se}} \end{pmatrix}) \in \mathbb{R}^{KN_{\mathrm{E}}} \times \mathbb{R}^{KN_{\mathrm{E}}} \mid$$

$$\left(\boldsymbol{F}_{\mathrm{C}}(\boldsymbol{x})\ \boldsymbol{F}_{\mathrm{R}}(\boldsymbol{x})\ \boldsymbol{F}_{\mathrm{Sf}}(\boldsymbol{x})\ \boldsymbol{F}_{\mathrm{Se}}(\boldsymbol{x})\right) \begin{pmatrix} -\boldsymbol{f}_{\mathrm{C}} \\ -\boldsymbol{f}_{\mathrm{R}} \\ \boldsymbol{f}_{\mathrm{Sf}} \\ \boldsymbol{f}_{\mathrm{Se}} \end{pmatrix}$$

$$+ \left(\boldsymbol{E}_{\mathrm{C}}(\boldsymbol{x})\ \boldsymbol{E}_{\mathrm{R}}(\boldsymbol{x})\ \boldsymbol{E}_{\mathrm{Sf}}(\boldsymbol{x})\ \boldsymbol{E}_{\mathrm{Se}}(\boldsymbol{x})\right) \begin{pmatrix} \boldsymbol{e}_{\mathrm{C}} \\ \boldsymbol{e}_{\mathrm{R}} \\ \boldsymbol{e}_{\mathrm{Sf}} \\ \boldsymbol{e}_{\mathrm{Se}} \end{pmatrix} = \boldsymbol{0}\}, \quad (3.82)$$

where $\boldsymbol{f}_{\alpha} = (\boldsymbol{f}_i) \in \mathbb{R}^{KN_{\alpha}}$, $\boldsymbol{e}_{\alpha} = (\boldsymbol{e}_i) \in \mathbb{R}^{KN_{\alpha}}$ *for all* $i \in \mathbb{V}_{\alpha}$ *and* $\boldsymbol{F}_{\alpha}(\boldsymbol{x}), \boldsymbol{E}_{\alpha}(\boldsymbol{x}) \in \mathbb{R}^{KN_{\mathrm{E}} \times KN_{\alpha}}$ *with* $\alpha \in \{\mathrm{C}, \mathrm{R}, \mathrm{Sf}, \mathrm{Se}\}$. *Let*

$$\mathrm{rank}\left(\boldsymbol{F}_{\mathrm{C}}(\boldsymbol{x})\ \boldsymbol{E}_{\mathrm{Sf}}(\boldsymbol{x})\ \boldsymbol{F}_{\mathrm{Se}}(\boldsymbol{x})\right) = K\left(N_{\mathrm{C}} + N_{\mathrm{Sf}} + N_{\mathrm{Se}}\right), \quad \forall \boldsymbol{x} \in \mathbb{X}. \quad (3.83)$$

hold (i.e., Assumption 3.27). Then, (3.82) can be formulated in an explicit representation

$$\mathbb{D}(\boldsymbol{x}) = \{(\begin{pmatrix} \boldsymbol{f}_{\mathrm{C}} \\ \boldsymbol{f}_{\mathrm{R}} \\ \boldsymbol{f}_{\mathrm{Sf}} \\ \boldsymbol{f}_{\mathrm{Se}} \end{pmatrix}, \begin{pmatrix} \boldsymbol{e}_{\mathrm{C}} \\ \boldsymbol{e}_{\mathrm{R}} \\ \boldsymbol{e}_{\mathrm{Sf}} \\ \boldsymbol{e}_{\mathrm{Se}} \end{pmatrix}) \in \mathbb{R}^{KN_{\mathrm{E}}} \times \mathbb{R}^{KN_{\mathrm{E}}} \mid \begin{pmatrix} \boldsymbol{y}_{\mathrm{C}} \\ \boldsymbol{y}_{\mathrm{R}} \\ \boldsymbol{y}_{\mathrm{P}} \end{pmatrix} = \boldsymbol{Z}(\boldsymbol{x}) \begin{pmatrix} \boldsymbol{u}_{\mathrm{C}} \\ \boldsymbol{u}_{\mathrm{R}} \\ \boldsymbol{u}_{\mathrm{P}} \end{pmatrix} \}, \quad (3.84)$$

where $\boldsymbol{Z}(\boldsymbol{x}) = -\boldsymbol{Z}^{\top}(\boldsymbol{x}) \in \mathbb{R}^{KN_{\mathrm{E}}}$ *for all* $\boldsymbol{x} \in \mathbb{X}$ *with* $\boldsymbol{u}_{\mathrm{C}} = \boldsymbol{e}_{\mathrm{C}}$, $\boldsymbol{u}_{\mathrm{R}} = (\boldsymbol{e}_{\mathrm{R},1}^{\top}\ \boldsymbol{f}_{\mathrm{R},2}^{\top})^{\top}$, $\boldsymbol{u}_{\mathrm{P}} = (\boldsymbol{f}_{\mathrm{Sf}}^{\top}\ \boldsymbol{e}_{\mathrm{Se}}^{\top})^{\top}$, $\boldsymbol{y}_{\mathrm{C}} = -\boldsymbol{f}_{\mathrm{C}}$, $\boldsymbol{y}_{\mathrm{R}} = (-\boldsymbol{f}_{\mathrm{R},1}^{\top}\ \boldsymbol{e}_{\mathrm{R},2}^{\top})^{\top}$, *and* $\boldsymbol{y}_{\mathrm{P}} = (\boldsymbol{e}_{\mathrm{Sf}}^{\top}\ \boldsymbol{f}_{\mathrm{Se}}^{\top})^{\top}$. *Suppose the resistive relations can be reorganized as* $\boldsymbol{u}_{\mathrm{R}} = -\tilde{\boldsymbol{R}}(\boldsymbol{x})\,\boldsymbol{y}_{\mathrm{R}}$ *with* $\tilde{\boldsymbol{R}}(\boldsymbol{x}) = \tilde{\boldsymbol{R}}(\boldsymbol{x})^{\top} \succeq 0$ *(i.e., Assumption 3.30). The bond graph can then be formulated as an explicit PHS (cf. Definition 2.23)*

$$\dot{\boldsymbol{x}} = (\boldsymbol{J}(\boldsymbol{x}) - \boldsymbol{R}(\boldsymbol{x}))\frac{\partial H}{\partial \boldsymbol{x}}(\boldsymbol{x}) + (\boldsymbol{G}(\boldsymbol{x}) - \boldsymbol{P}(\boldsymbol{x}))\,\boldsymbol{u}, \quad (3.85a)$$

$$\boldsymbol{y} = (\boldsymbol{G}(\boldsymbol{x}) + \boldsymbol{P}(\boldsymbol{x}))^{\top}\frac{\partial H}{\partial \boldsymbol{x}}(\boldsymbol{x}) + (\boldsymbol{M}(\boldsymbol{x}) + \boldsymbol{S}(\boldsymbol{x}))\,\boldsymbol{u}, \quad (3.85b)$$

that has Property 3.8. Assumption 3.27 and Assumption 3.30 together form a sufficient condition for formulating the bond graph as an explicit PHS (3.85) with Property 3.8. A necessary condition for the existence of such a model is given by $\mathrm{rank}\,(\boldsymbol{E}_{\mathrm{Sf}}(\boldsymbol{x})\ \boldsymbol{F}_{\mathrm{Se}}(\boldsymbol{x})) = K(N_{\mathrm{Sf}} + N_{\mathrm{Se}})$ *for all* $\boldsymbol{x} \in \mathbb{X}$ *(i.e., Assumption 3.22).*

Proof:
Consider a K-dimensional bond graph as in Definition 3.5. First, consider the set of interior elements $\mathbb{V}_{\mathrm{I}}$ with $|\mathbb{V}_{\mathrm{I}}| = N_{\mathrm{I}}$. According to Lemma 3.12, the equations of the elements of $\mathbb{V}_{\mathrm{I}}$ can be described by a corresponding set of Dirac structures $\mathbb{DS}$ with $|\mathbb{DS}| = N_{\mathrm{I}}$. From Lemma 3.17 it follows that the N_{I} elements of $\mathbb{DS}$ can be composed into a single Dirac structure in implicit form (3.82). Now let Assumption 3.27 hold. Then, (3.82) can be formulated in an explicit representation (3.84). Consider the constitutive relations (3.65) of C-type elements. Let Assumption 3.30 hold for $\boldsymbol{u}_{\mathrm{R}}, \boldsymbol{y}_{\mathrm{R}}$ as in (3.84), i.e., the constitutive relation of R-type elements can be written as (3.66). According to Lemma 3.31, the equations (3.62), (3.65), and (3.66) can then be written as PHS (3.85). From

Lemma 3.31, it follows that $\boldsymbol{u} = \boldsymbol{u}_\mathrm{P}$ and $\boldsymbol{y} = \boldsymbol{y}_\mathrm{P}$ which is why the explicit PHS has Property 3.8.
Next, we prove sufficiency and necessity for the respective assumptions. Sufficiency of assumptions 3.27 and 3.30 follows directly from the above considerations. Now for the necessity of Assumption 3.22. Property 3.8 implies Property 3.21. In Proposition 3.26 we show that Assumption 3.22 is necessary (and sufficient) for formulating the junction structure equations as explicit Dirac structure satisfying Property 3.21. In Lemma 3.31 it is shown, that the inputs and outputs of the explicit Dirac structure directly translate into the inputs and outputs of the explicit PHS. Thus, under Property 3.8, the necessity of Assumption 3.22 from Proposition 3.26 also accounts for the subsequent derivation of an explicit PHS. This concludes the proof. □

Theorem 3.35 provides two conditions, one sufficient and one necessary, for the existence of an explicit port-Hamiltonian formulation of a bond graph. The sufficient condition is composed of two subconditions, viz. (i) that the system contains no dependent sources and no dependent storages (i.e., Assumption 3.27) and (ii) that the constitutive relations of energy-dissipating elements can be formulated in a suitable input-output representation (i.e., Assumption 3.30). The necessary existence condition requires the system to contain no dependent sources (i.e., Assumption 3.22).

The methods associated to Theorem 3.35 constitute an algorithm which can be used to generate an explicit PHS from a given bond graph. A pseudo code listing is given in Algorithm 3.36. The input of the algorithm is a formal bond graph as defined in Definition 3.5. The output is a complete specification of the resulting port-Hamiltonian model, viz. the matrices, Hamiltonian, state vector, input vector, and output vector of the PHS.

It is straightforward to implement Algorithm 3.36 in a computer algebra system. Based on such an implementation, one can compute an explicit PHS from a bond graph. This computation runs fully automatic, i.e., except for the specification of the bond graph, it requires no action from the user. Thus, provided the existence conditions from Theorem 3.35 are satisfied, we can automatically generate an explicit PHS from a bond graph.

A limitation of Algorithm 3.36 is that it cannot be applied to systems with dependent storages as such systems violate Assumption 3.27. Dependent storages occur for example in multi-phase power systems with Y-connected inductive loads and in mechanic system with rigidly coupled masses. Of course, we are also interested in an automated modeling of such systems. Hence, in the following subsection, we relax Assumption 3.27 in order to enable a generation of PHSs from bond graphs possibly containing dependent storages.

Algorithm 3.36 Port-Hamiltonian formulation of a bond graph

Input: K-dimensional bond graph (3.12)
// Methods from Subsection 3.2.3
for all $i \in \mathbb{V}_\mathrm{I}$ **do**
    compute $\boldsymbol{F}_i(\boldsymbol{x}), \boldsymbol{E}_i(\boldsymbol{x})$ according to (3.20)
    construct $\mathbb{D}_i(\boldsymbol{x})$ as in (3.19)
end for
// Methods from Subsection 3.2.4
for all $i \in \mathbb{V}_\mathrm{I}$ **do**
    bring $\mathbb{D}_i(\boldsymbol{x})$ to the form (3.23)
    compute $\mathbb{D}^\mathrm{IC}$ according to (3.25)
    compute $\boldsymbol{\Gamma}_i^\top(\boldsymbol{x})$ according to (3.28)
end for
$\boldsymbol{\Gamma}^\top(\boldsymbol{x}) \leftarrow (\boldsymbol{\Gamma}_i^\top(\boldsymbol{x}))$ for all $i \in \mathbb{V}_\mathrm{I}$
$\boldsymbol{\Lambda}^\top(\boldsymbol{x}) \leftarrow \ker(\boldsymbol{\Gamma}^\top(\boldsymbol{x}))$
write $\boldsymbol{\Lambda}^\top(\boldsymbol{x})$ as $(\boldsymbol{\Lambda}_i^\top(\boldsymbol{x}))$ for all $i \in \mathbb{V}_\mathrm{I}$
compute $\mathbb{D}(\boldsymbol{x})$ according to (3.29)
bring $\mathbb{D}(\boldsymbol{x})$ to the form (3.27)
// Methods from Subsection 3.2.5
if Assumption 3.22 is violated **then**
    print "Bond graph contains dependent sources. No PHS can be computed!"
    terminate
end if
if Assumption 3.27 is violated **then**
    print "Bond graph contains dependent storages or
        storages determined by sources. Algorithm terminates!"
    terminate
end if
split $\boldsymbol{F}_\mathrm{R}(\boldsymbol{x})$ such that (3.39) is fulfilled
split $\boldsymbol{E}_\mathrm{R}(\boldsymbol{x}), \boldsymbol{f}_\mathrm{R}, \boldsymbol{e}_\mathrm{R}$ in same parts as $\boldsymbol{F}_\mathrm{R}(\boldsymbol{x})$
compute $\boldsymbol{Z}(\boldsymbol{x})$ according to (3.38) with $\boldsymbol{F}_{\mathrm{C},1}(\boldsymbol{x}) = \boldsymbol{F}_\mathrm{C}(\boldsymbol{x}), \boldsymbol{E}_{\mathrm{C},1}(\boldsymbol{x}) = \boldsymbol{E}_\mathrm{C}(\boldsymbol{x})$
compute $\boldsymbol{u}_i, \boldsymbol{y}_i$ according to (3.63), $i \in \{\mathrm{C}, \mathrm{P}, \mathrm{R}\}$
compute $\mathbb{D}(\boldsymbol{x})$ as in (3.62)
// Methods from Subsection 3.2.6
if Assumption 3.30 is violated **then**
    print "No suitable input-output splitting of
        R-type elements exists. Algorithm terminates!"
    terminate
end if
bring resistive relation to form (3.66)
compute PHS matrices with (3.67)
$\boldsymbol{x} \leftarrow (\boldsymbol{x}_i), \forall i \in \mathbb{V}_\mathrm{C}$ and $H(\boldsymbol{x}) \leftarrow V(\boldsymbol{x})$
$\boldsymbol{u} \leftarrow \boldsymbol{u}_\mathrm{P}, \boldsymbol{y} \leftarrow \boldsymbol{y}_\mathrm{P}$
return explicit PHS (3.85)

3.2.8 Systems with Dependent Storages

The aim of this subsection is to develop a method which derives an explicit PHS from bond graphs with dependent storages. The leading idea is to express variables according to dependent storages as functions of variables only according to independent storages. Based hereon, we formulate the port-Hamiltonian dynamics and output in terms of the system inputs and independent storages. Portions of this section have previously been published in Pfeifer et al. [2020b].

Consider a bond graph as from Definition 3.5. In the following, we distinguish between two types of storages elements, viz. independent storages $\mathrm{C_i}$ and dependent storages $\mathrm{C_d}$.[14] Correspondingly, let us decompose the set of storages $\mathbb{V}_\mathrm{C}$ into $\mathbb{V}_{\mathrm{C_i}}$ and $\mathbb{V}_{\mathrm{C_d}}$. We then have $N_\mathrm{C} = N_{\mathrm{C_i}} + N_{\mathrm{C_d}}$, where $N_{\mathrm{C_i}} = |\mathbb{V}_{\mathrm{C_i}}|$ and $N_{\mathrm{C_d}} = |\mathbb{V}_{\mathrm{C_d}}|$. Recall Lemma 3.23: provided there are no dependent sources, the junction structure of the bond graph can always be described as a Dirac structure in input-output representation (3.36). Using the composition $\mathbb{V}_\mathrm{C} = \mathbb{V}_{\mathrm{C_i}} \cup \mathbb{V}_{\mathrm{C_d}}$, the Dirac structure (3.36) reads:

$$\mathbb{D} = \{ (\begin{pmatrix} \boldsymbol{f}_\mathrm{C} \\ \boldsymbol{f}_\mathrm{R} \\ \boldsymbol{f}_\mathrm{Sf} \\ \boldsymbol{f}_\mathrm{Se} \end{pmatrix}, \begin{pmatrix} \boldsymbol{e}_\mathrm{C} \\ \boldsymbol{e}_\mathrm{R} \\ \boldsymbol{e}_\mathrm{Sf} \\ \boldsymbol{e}_\mathrm{Se} \end{pmatrix}) \in \mathbb{R}^{KN_\mathrm{E}} \times \mathbb{R}^{KN_\mathrm{E}} \mid \begin{pmatrix} \boldsymbol{y}_{\mathrm{C_i}} \\ \boldsymbol{y}_{\mathrm{C_d}} \\ \boldsymbol{y}_\mathrm{R} \\ \boldsymbol{y}_\mathrm{P} \end{pmatrix} = \\ \begin{pmatrix} \boldsymbol{Z}_{\mathrm{C_iC_i}}(\boldsymbol{x}) & -\boldsymbol{Z}_{\mathrm{C_iC_d}}(\boldsymbol{x}) & -\boldsymbol{Z}_{\mathrm{C_iR}}(\boldsymbol{x}) & -\boldsymbol{Z}_{\mathrm{C_iP}}(\boldsymbol{x}) \\ \boldsymbol{Z}_{\mathrm{C_iC_d}}^\top(\boldsymbol{x}) & \boldsymbol{Z}_{\mathrm{C_dC_d}}(\boldsymbol{x}) & -\boldsymbol{Z}_{\mathrm{C_dR}}(\boldsymbol{x}) & -\boldsymbol{Z}_{\mathrm{C_dP}}(\boldsymbol{x}) \\ \boldsymbol{Z}_{\mathrm{C_iR}}^\top(\boldsymbol{x}) & \boldsymbol{Z}_{\mathrm{C_dR}}^\top(\boldsymbol{x}) & \boldsymbol{Z}_\mathrm{RR}(\boldsymbol{x}) & -\boldsymbol{Z}_\mathrm{RP}(\boldsymbol{x}) \\ \boldsymbol{Z}_{\mathrm{C_iP}}^\top(\boldsymbol{x}) & \boldsymbol{Z}_{\mathrm{C_dP}}^\top(\boldsymbol{x}) & \boldsymbol{Z}_\mathrm{RP}^\top(\boldsymbol{x}) & \boldsymbol{Z}_\mathrm{PP}(\boldsymbol{x}) \end{pmatrix} \begin{pmatrix} \boldsymbol{u}_{\mathrm{C_i}} \\ \boldsymbol{u}_{\mathrm{C_d}} \\ \boldsymbol{u}_\mathrm{R} \\ \boldsymbol{u}_\mathrm{P} \end{pmatrix} \}, \tag{3.86}$$

where

$$\boldsymbol{u}_{\mathrm{C_i}} = \boldsymbol{e}_{\mathrm{C_i}}, \qquad \boldsymbol{u}_{\mathrm{C_d}} = -\boldsymbol{f}_{\mathrm{C_d}}, \tag{3.87a}$$

$$\boldsymbol{y}_{\mathrm{C_i}} = -\boldsymbol{f}_{\mathrm{C_i}}, \qquad \boldsymbol{y}_{\mathrm{C_d}} = \boldsymbol{e}_{\mathrm{C_d}}, \tag{3.87b}$$

[14] The index i refers to *independent* and is not to be confused with the index I which stands for *interior*.

and the remaining vectors as in (3.37). It is well known that power variables of dependent storages are functions only of power variables of independent storages and inputs from sources [Wellstead, 1979, p. 199]. Thus, without loss of generality we may assume $\boldsymbol{Z}_{\mathrm{C_dC_d}}(\boldsymbol{x}) = \boldsymbol{0}$, $\boldsymbol{Z}_{\mathrm{C_dR}}(\boldsymbol{x}) = \boldsymbol{0}$ and write (3.86) as

$$\mathbb{D} = \{ (\begin{pmatrix} \boldsymbol{f}_{\mathrm{C}} \\ \boldsymbol{f}_{\mathrm{R}} \\ \boldsymbol{f}_{\mathrm{Sf}} \\ \boldsymbol{f}_{\mathrm{Se}} \end{pmatrix}, \begin{pmatrix} \boldsymbol{e}_{\mathrm{C}} \\ \boldsymbol{e}_{\mathrm{R}} \\ \boldsymbol{e}_{\mathrm{Sf}} \\ \boldsymbol{e}_{\mathrm{Se}} \end{pmatrix}) \in \mathbb{R}^{KN_{\mathrm{E}}} \times \mathbb{R}^{KN_{\mathrm{E}}} \mid \begin{pmatrix} \boldsymbol{y}_{\mathrm{C_i}} \\ \boldsymbol{y}_{\mathrm{C_d}} \\ \boldsymbol{y}_{\mathrm{R}} \\ \boldsymbol{y}_{\mathrm{P}} \end{pmatrix} = \begin{pmatrix} \boldsymbol{Z}_{\mathrm{C_iC_i}}(\boldsymbol{x}) & -\boldsymbol{Z}_{\mathrm{C_iC_d}}(\boldsymbol{x}) & -\boldsymbol{Z}_{\mathrm{C_iR}}(\boldsymbol{x}) & -\boldsymbol{Z}_{\mathrm{C_iP}}(\boldsymbol{x}) \\ \boldsymbol{Z}^{\top}_{\mathrm{C_iC_d}}(\boldsymbol{x}) & \boldsymbol{0} & \boldsymbol{0} & -\boldsymbol{Z}_{\mathrm{C_dP}}(\boldsymbol{x}) \\ \boldsymbol{Z}^{\top}_{\mathrm{C_iR}}(\boldsymbol{x}) & \boldsymbol{0} & \boldsymbol{Z}_{\mathrm{RR}}(\boldsymbol{x}) & -\boldsymbol{Z}_{\mathrm{RP}}(\boldsymbol{x}) \\ \boldsymbol{Z}^{\top}_{\mathrm{C_iP}}(\boldsymbol{x}) & \boldsymbol{Z}^{\top}_{\mathrm{C_dP}}(\boldsymbol{x}) & \boldsymbol{Z}^{\top}_{\mathrm{RP}}(\boldsymbol{x}) & \boldsymbol{Z}_{\mathrm{PP}}(\boldsymbol{x}) \end{pmatrix} \begin{pmatrix} \boldsymbol{u}_{\mathrm{C_i}} \\ \boldsymbol{u}_{\mathrm{C_d}} \\ \boldsymbol{u}_{\mathrm{R}} \\ \boldsymbol{u}_{\mathrm{P}} \end{pmatrix} \}. \tag{3.88}$$

As can be seen in (3.88), dependent storages can be (i) dependent on independent storages (i.e., $\boldsymbol{Z}_{\mathrm{C_iC_d}}(\boldsymbol{x}) \neq 0$) and/or (ii) determined by sources (i.e., $\boldsymbol{Z}_{\mathrm{C_dP}}(\boldsymbol{x}) \neq 0$). Wellstead [1979, p. 200] showed that case (ii) leads to mathematical models of the form $\dot{\boldsymbol{x}} = \boldsymbol{f}(\boldsymbol{x}, \boldsymbol{u}, \dot{\boldsymbol{u}}, t)$. Due to the dependence on $\dot{\boldsymbol{u}}$, no explicit PHS of the form (2.12) can be obtained for such systems. Hence, we exclude case (ii) in the following considerations. Moreover, we assume $\boldsymbol{Z}_{\mathrm{C_iC_d}}(\boldsymbol{x})$ to be constant. The following assumption summarizes the preconditions on (3.88):

Assumption 3.37 (Matrix blocks in (3.88))
In (3.88), we have $\boldsymbol{Z}_{\mathrm{C_dP}}(\boldsymbol{x}) = \boldsymbol{0}$ for all $\boldsymbol{x} \in \mathbb{X}$ and $\boldsymbol{Z}_{\mathrm{C_iC_d}}(\boldsymbol{x}) = \boldsymbol{Z}_{\mathrm{C_iC_d}} = const.$

Next, we elaborate the consequence of the decomposition $\mathbb{V}_{\mathrm{C}} = \mathbb{V}_{\mathrm{C_i}} \cup \mathbb{V}_{\mathrm{C_d}}$ to the constitutive relations of storages. With the inputs and outputs from (3.88), we can rewrite the constitutive relations (3.6) as

$$\begin{pmatrix} \boldsymbol{f}_{\mathrm{C_i}} \\ \boldsymbol{f}_{\mathrm{C_d}} \end{pmatrix} \overset{(3.87)}{=} - \begin{pmatrix} \boldsymbol{y}_{\mathrm{C_i}} \\ \boldsymbol{u}_{\mathrm{C_d}} \end{pmatrix} = \begin{pmatrix} \dot{\boldsymbol{x}}_{\mathrm{i}} \\ \dot{\boldsymbol{x}}_{\mathrm{d}} \end{pmatrix}, \tag{3.89a}$$

$$\begin{pmatrix} \boldsymbol{e}_{\mathrm{C_i}} \\ \boldsymbol{e}_{\mathrm{C_d}} \end{pmatrix} \overset{(3.87)}{=} \begin{pmatrix} \boldsymbol{u}_{\mathrm{C_i}} \\ \boldsymbol{y}_{\mathrm{C_d}} \end{pmatrix} = \begin{pmatrix} \frac{\partial V_{\mathrm{i}}}{\partial \boldsymbol{x}_{\mathrm{i}}}(\boldsymbol{x}_{\mathrm{i}}) \\ \frac{\partial V_{\mathrm{d}}}{\partial \boldsymbol{x}_{\mathrm{d}}}(\boldsymbol{x}_{\mathrm{d}}) \end{pmatrix}, \tag{3.89b}$$

where $\boldsymbol{x}_\mathrm{i} \in \mathbb{R}^{KN_{\mathrm{C_i}}}$, $\boldsymbol{x}_\mathrm{d} \in \mathbb{R}^{KN_{\mathrm{C_d}}}$ are the energy states and $V_\mathrm{i}\colon \mathbb{R}^{KN_{\mathrm{C_i}}} \to \mathbb{R}$, $V_\mathrm{d}\colon \mathbb{R}^{KN_{\mathrm{C_d}}} \to \mathbb{R}$ are the storage functions of the independent and dependent storage elements, respectively. The overall energy in the system is then given by the composite storage function $V(\boldsymbol{x}_\mathrm{i}, \boldsymbol{x}_\mathrm{d}) = V_\mathrm{i}(\boldsymbol{x}_\mathrm{i}) + V_\mathrm{d}(\boldsymbol{x}_\mathrm{d})$. The storage functions $V_\mathrm{i}(\boldsymbol{x}_\mathrm{i})$ and $V_\mathrm{d}(\boldsymbol{x}_\mathrm{d})$ are bounded from below. Moreover, so far, they are allowed to be any differentiable linear or nonlinear function. In the latter case, however, it is (in general) impossible to solve (3.88) and (3.65) for an explicit expression which relates the states of the dependent storages as a function of the states of the independent storages. As a remedy, we restrict our attention to the special case of linear storages, i.e., the case of quadratic positive-definite storage functions.

Assumption 3.38 (Quadratic storage functions)
The storage functions in (3.65) are of the form $V_\mathrm{i}(\boldsymbol{x}_\mathrm{i}) = \frac{1}{2}\boldsymbol{x}_\mathrm{i}^\top \boldsymbol{Q}_\mathrm{i}\boldsymbol{x}_\mathrm{i}$ *and* $V_\mathrm{d}(\boldsymbol{x}_\mathrm{d}) = \frac{1}{2}\boldsymbol{x}_\mathrm{d}^\top \boldsymbol{Q}_\mathrm{d}\boldsymbol{x}_\mathrm{d}$ *where* $\boldsymbol{Q}_\mathrm{i} = \boldsymbol{Q}_\mathrm{i}^\top \succ 0$ *and* $\boldsymbol{Q}_\mathrm{d} = \boldsymbol{Q}_\mathrm{d}^\top \succ 0$.

Based on assumptions 3.37 and 3.38, we express the variables of dependent storages by means of variables of the independent storages in the following lemma. This is the key lemma of this subsection.

Lemma 3.39 (Port-Hamiltonian formulation in case of dependent storages)
Given a Dirac structure in explicit representation (3.88) *that fulfills Assumption 3.37. Moreover, suppose constitutive relations of storages* (3.65) *which satisfy Assumption 3.38. For the constitutive relations of resistors, let Assumption 3.30 hold, which enables us to write them as* (3.66)*. Then,* (3.88)*,* (3.65)*, and* (3.66) *can be written as explicit input-state-output PHS of the form* (2.12)*. The input, state, and output of the PHS are given as* $\boldsymbol{u} = \boldsymbol{u}_\mathrm{P}$, $\boldsymbol{x} = \boldsymbol{x}_\mathrm{i}$, *and* $\boldsymbol{y} = \boldsymbol{y}_\mathrm{P}$*, respectively. The Hamiltonian is* $H(\boldsymbol{x}) = V_\mathrm{i}(\boldsymbol{x}_\mathrm{i}) = \frac{1}{2}\boldsymbol{x}_\mathrm{i}^\top \boldsymbol{Q}_\mathrm{i}\boldsymbol{x}_\mathrm{i}$

where $\boldsymbol{Q}_{\mathrm{i}} = \boldsymbol{Q}_{\mathrm{i}}^{\top} \succ 0$. *The matrices of the PHS are calculated as*

$$\boldsymbol{J}(\boldsymbol{x}) = \frac{1}{2}\left(\boldsymbol{Z}_{\mathrm{C_iR}}(\boldsymbol{x})\tilde{\boldsymbol{R}}(\boldsymbol{x})\tilde{\boldsymbol{K}}^{\top}(\boldsymbol{x})\boldsymbol{Z}_{\mathrm{C_iR}}^{\top}(\boldsymbol{x})\tilde{\boldsymbol{L}}^{\top}\right) - \frac{1}{2}\left(\tilde{\boldsymbol{L}}\boldsymbol{Z}_{\mathrm{C_iR}}(\boldsymbol{x})\tilde{\boldsymbol{K}}(\boldsymbol{x})\tilde{\boldsymbol{R}}(\boldsymbol{x})\boldsymbol{Z}_{\mathrm{C_iR}}^{\top}(\boldsymbol{x}) + \boldsymbol{Z}_{\mathrm{C_iC_i}}(\boldsymbol{x})\tilde{\boldsymbol{L}}^{\top} + \tilde{\boldsymbol{L}}\boldsymbol{Z}_{\mathrm{C_iC_i}}(\boldsymbol{x})\right), \tag{3.90a}$$

$$\boldsymbol{R}(\boldsymbol{x}) = \frac{1}{2}\left(\boldsymbol{Z}_{\mathrm{C_iR}}(\boldsymbol{x})\tilde{\boldsymbol{R}}(\boldsymbol{x})\tilde{\boldsymbol{K}}^{\top}(\boldsymbol{x})\boldsymbol{Z}_{\mathrm{C_iR}}^{\top}(\boldsymbol{x})\tilde{\boldsymbol{L}}^{\top}\right) + \frac{1}{2}\left(\tilde{\boldsymbol{L}}\boldsymbol{Z}_{\mathrm{C_iR}}(\boldsymbol{x})\tilde{\boldsymbol{K}}(\boldsymbol{x})\tilde{\boldsymbol{R}}(\boldsymbol{x})\boldsymbol{Z}_{\mathrm{C_iR}}^{\top}(\boldsymbol{x}) - \boldsymbol{Z}_{\mathrm{C_iC_i}}(\boldsymbol{x})\tilde{\boldsymbol{L}}^{\top} + \tilde{\boldsymbol{L}}\boldsymbol{Z}_{\mathrm{C_iC_i}}(\boldsymbol{x})\right), \tag{3.90b}$$

$$\boldsymbol{G}(\boldsymbol{x}) = \frac{1}{2}\left(\boldsymbol{I} + \tilde{\boldsymbol{L}}\right)\boldsymbol{Z}_{\mathrm{C_iP}}(\boldsymbol{x}) - \frac{1}{2}\left(\boldsymbol{Z}_{\mathrm{C_iR}}(\boldsymbol{x})\tilde{\boldsymbol{R}}(\boldsymbol{x})\tilde{\boldsymbol{K}}^{\top}(\boldsymbol{x}) - \tilde{\boldsymbol{L}}\boldsymbol{Z}_{\mathrm{C_iR}}(\boldsymbol{x})\tilde{\boldsymbol{K}}(\boldsymbol{x})\tilde{\boldsymbol{R}}(\boldsymbol{x})\right)\boldsymbol{Z}_{\mathrm{RP}}(\boldsymbol{x}), \tag{3.90c}$$

$$\boldsymbol{P}(\boldsymbol{x}) = \frac{1}{2}\left(\boldsymbol{I} - \tilde{\boldsymbol{L}}\right)\boldsymbol{Z}_{\mathrm{C_iP}}(\boldsymbol{x}) - \frac{1}{2}\left(\boldsymbol{Z}_{\mathrm{C_iR}}(\boldsymbol{x})\tilde{\boldsymbol{R}}(\boldsymbol{x})\tilde{\boldsymbol{K}}^{\top}(\boldsymbol{x}) + \tilde{\boldsymbol{L}}\boldsymbol{Z}_{\mathrm{C_iR}}(\boldsymbol{x})\tilde{\boldsymbol{K}}(\boldsymbol{x})\tilde{\boldsymbol{R}}(\boldsymbol{x})\right)\boldsymbol{Z}_{\mathrm{RP}}(\boldsymbol{x}), \tag{3.90d}$$

$$\boldsymbol{M}(\boldsymbol{x}) = \frac{1}{2}\boldsymbol{Z}_{\mathrm{RP}}^{\top}(\boldsymbol{x})\left(\tilde{\boldsymbol{K}}(\boldsymbol{x})\tilde{\boldsymbol{R}}(\boldsymbol{x}) - \tilde{\boldsymbol{R}}(\boldsymbol{x})\tilde{\boldsymbol{K}}^{\top}(\boldsymbol{x})\right)\boldsymbol{Z}_{\mathrm{RP}}(\boldsymbol{x}) + \boldsymbol{Z}_{\mathrm{PP}}(\boldsymbol{x}), \tag{3.90e}$$

$$\boldsymbol{S}(\boldsymbol{x}) = \frac{1}{2}\boldsymbol{Z}_{\mathrm{RP}}^{\top}(\boldsymbol{x})\left(\tilde{\boldsymbol{K}}(\boldsymbol{x})\tilde{\boldsymbol{R}}(\boldsymbol{x}) + \tilde{\boldsymbol{R}}(\boldsymbol{x})\tilde{\boldsymbol{K}}^{\top}(\boldsymbol{x})\right)\boldsymbol{Z}_{\mathrm{RP}}(\boldsymbol{x}), \tag{3.90f}$$

where $\tilde{\boldsymbol{K}}(\boldsymbol{x}) = \left(\boldsymbol{I} + \tilde{\boldsymbol{R}}(\boldsymbol{x})\boldsymbol{Z}_{\mathrm{RR}}(\boldsymbol{x})\right)^{-1}$ *and* $\tilde{\boldsymbol{L}} = \left(\boldsymbol{I} + \boldsymbol{Z}_{\mathrm{C_iC_d}}\boldsymbol{Q}_{\mathrm{d}}^{-1}\boldsymbol{Z}_{\mathrm{C_iC_d}}^{\top}\boldsymbol{Q}_{\mathrm{i}}\right)^{-1}$.

Remark 3.40 (Existence of $\tilde{\boldsymbol{K}}$ and $\tilde{\boldsymbol{L}}$). In Lemma 3.33, it has been shown that the matrix $\tilde{\boldsymbol{K}}(\boldsymbol{x})$ always exists for all $x \in \mathbb{X}$. The existence of the matrix $\tilde{\boldsymbol{L}}$ will be discussed later in Lemma 3.44.

Proof:

The basic approach of the proof is similar as in the proof of Lemma 3.31, i.e., in the case without dependent storages. In the proof of Lemma 3.31, the first step was to eliminate variables that belong to resistive elements. In the proof at hand, however, we will eliminate both, variables that belong to resistive elements and variables that belong to dependent storages. In this context, a cornerstone is to derive an expression which relates the variables of interest without using variables that belong to dependent storages—which is step (i) of this proof. In step (ii), we decompose the equation system obtained from (i) into a symmetric and a skew-symmetric part and insert the constitutive relations of the independent storages from (3.65). In the third step (iii), we prove the definiteness condition (2.13). Compared to the proof of Lemma 3.31, step (iii) is more delicate as we have to consider the dependent storages in the power balance of the system. For the sake of releasing notational burden, we will suppress the argument $\boldsymbol{x}$ to the matrices during the proof.

(i) Let assumption 3.37 hold. Substituting the third line from the equation system of (3.88) into (3.66) gives

$$\begin{aligned} \boldsymbol{u}_{\mathrm{R}} &= -\tilde{\boldsymbol{R}}\left(\boldsymbol{Z}_{\mathrm{C_iR}}^{\top}\boldsymbol{u}_{\mathrm{C_i}} + \boldsymbol{Z}_{\mathrm{RR}}\boldsymbol{u}_{\mathrm{R}} - \boldsymbol{Z}_{\mathrm{RP}}\boldsymbol{u}_{\mathrm{P}}\right) \\ \Leftrightarrow \boldsymbol{u}_{\mathrm{R}} &= -\tilde{\boldsymbol{K}}\tilde{\boldsymbol{R}}\boldsymbol{Z}_{\mathrm{C_iR}}^{\top}\boldsymbol{u}_{\mathrm{C_i}} + \tilde{\boldsymbol{K}}\tilde{\boldsymbol{R}}\boldsymbol{Z}_{\mathrm{RP}}\boldsymbol{u}_{\mathrm{P}}, \end{aligned} \tag{3.91}$$

with $\tilde{\boldsymbol{K}} = \left(\boldsymbol{I} + \tilde{\boldsymbol{R}}\boldsymbol{Z}_{\mathrm{RR}}\right)^{-1}$. According to Lemma 3.33, $\tilde{\boldsymbol{K}}$ always exists. Now we use an idea of Wellstead [1979, pp. 199-200] to eliminate the variables that belong to dependent storages. In addition to Assumption 3.37, let Assumption 3.38 hold. With (3.89b), the second line of the equation system of (3.88) reads:

$$\begin{aligned} \boldsymbol{y}_{\mathrm{C_d}} &= \boldsymbol{Z}_{\mathrm{C_iC_d}}^{\top}\boldsymbol{u}_{\mathrm{C_i}} \\ \Leftrightarrow \frac{\partial V_{\mathrm{d}}}{\partial \boldsymbol{x}_{\mathrm{d}}}(\boldsymbol{x}_{\mathrm{d}}) &= \boldsymbol{Z}_{\mathrm{C_iC_d}}^{\top}\left(\frac{\partial V_{\mathrm{i}}}{\partial \boldsymbol{x}_i}(\boldsymbol{x}_i)\right) \\ \Leftrightarrow \boldsymbol{Q}_{\mathrm{d}}\boldsymbol{x}_{\mathrm{d}} &= \boldsymbol{Z}_{\mathrm{C_iC_d}}^{\top}\boldsymbol{Q}_{\mathrm{i}}\boldsymbol{x}_{\mathrm{i}} \\ \Leftrightarrow \boldsymbol{x}_{\mathrm{d}} &= \boldsymbol{Q}_{\mathrm{d}}^{-1}\boldsymbol{Z}_{\mathrm{C_iC_d}}^{\top}\boldsymbol{Q}_{\mathrm{i}}\boldsymbol{x}_{\mathrm{i}}. \end{aligned} \tag{3.92}$$

By differentiating (3.92) with respect to time and using (3.89a), we obtain

$$\begin{aligned} \dot{\boldsymbol{x}}_{\mathrm{d}} &= \boldsymbol{Q}_{\mathrm{d}}^{-1}\boldsymbol{Z}_{\mathrm{C_iC_d}}^{\top}\boldsymbol{Q}_{\mathrm{i}}\dot{\boldsymbol{x}}_{\mathrm{i}} \\ \Leftrightarrow \boldsymbol{0} &= -\boldsymbol{u}_{\mathrm{C_d}} + \boldsymbol{Q}_{\mathrm{d}}^{-1}\boldsymbol{Z}_{\mathrm{C_iC_d}}^{\top}\boldsymbol{Q}_{\mathrm{i}}\boldsymbol{y}_{\mathrm{C_i}}. \end{aligned} \tag{3.93}$$

Insertion of (3.91) into the first line of the equation system in (3.88) gives

$$y_{C_i} = \left(Z_{C_iC_i} + Z_{C_iR}\tilde{K}\tilde{R}Z_{C_iR}^{\top}\right)u_{C_i} - Z_{C_iC_d}u_{C_d} + \left(-Z_{C_iP} - Z_{C_iR}\tilde{K}\tilde{R}Z_{RP}\right)u_P. \tag{3.94}$$

Equations (3.93) and (3.94) can be written in matrix-vector form:

$$\begin{pmatrix} I & Z_{C_iC_d} \\ -Q_d^{-1}Z_{C_iC_d}^{\top}Q_i & I \end{pmatrix}\begin{pmatrix} y_{C_i} \\ u_{C_d} \end{pmatrix} = \begin{pmatrix} Z_{C_iC_i} + Z_{C_iR}\tilde{K}\tilde{R}Z_{C_iR}^{\top} & -Z_{C_iP} - Z_{C_iR}\tilde{K}\tilde{R}Z_{RP} \\ 0 & 0 \end{pmatrix}\begin{pmatrix} u_{C_i} \\ u_P \end{pmatrix}. \tag{3.95}$$

Next, we use the inversion rules for 2×2 block matrices from Lu and Shiou [2002] to invert the matrix on the left-hand side of equation (3.95). The Schur complement of this matrix is given by:

$$L := I + Z_{C_iC_d}Q_d^{-1}Z_{C_iC_d}^{\top}Q_i. \tag{3.96}$$

The regularity of the matrix L will be discussed in Lemma 3.44. Provisionally, let us suppose that the inverse of L exists and is given by $\tilde{L} := L^{-1}$. By applying the inversion rules from Lu and Shiou [2002], we can write (3.95) equivalently as

$$\begin{pmatrix} y_{C_i} \\ u_{C_d} \end{pmatrix} = \begin{pmatrix} Z_1 & Z_2 \\ Z_3 & Z_4 \end{pmatrix}\begin{pmatrix} u_{C_i} \\ u_P \end{pmatrix}, \tag{3.97}$$

with

$$Z_1 = \tilde{L}\left(Z_{C_iC_i} + Z_{C_iR}\tilde{K}\tilde{R}Z_{C_iR}^{\top}\right), \tag{3.98a}$$

$$Z_2 = \tilde{L}\left(-Z_{C_iP} - Z_{C_iR}\tilde{K}\tilde{R}Z_{RP}\right), \tag{3.98b}$$

$$Z_3 = Q_d^{-1}Z_{C_iC_d}^{\top}Q_i\tilde{L}\left(Z_{C_iC_i} - Z_{C_iR}\tilde{K}\tilde{R}Z_{C_iR}^{\top}\right), \tag{3.98c}$$

$$Z_4 = Q_d^{-1}Z_{C_iC_d}^{\top}Q_i\tilde{L}\left(-Z_{C_iP} - Z_{C_iR}\tilde{K}\tilde{R}Z_{RP}\right). \tag{3.98d}$$

The first line of (3.97) will pave the way to the state differential equation (2.12a) of the PHS. To obtain an expression for the output equation (2.12b), we insert (3.91) into the fourth line of the equation system of (3.88) and obtain

$$y_{\mathrm{P}} = \left(\boldsymbol{Z}_{\mathrm{C_iP}}^{\top} - \boldsymbol{Z}_{\mathrm{RP}}^{\top}\tilde{\boldsymbol{K}}\tilde{\boldsymbol{R}}\boldsymbol{Z}_{\mathrm{C_iR}}^{\top}\right)\boldsymbol{u}_{\mathrm{C_i}} + \left(\boldsymbol{Z}_{\mathrm{PP}} + \boldsymbol{Z}_{\mathrm{RP}}^{\top}\tilde{\boldsymbol{K}}\tilde{\boldsymbol{R}}\boldsymbol{Z}_{\mathrm{RP}}\right)\boldsymbol{u}_{\mathrm{P}}. \tag{3.99}$$

The first line of the equation system in (3.97) and equation (3.99) can be written together in matrix-vector form as

$$\begin{pmatrix} y_{\mathrm{C_i}} \\ y_{\mathrm{P}} \end{pmatrix} = \begin{pmatrix} \tilde{\boldsymbol{L}}\left(\boldsymbol{Z}_{\mathrm{C_iC_i}} + \boldsymbol{Z}_{\mathrm{C_iR}}\tilde{\boldsymbol{K}}\tilde{\boldsymbol{R}}\boldsymbol{Z}_{\mathrm{C_iR}}^{\top}\right) & -\tilde{\boldsymbol{L}}\left(\boldsymbol{Z}_{\mathrm{C_iP}} + \boldsymbol{Z}_{\mathrm{C_iR}}\tilde{\boldsymbol{K}}\tilde{\boldsymbol{R}}\boldsymbol{Z}_{\mathrm{RP}}\right) \\ \boldsymbol{Z}_{\mathrm{C_iP}}^{\top} - \boldsymbol{Z}_{\mathrm{RP}}^{\top}\tilde{\boldsymbol{K}}\tilde{\boldsymbol{R}}\boldsymbol{Z}_{\mathrm{C_iR}}^{\top} & \boldsymbol{Z}_{\mathrm{PP}} + \boldsymbol{Z}_{\mathrm{RP}}^{\top}\tilde{\boldsymbol{K}}\tilde{\boldsymbol{R}}\boldsymbol{Z}_{\mathrm{RP}} \end{pmatrix} \begin{pmatrix} \boldsymbol{u}_{\mathrm{C_i}} \\ \boldsymbol{u}_{\mathrm{P}} \end{pmatrix}. \tag{3.100}$$

Note that (3.100) is independent of variables that belong to resistive elements and dependent storages.

(ii) The matrix in (3.100) can be decomposed into a skew-symmetric and a symmetric part. Using this decomposition, (3.100) can be equivalently written as

$$\begin{pmatrix} y_{\mathrm{C_i}} \\ y_{\mathrm{P}} \end{pmatrix} = \Bigg[\underbrace{\begin{pmatrix} -\boldsymbol{J} & -\boldsymbol{G} \\ \boldsymbol{G}^{\top} & \boldsymbol{M} \end{pmatrix}}_{=\boldsymbol{\Psi}} + \underbrace{\begin{pmatrix} \boldsymbol{R} & \boldsymbol{P} \\ \boldsymbol{P}^{\top} & \boldsymbol{S} \end{pmatrix}}_{=\boldsymbol{\Theta}}\Bigg] \begin{pmatrix} \boldsymbol{u}_{\mathrm{C_i}} \\ \boldsymbol{u}_{\mathrm{P}} \end{pmatrix}, \tag{3.101}$$

with $\boldsymbol{J}$, $\boldsymbol{R}$, $\boldsymbol{G}$, $\boldsymbol{P}$, $\boldsymbol{M}$, $\boldsymbol{S}$ as in (3.90) and $\boldsymbol{\Psi} = -\boldsymbol{\Psi}^{\top}$, $\boldsymbol{\Theta} = \boldsymbol{\Theta}^{\top}$. By inserting the identities of the independent variables from (3.65) into (3.101), we finally obtain an explicit PHS (2.12) with $\boldsymbol{u} = \boldsymbol{u}_{\mathrm{P}}$, $\boldsymbol{x} = \boldsymbol{x}_{\mathrm{i}}$, $\boldsymbol{y} = y_{\mathrm{P}}$ and Hamiltonian $H(\boldsymbol{x}) = V_{\mathrm{i}}(\boldsymbol{x}_{\mathrm{i}})$.

(iii) In the last step, we show that the definiteness condition (2.13) holds. Let us merge (3.101) with the second line of the equation system in (3.88):

$$\begin{pmatrix} y_{\mathrm{C_i}} \\ y_{\mathrm{P}} \\ y_{\mathrm{C_d}} \end{pmatrix} = \Bigg[\underbrace{\begin{pmatrix} -\boldsymbol{J} & -\boldsymbol{G} & 0 \\ \boldsymbol{G}^{\top} & \boldsymbol{M} & 0 \\ 0 & 0 & 0 \end{pmatrix}}_{=:\tilde{\boldsymbol{\Psi}}} + \underbrace{\begin{pmatrix} \boldsymbol{R} & \boldsymbol{P} & 0 \\ \boldsymbol{P}^{\top} & \boldsymbol{S} & 0 \\ \boldsymbol{Z}_{\mathrm{C_iC_d}}^{\top} & 0 & 0 \end{pmatrix}}_{=:\tilde{\boldsymbol{\Theta}}}\Bigg] \begin{pmatrix} \boldsymbol{u}_{\mathrm{C_i}} \\ \boldsymbol{u}_{\mathrm{P}} \\ \boldsymbol{u}_{\mathrm{C_d}} \end{pmatrix}. \tag{3.102}$$

Note that $\tilde{\boldsymbol{\Psi}}$ is skew-symmetric but $\tilde{\boldsymbol{\Theta}}$ is not symmetric. Nevertheless, $\tilde{\boldsymbol{\Theta}} \succeq 0$ implies $\boldsymbol{\Theta} \succeq 0$. In the following, we show that $\tilde{\boldsymbol{\Theta}} \succeq 0$:

$$\begin{aligned}
\begin{pmatrix} \boldsymbol{u}_{\mathrm{C_i}}^\top & \boldsymbol{u}_{\mathrm{P}}^\top & \boldsymbol{u}_{\mathrm{C_d}}^\top \end{pmatrix} \tilde{\boldsymbol{\Theta}} \begin{pmatrix} \boldsymbol{u}_{\mathrm{C_i}} \\ \boldsymbol{u}_{\mathrm{P}} \\ \boldsymbol{u}_{\mathrm{C_d}} \end{pmatrix} &= \begin{pmatrix} \boldsymbol{u}_{\mathrm{C_i}}^\top & \boldsymbol{u}_{\mathrm{P}}^\top & \boldsymbol{u}_{\mathrm{C_d}}^\top \end{pmatrix} (\tilde{\boldsymbol{\Psi}} + \tilde{\boldsymbol{\Theta}}) \begin{pmatrix} \boldsymbol{u}_{\mathrm{C_i}} \\ \boldsymbol{u}_{\mathrm{P}} \\ \boldsymbol{u}_{\mathrm{C_d}} \end{pmatrix} \\
&\overset{(3.102)}{=} \begin{pmatrix} \boldsymbol{u}_{\mathrm{C_i}}^\top & \boldsymbol{u}_{\mathrm{P}}^\top & \boldsymbol{u}_{\mathrm{C_d}}^\top \end{pmatrix} \begin{pmatrix} \boldsymbol{y}_{\mathrm{C_i}} \\ \boldsymbol{y}_{\mathrm{P}} \\ \boldsymbol{y}_{\mathrm{C_d}} \end{pmatrix} \\
&\overset{(2.9)}{=} -\boldsymbol{y}_{\mathrm{R}}^\top \boldsymbol{u}_{\mathrm{R}} \overset{(3.66)}{=} \boldsymbol{y}_{\mathrm{R}}^\top \tilde{\boldsymbol{R}} \boldsymbol{y}_{\mathrm{R}} \geq 0.
\end{aligned} \tag{3.103}$$

Hence, we have $\tilde{\boldsymbol{\Theta}} \succeq 0$, which implies $\boldsymbol{\Theta} \succeq 0$. This concludes the proof. □

Remark 3.41 (Dependent states as a function of independent states). The following expression (i.e., (3.92)) relates the dependent states as functions of the independent states:

$$\boldsymbol{x}_{\mathrm{d}} = \boldsymbol{Q}_{\mathrm{d}}^{-1} \boldsymbol{Z}_{\mathrm{C_i C_d}}^\top \boldsymbol{Q}_{\mathrm{i}} \boldsymbol{x}_{\mathrm{i}}. \tag{3.104}$$

Equation (3.104) can be used, for example, to express the total energy in the system as a function only of $\boldsymbol{x}_{\mathrm{i}}$:

$$\begin{aligned}
V(\boldsymbol{x}_{\mathrm{i}}, \boldsymbol{x}_{\mathrm{d}}) &= V_{\mathrm{i}}(\boldsymbol{x}_{\mathrm{i}}) + V_{\mathrm{d}}(\boldsymbol{x}_{\mathrm{d}}) \\
&= \tfrac{1}{2} \boldsymbol{x}_{\mathrm{i}}^\top \boldsymbol{Q}_{\mathrm{i}} \boldsymbol{x}_{\mathrm{i}} + \tfrac{1}{2} \boldsymbol{x}_{\mathrm{d}}^\top \boldsymbol{Q}_{\mathrm{d}} \boldsymbol{x}_{\mathrm{d}} \\
&\overset{(3.104)}{=} \tfrac{1}{2} \boldsymbol{x}_{\mathrm{i}}^\top \boldsymbol{Q}_{\mathrm{i}} \left(\boldsymbol{I} + \boldsymbol{Z}_{\mathrm{C_i C_d}} \boldsymbol{Q}_{\mathrm{d}}^{-1} \boldsymbol{Z}_{\mathrm{C_i C_d}}^\top \boldsymbol{Q}_{\mathrm{i}} \right) \boldsymbol{x}_{\mathrm{i}} \\
&\overset{(3.96)}{=} \tfrac{1}{2} \boldsymbol{x}_{\mathrm{i}}^\top \boldsymbol{Q}_{\mathrm{i}} \boldsymbol{L} \boldsymbol{x}_{\mathrm{i}} = V(\boldsymbol{x}_{\mathrm{i}}).
\end{aligned} \tag{3.105}$$

Remark 3.42 (Dynamics of the dependent states). By inserting (3.65) into the second line of (3.97), we yield an explicit expression for the dynamics of the dependent states:

$$\begin{aligned}\dot{\boldsymbol{x}}_{\mathrm{d}} = \boldsymbol{Q}_{\mathrm{d}}^{-1}\boldsymbol{Z}_{\mathrm{C_iC_d}}^{\top}\boldsymbol{Q}_{\mathrm{i}}\tilde{\boldsymbol{L}}\left(\boldsymbol{Z}_{\mathrm{C_iC_i}} - \boldsymbol{Z}_{\mathrm{C_iR}}\tilde{\boldsymbol{K}}\tilde{\boldsymbol{R}}\boldsymbol{Z}_{\mathrm{C_iR}}^{\top}\right)\boldsymbol{Q}_{\mathrm{i}}\boldsymbol{x}_{\mathrm{i}} \\ -\boldsymbol{Q}_{\mathrm{d}}^{-1}\boldsymbol{Z}_{\mathrm{C_iC_d}}^{\top}\boldsymbol{Q}_{\mathrm{i}}\tilde{\boldsymbol{L}}\left(\boldsymbol{Z}_{\mathrm{C_iP}} + \boldsymbol{Z}_{\mathrm{C_iR}}\tilde{\boldsymbol{K}}\tilde{\boldsymbol{R}}\boldsymbol{Z}_{\mathrm{RP}}\right)\boldsymbol{u}_{\mathrm{P}}.\end{aligned} \tag{3.106}$$

A central prerequisite for the proof of Lemma 3.39 is the existence of the matrix $\tilde{\boldsymbol{L}} = \boldsymbol{L}^{-1}$, where $\boldsymbol{L} = \left(\boldsymbol{I} + \boldsymbol{Z}_{\mathrm{C_iC_d}}\boldsymbol{Q}_{\mathrm{d}}^{-1}\boldsymbol{Z}_{\mathrm{C_iC_d}}^{\top}\boldsymbol{Q}_{\mathrm{i}}\right)$ with $\boldsymbol{Q}_{\mathrm{i}}$, $\boldsymbol{Q}_{\mathrm{d}}$ symmetric, positive-definite and $\boldsymbol{Z}_{\mathrm{C_iC_d}}$ a matrix of proper size. In the sequel, we prove that this prerequisite is indeed fulfilled, i.e., that $\boldsymbol{L}$ is always regular. To this end, we will make use of the following mathematical concepts (see Deuflhard and Hohmann [1995, p. 273]):

Definition 3.43 (Rayleigh quotient and numerical range)
Given a quadratic matrix $\boldsymbol{A} \in \mathbb{R}^{n\times n}$ and a non-zero vector $\boldsymbol{x} \in \mathbb{R}^{n}$. The Rayleigh quotient $\rho(\boldsymbol{A},\boldsymbol{x})$ is defined as

$$\rho(\boldsymbol{A},\boldsymbol{x}) := \frac{\boldsymbol{x}^{\top}\boldsymbol{A}\boldsymbol{x}}{\boldsymbol{x}^{\top}\boldsymbol{x}}. \tag{3.107}$$

The set of all Rayleigh quotients over non-zero vectors

$$W(\boldsymbol{A}) := \{\rho(\boldsymbol{A},\boldsymbol{x})\,|\,\boldsymbol{x} \in \mathbb{R}^{n} \setminus \{\boldsymbol{0}\}\} \tag{3.108}$$

is called numerical range of $\boldsymbol{A}$.

We have $\mathrm{Spec}(\boldsymbol{A}) \subseteq W(\boldsymbol{A})$ and for a symmetric matrix $\boldsymbol{A}$ we have $W(\boldsymbol{A}) = [\lambda_{\mathrm{min}}, \lambda_{\mathrm{max}}]$ by the min-max Theorem (also known as Courant-Fischer Theorem), where λ_{min} and λ_{max} are the smallest and the largest eigenvalues of $\boldsymbol{A}$ [Deuflhard and Hohmann, 1995, Lemma 8.29, p. 273].

The following lemma is the key step to show that $\tilde{\boldsymbol{L}}$ always exists:

Lemma 3.44 (Regularity of a matrix)
Let $\boldsymbol{X} \in \mathbb{R}^{p \times p}$ be a symmetric positive-definite matrix and $\boldsymbol{Y} \in \mathbb{R}^{p \times p}$ be any matrix whose numerical range $W(\boldsymbol{Y})$ is contained in $[0, \infty)$, i.e., $\boldsymbol{Y}$ has only non-negative eigenvalues. Then the matrix $\boldsymbol{I} + \boldsymbol{Y}\boldsymbol{X}$ is regular.

Proof:
Without loss of generality we may assume $\boldsymbol{X}$ to be diagonal.[15] The regularity of $\boldsymbol{I} + \boldsymbol{Y}\boldsymbol{X}$ will be proven by showing that $\boldsymbol{Y}\boldsymbol{X}$ has only eigenvalues in $[0, \infty)$, i.e., that -1 is not an eigenvalue of $\boldsymbol{Y}\boldsymbol{X}$. Let $\sqrt{\boldsymbol{X}}$ be the diagonal matrix which is a square root of $\boldsymbol{X}$, i.e., $\sqrt{\boldsymbol{X}}\sqrt{\boldsymbol{X}} = \sqrt{\boldsymbol{X}}\sqrt{\boldsymbol{X}}^\top = \boldsymbol{X}$. Such a matrix exists and is invertible since $\boldsymbol{X}$ is diagonal and positive-definite. Because the spectrum of a matrix is invariant under conjugation, we have

$$\begin{aligned}
\operatorname{Spec}(\boldsymbol{Y}\boldsymbol{X}) &= \operatorname{Spec}\left(\sqrt{\boldsymbol{X}}\boldsymbol{Y}\boldsymbol{X}\sqrt{\boldsymbol{X}}^{-1}\right) \\
&= \operatorname{Spec}\left(\sqrt{\boldsymbol{X}}\boldsymbol{Y}\sqrt{\boldsymbol{X}}\right) = \operatorname{Spec}\left(\sqrt{\boldsymbol{X}}\boldsymbol{Y}\sqrt{\boldsymbol{X}}^\top\right) \\
&\subseteq W\left(\sqrt{\boldsymbol{X}}\boldsymbol{Y}\sqrt{\boldsymbol{X}}^\top\right) \subseteq W(\boldsymbol{Y}) \cdot (0, \infty) \subseteq [0, \infty).
\end{aligned} \tag{3.109}$$

In (3.109), the second to last inclusion holds since

$$\frac{\boldsymbol{x}^\top \sqrt{\boldsymbol{X}}\boldsymbol{Y}\sqrt{\boldsymbol{X}}^\top \boldsymbol{x}}{\boldsymbol{x}^\top \boldsymbol{x}} = \underbrace{\frac{\boldsymbol{x}^\top \sqrt{\boldsymbol{X}}\boldsymbol{Y}\sqrt{\boldsymbol{X}}^\top \boldsymbol{x}}{\boldsymbol{x}^\top \sqrt{\boldsymbol{X}}\sqrt{\boldsymbol{X}}^\top \boldsymbol{x}}}_{\in W(\boldsymbol{Y})} \cdot \underbrace{\frac{\boldsymbol{x}^\top \sqrt{\boldsymbol{X}}\sqrt{\boldsymbol{X}}^\top \boldsymbol{x}}{\boldsymbol{x}^\top \boldsymbol{x}}}_{\in W(\boldsymbol{X})} \tag{3.110}$$

and $W(\boldsymbol{X}) \subseteq (0, \infty)$ by the min-max Theorem. Thus, -1 is not an eigenvalue of $\boldsymbol{Y}\boldsymbol{X}$ and $\boldsymbol{I} + \boldsymbol{Y}\boldsymbol{X}$ is invertible. □

[15] cf. the proof of Lemma 3.33 and note that for $\boldsymbol{T} \in \mathbb{O}(p)$ we have $W(\boldsymbol{T}\boldsymbol{Y}\boldsymbol{T}^\top) \in [0, \infty)$.

Corollary 3.45
The matrix $\tilde{\boldsymbol{L}}(\boldsymbol{x})$ from Lemma 3.39 always exists.

Proof:
Recall that $\boldsymbol{Q}_{\mathrm{d}} \succ 0$ and thus $\boldsymbol{Q}_{\mathrm{d}}^{-1} \succ 0$, which also implies $\boldsymbol{Q}_{\mathrm{d}}^{-1} \succeq 0$. Hence, $\boldsymbol{Z}_{\mathrm{C_iC_d}}\boldsymbol{Q}_{\mathrm{d}}^{-1}\boldsymbol{Z}_{\mathrm{C_iC_d}}^{\top}$ is positive semi-definite. The Rayleight quotient of a positive semi-definite matrix is always ≥ 0, i.e., $W(\boldsymbol{Z}_{\mathrm{C_iC_d}}\boldsymbol{Q}_{\mathrm{d}}^{-1}\boldsymbol{Z}_{\mathrm{C_iC_d}}^{\top}) \subseteq [0, \infty)$. As $\boldsymbol{Q}_{\mathrm{i}}$ is symmetric and positive-definite, the claim follows from Lemma 3.44. □

Remark 3.46 (Alternative proof of Lemma 3.33). By interchanging $\boldsymbol{Y}$ and $\boldsymbol{X}$, the proof of Lemma 3.44 can be easily extended to also show regularity of $\boldsymbol{I}+\boldsymbol{X}\boldsymbol{Y}$. By this, we obtain an alternative proof of Lemma 3.33 for the existence of matrix $\tilde{\boldsymbol{K}}(\boldsymbol{x})$ (cf. Remark 3.40).

Now we have everything prepared to collect the insights from the previous lemmas in a summarizing theorem about the existence of an explicit port-Hamiltonian formulation—similar to Theorem 3.35. In contrast to Theorem 3.35, however, the following theorem focuses on bond graphs that contain dependent storages.

Theorem 3.47 (PHSs from Bond Graphs with Dependent Storages)
Given a K-dimensional bond graph as in Definition 3.5 with linear storages (i.e., Assumption 3.38). The junction structure of the bond graph can be described by a Dirac structure in implicit form (3.82). Let $\operatorname{rank}(\boldsymbol{E}_{\mathrm{Sf}}(\boldsymbol{x})\ \boldsymbol{F}_{\mathrm{Se}}(\boldsymbol{x})) = K(N_{\mathrm{Sf}}+N_{\mathrm{Se}})$ *hold for all* $\boldsymbol{x} \in \mathbb{X}$ *(i.e., Assumption 3.22). Then, (3.82) can be formulated in an explicit form*

$$\mathbb{D} = \{(\begin{pmatrix} \boldsymbol{f}_{\mathrm{C}} \\ \boldsymbol{f}_{\mathrm{R}} \\ \boldsymbol{f}_{\mathrm{Sf}} \\ \boldsymbol{f}_{\mathrm{Se}} \end{pmatrix}, \begin{pmatrix} \boldsymbol{e}_{\mathrm{C}} \\ \boldsymbol{e}_{\mathrm{R}} \\ \boldsymbol{e}_{\mathrm{Sf}} \\ \boldsymbol{e}_{\mathrm{Se}} \end{pmatrix}) \in \mathbb{R}^{KN_{\mathrm{E}}} \times \mathbb{R}^{KN_{\mathrm{E}}} \mid \begin{pmatrix} \boldsymbol{y}_{\mathrm{C_i}} \\ \boldsymbol{y}_{\mathrm{C_d}} \\ \boldsymbol{y}_{\mathrm{R}} \\ \boldsymbol{y}_{\mathrm{P}} \end{pmatrix} = \underbrace{\begin{pmatrix} \boldsymbol{Z}_{\mathrm{C_iC_i}}(\boldsymbol{x}) & -\boldsymbol{Z}_{\mathrm{C_iC_d}}(\boldsymbol{x}) & -\boldsymbol{Z}_{\mathrm{C_iR}}(\boldsymbol{x}) & -\boldsymbol{Z}_{\mathrm{C_iP}}(\boldsymbol{x}) \\ \boldsymbol{Z}^{\top}_{\mathrm{C_iC_d}}(\boldsymbol{x}) & \boldsymbol{0} & \boldsymbol{0} & -\boldsymbol{Z}_{\mathrm{C_dP}}(\boldsymbol{x}) \\ \boldsymbol{Z}^{\top}_{\mathrm{C_iR}}(\boldsymbol{x}) & \boldsymbol{0} & \boldsymbol{Z}_{\mathrm{RR}}(\boldsymbol{x}) & -\boldsymbol{Z}_{\mathrm{RP}}(\boldsymbol{x}) \\ \boldsymbol{Z}^{\top}_{\mathrm{C_iP}}(\boldsymbol{x}) & \boldsymbol{Z}^{\top}_{\mathrm{C_dP}}(\boldsymbol{x}) & \boldsymbol{Z}^{\top}_{\mathrm{RP}}(\boldsymbol{x}) & \boldsymbol{Z}_{\mathrm{PP}}(\boldsymbol{x}) \end{pmatrix}}_{\boldsymbol{Z}(\boldsymbol{x})} \begin{pmatrix} \boldsymbol{u}_{\mathrm{C_i}} \\ \boldsymbol{u}_{\mathrm{C_d}} \\ \boldsymbol{u}_{\mathrm{R}} \\ \boldsymbol{u}_{\mathrm{P}} \end{pmatrix} \}, \tag{3.111}$$

where $\boldsymbol{Z}(\boldsymbol{x}) = -\boldsymbol{Z}^{\top}(\boldsymbol{x})$ *for all* $\boldsymbol{x} \in \mathbb{X}$ *and* $\boldsymbol{u}_{\mathrm{C_i}} = \boldsymbol{e}_{\mathrm{C_i}}$, $\boldsymbol{u}_{\mathrm{C_d}} = -\boldsymbol{f}_{\mathrm{C_d}}$, $\boldsymbol{u}_{\mathrm{R}} = (\boldsymbol{e}^{\top}_{\mathrm{R},1}\ -\boldsymbol{f}^{\top}_{\mathrm{R},2})^{\top}$, $\boldsymbol{u}_{\mathrm{P}} = (\boldsymbol{f}^{\top}_{\mathrm{Sf}}\ \boldsymbol{e}_{\mathrm{Se}})^{\top}$, $\boldsymbol{y}_{\mathrm{C_i}} = -\boldsymbol{f}_{\mathrm{C_i}}$, $\boldsymbol{y}_{\mathrm{C_d}} = \boldsymbol{e}_{\mathrm{C_d}}$, $\boldsymbol{y}_{\mathrm{R}} = (-\boldsymbol{f}^{\top}_{\mathrm{R},1}\ \boldsymbol{e}^{\top}_{\mathrm{R},2})^{\top}$, $\boldsymbol{y}_{\mathrm{P}} = (\boldsymbol{e}^{\top}_{\mathrm{Sf}}\ \boldsymbol{f}^{\top}_{\mathrm{Se}})^{\top}$. *Suppose the resistive relations can be written as* $\boldsymbol{u}_{\mathrm{R}} = \tilde{\boldsymbol{R}}(\boldsymbol{x})\,\boldsymbol{y}_{\mathrm{R}}$ *with* $\tilde{\boldsymbol{R}}(\boldsymbol{x}) = \tilde{\boldsymbol{R}}(\boldsymbol{x})^{\top} \succeq 0$ *(i.e., Assumption 3.30). Moreover, let* $\boldsymbol{Z}_{\mathrm{C_dP}}(\boldsymbol{x}) = \boldsymbol{0}$ *for all* $\boldsymbol{x} \in \mathbb{X}$ *and* $\boldsymbol{Z}_{\mathrm{C_iC_d}}(\boldsymbol{x}) = \boldsymbol{Z}_{\mathrm{C_iC_d}} =$ *const. (i.e., Assumption 3.37). The bond graph can then be formulated as an explicit PHS as from Definition 2.23:*

$$\dot{\boldsymbol{x}} = (\boldsymbol{J}(\boldsymbol{x}) - \boldsymbol{R}(\boldsymbol{x}))\frac{\partial H}{\partial \boldsymbol{x}}(\boldsymbol{x}) + (\boldsymbol{G}(\boldsymbol{x}) - \boldsymbol{P}(\boldsymbol{x}))\,\boldsymbol{u}, \tag{3.112a}$$

$$\boldsymbol{y} = (\boldsymbol{G}(\boldsymbol{x}) + \boldsymbol{P}(\boldsymbol{x}))^{\top}\frac{\partial H}{\partial \boldsymbol{x}}(\boldsymbol{x}) + (\boldsymbol{M}(\boldsymbol{x}) + \boldsymbol{S}(\boldsymbol{x}))\,\boldsymbol{u}, \tag{3.112b}$$

that has Property 3.8. The input, state, and output of the PHS are given as $\boldsymbol{u} = \boldsymbol{u}_{\mathrm{P}}$, $\boldsymbol{x} = \boldsymbol{x}_{\mathrm{i}}$, *and* $\boldsymbol{y} = \boldsymbol{y}_{\mathrm{P}}$, *respectively, with the Hamiltonian* $H(\boldsymbol{x}) = V_{\mathrm{i}}(\boldsymbol{x}_{\mathrm{i}})$.

Proof:
Consider a K-dimensional bond graph as in Definition 3.5 which fulfills Assumption 3.38. From Theorem 3.35 it is known that the junction structure of the bond graph can be described by a Dirac structure of the form (3.82). Let Assumption 3.22 hold. From Lemma 3.23 it follows that (3.82) can be written as (3.111) where the zero blocks are due to the fact that variables of dependent storages are functions only of power variables of independent storages and inputs from sources [Wellstead, 1979, p. 199]. Let assumptions 3.30 and 3.37 hold. It follows from Lemma 3.39 that the bond graph can then be formulated as an explicit PHS (3.112) with inputs, states, outputs, and Hamiltonian as stated. By $\boldsymbol{u} = \boldsymbol{u}_{\mathrm{P}}$ and $\boldsymbol{y} = \boldsymbol{y}_{\mathrm{P}}$, the PHS has Property 3.8. □

Theorem 3.47 provides a sufficient condition for the existence of an explicit port-Hamiltonian formulation of bond graphs with dependent storages. The condition is composed of four subconditions, viz. (i) that the system contains no dependent sources (i.e., Assumption 3.22); (ii) that the constitutive relation of energy-dissipating elements can be formulated in a suitable input-output representation (i.e., Assumption 3.30); (iii) that the variables from dependent storages are static functions of only variables from independent storages (i.e., Assumption 3.37); and (iv) that the storage elements are linear (i.e., Assumption 3.38). By these conditions, Theorem 3.47 generalizes Theorem 3.35 with respect to the occurence of dependent storages but is less general concerning the permitted storage functions.

Algorithm 3.48 collects the methods associated to Theorem 3.47 in a pseudo code listing to calculate an explicit PHS from a given bond graph. It is straightforward to implement the algorithm in a computer algebra system. Such an implementation enables the automated generation of an explicit PHS based on a bond graph system with dependent storages.

Algorithm 3.48 Port-Hamiltonian formulation of a bond graph with dependent storages

Input: K-dimensional bond graph (3.12) with linear storages
// Methods from subsections 3.2.3 and 3.2.4 are as in Algorithm 3.36
execute line 1 to 14 from Algorithm 3.36
// Methods from Subsection 3.2.5
if Assumption 3.22 is violated **then**
    print "Bond graph contains dependent sources. No PHS can be computed!"
    terminate
end if
split $\boldsymbol{F}_{\mathrm{C}}(\boldsymbol{x})$ and $\boldsymbol{F}_{\mathrm{R}}(\boldsymbol{x})$ such that (3.39) is fulfilled
split $\boldsymbol{E}_{\mathrm{C}}(\boldsymbol{x})$, $\boldsymbol{f}_{\mathrm{C}}$, $\boldsymbol{e}_{\mathrm{C}}$ in same parts as $\boldsymbol{F}_{\mathrm{C}}(\boldsymbol{x})$
split $\boldsymbol{E}_{\mathrm{R}}(\boldsymbol{x})$, $\boldsymbol{f}_{\mathrm{R}}$, $\boldsymbol{e}_{\mathrm{R}}$ in same parts as $\boldsymbol{F}_{\mathrm{R}}(\boldsymbol{x})$
compute $\boldsymbol{Z}(\boldsymbol{x})$ according to (3.38)
compute $\boldsymbol{u}_i, \boldsymbol{y}_i$ according to (3.37), $i \in \{\mathrm{C}, \mathrm{P}, \mathrm{R}\}$
compute $\mathbb{D}(\boldsymbol{x})$ as in (3.111)
// Methods from Subsection 3.2.8
if Assumption 3.30 or 3.37 is violated **then**
    print "Existence conditions are violated. Algorithm terminates!"
    terminate
end if
bring resistive relation to form (3.66)
compute PHS matrices with (3.90)
$\boldsymbol{x} \leftarrow \boldsymbol{x}_{\mathrm{i}}$ and $H(\boldsymbol{x}) \leftarrow V_{\mathrm{i}}(\boldsymbol{x}_{\mathrm{i}})$
$\boldsymbol{u} \leftarrow \boldsymbol{u}_{\mathrm{P}}$, $\boldsymbol{y} \leftarrow \boldsymbol{y}_{\mathrm{P}}$
return explicit PHS (3.112)

To illustrate the automated model generation based on Algorithm 3.48, let us consider a new example.

Example 3.49:

Consider the single-bond graph ($K = 1$) in Figure 3.4.

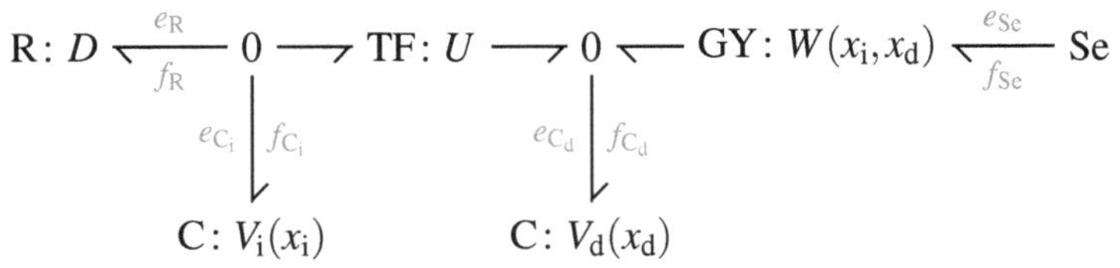

Figure 3.4: Exemplary bond graph with dependent storages

The C*-type element with storage function $V_{\mathrm{i}}(x_{\mathrm{i}})$ is considered as independent storage element; the* C*-type element with storage function $V_{\mathrm{d}}(x_{\mathrm{d}})$ is a dependent storage element. The respective storage functions are given by $V_{\mathrm{i}}(x_{\mathrm{i}}) = x_{\mathrm{i}}^2/(2c_{\mathrm{i}})$ and $V_{\mathrm{d}}(x_{\mathrm{d}}) = x_{\mathrm{d}}^2/(2c_{\mathrm{d}})$ where $c_{\mathrm{i}}, c_{\mathrm{d}} > 0$. The constitutive relation of the* R*-type element is specified by $f_{\mathrm{R}} = De_{\mathrm{R}}$ where $D > 0$. The* *transformer* TF *has a constant transformation ratio* $U > 0$ *and the gyrator* GY *is state-modulated with an arbitrary gyration ratio* $W(x_\mathrm{i}, x_\mathrm{d}) > 0$ *for all* $x_\mathrm{i}, x_\mathrm{d} \in \mathbb{R}$.

By executing the first two lines of Algorithm 3.48, we obtain a Dirac structure of the form (3.111):

$$\mathbb{D} = \{(\begin{pmatrix} f_{\mathrm{C_i}} \\ f_{\mathrm{C_d}} \\ f_\mathrm{R} \\ f_\mathrm{Se} \end{pmatrix}, \begin{pmatrix} e_{\mathrm{C_i}} \\ e_{\mathrm{C_d}} \\ e_\mathrm{R} \\ e_\mathrm{Se} \end{pmatrix}) \in \mathbb{R}^4 \times \mathbb{R}^4 \mid$$
$$\begin{pmatrix} -f_{\mathrm{C_i}} \\ e_{\mathrm{C_d}} \\ e_\mathrm{R} \\ f_\mathrm{Se} \end{pmatrix} = \begin{pmatrix} 0 & -U & -1 & -\frac{U}{W(x_\mathrm{i}, x_\mathrm{d})} \\ U & 0 & 0 & 0 \\ 1 & 0 & 0 & 0 \\ \frac{U}{W(x_\mathrm{i}, x_\mathrm{d})} & 0 & 0 & 0 \end{pmatrix} \begin{pmatrix} e_{\mathrm{C_i}} \\ -f_{\mathrm{C_d}} \\ -f_\mathrm{R} \\ e_\mathrm{Se} \end{pmatrix} \}. \tag{3.113}$$

Assumptions 3.30 and 3.37 are satisfied for (3.113). Hence, from Theorem 3.47 we know that the bond graph from Figure 3.4 permits an explicit port-Hamiltonian formulation of the form (3.112). By executing the remainder of Algorithm 3.48, we obtain the following explicit PHS:

$$\dot{x} = -\underbrace{\left(\frac{D c_\mathrm{i}}{c_\mathrm{i} + U^2 c_\mathrm{d}}\right)}_{= \boldsymbol{R}(x) = \boldsymbol{R}} \frac{\partial H}{\partial x}(x) + \underbrace{\left(\frac{U c_\mathrm{i}}{W(x)(c_\mathrm{i} + U^2 c_\mathrm{d})}\right)}_{= \boldsymbol{G}(x) - \boldsymbol{P}(x)} u, \tag{3.114a}$$

$$y = \underbrace{\left(\frac{U}{W(x)}\right)}_{= (\boldsymbol{G}(x) + \boldsymbol{P}(x))^\top} \frac{\partial H}{\partial x}(x), \tag{3.114b}$$

with $u = e_\mathrm{Se}$, $x = x_\mathrm{i}$, $y = f_\mathrm{Se}$, *and* $H(x) = V_\mathrm{i}(x_\mathrm{i})$.[16] *In this example, the expressions for* $\boldsymbol{J}$, $\boldsymbol{M}$, *and* $\boldsymbol{S}$ *are calculated as zero. Note that* $\boldsymbol{R} > 0$. *Moreover, the inputs and outputs are such that the PHS has Property 3.8.*

Example 3.49 illustrates that one can obtain an explicit PHS also for systems with linear dependent storages. In the next section, we discuss the main results

[16] The function $W(x) = W(x_\mathrm{i})$ in (3.114) can be obtained from the gyration ratio $W(x_\mathrm{i}, x_\mathrm{d})$ by using the substitution rule for x_d from (3.104) (cf. Remark 3.41).

presented in subsections 3.2.7 and 3.2.8 with respect to the overall objectives of this thesis and previous results from the related literature.

3.3 Discussion

Theorems 3.35 and 3.47 are the main theoretical results of this chapter. The two theorems are the first to give rigorous existence conditions for the complete transfer from a bond graph to an explicit PHS. In the literature, there exist some existence conditions for intermediate steps of this transfer. Golo et al. [2000] and Golo et al. [2003] show that the junction structure of a bond graph can always be related to a Dirac structure in implicit form which is in line with the findings from Theorem 3.35. Donaire and Junco [2009] provide sufficient conditions for the transfer from a Dirac structure in explicit form to a non-feedthrough PHS. Verbally formulated sufficient conditions for the existence of an explicit PHS have been given by van der Schaft [2009, p. 70] and van der Schaft and Jeltsema [2014, p. 53]. Theorems 3.35 and 3.47 put the conditions of Golo et al. [2000], Golo et al. [2003], Donaire and Junco [2009], van der Schaft [2009], and van der Schaft and Jeltsema [2014] into the perspective of an explicit port-Hamiltonian formulation of bond graphs.[17] Therewith, the conditions from the two theorems relate to, link, and generalize existing conditions and knowledge and streamline the present body of literature towards a fully automated model generation for interconnected systems.

From theorems 3.35 and 3.47 it follows that two important classes of bond graphs permit an explicit port-Hamiltonian formulation, viz. (i) bond graphs with nonlinear independent storages and (ii) bond graphs with linear dependent storages. Both cases, (i) and (ii), allow for nonlinearities in the interconnection structure arising from a state-modulation of energy-routing or energy-dissipating elements.

The independence of sources (i.e., Assumption 3.22) is shown to be a necessary condition for the existence of an explicit PHS. This condition is not strict as, in real physical systems, dependent sources are impossible to occur. It is noteworthy that Assumption 3.22 is also necessary for a bond graph to be well-posed in the sense of Golo et al. [2003, Def. 2].

Another condition from theorems 3.35 and 3.47 for the existence of a PHS is that the resistive relations can be rewritten in an input-output form (i.e., As-

[17] A more detailed discussion of theorems 3.35 and 3.47 in the light of previous work is out of the scope of this section but can be found in the literature notes provided in Appendix B.5.

sumption 3.30). This assumption is well-known from the literature (see, e.g., van der Schaft and Jeltsema [2014, p. 53] or van der Schaft [2009, p. 69] and is also mild. In Appendix B.3, it is shown that Assumption 3.30 is always fulfilled if the resistive elements are truly dissipative. This is satisfied by most multi-bond graphs and, in particular, for all single-bond graphs.

A limitation of theorems 3.35 and 3.47 is given for systems with nonlinear dependent storages. For such systems, however, there exist different strategies to eliminate dependent storages by adding or removing certain bond graph elements. The interested reader is referred to Borutzky [2010] and Karnopp et al. [2012].

As an interim conclusion, explicit PHSs are capable of describing a large class of bond graphs according to Definition 3.5. So it is left to discuss the capabilities and restrictions of Definition 3.5.

Definition 3.5 encompasses a comprehensive class of bond graphs as it allows for multi-port systems with nonlinearities in the energy-storing elements and state-modulations in the energy-routing and energy-dissipating elements. On the other hand, Definition 3.5 is limited to resistive relations that are linear in the power variables (i.e., Assumption 3.4). This represents the most relevant case of resistive relations (cf. van der Schaft and Jeltsema [2014, pp. 53–54] and Borutzky [2010, p. 364]), but excludes some phenomena that are of practical interest as, e.g., the dissipative relation describing the pressure drop in gas networks (cf. Strehle et al. [2018]) or the dissipative relation of nonlinear loads in AC power systems (cf. Strehle et al. [2020]). Hence, Appendix B.4 relaxes Assumption 3.4 which allows to consider bond graphs containing nonlinear resistive relations. Based hereon, Theorem 3.35 is generalized to systems with nonlinear dissipation. This leads to a novel class of explicit PHSs with nonlinear dissipation and feedthrough, which has, to the best of our knowledge, not been presented in the literature so far.[18] This generalization is straightforward which is why Assumption 3.4 is uncritical.[19]

Provided the existence conditions from theorems 3.35 and 3.47 are fulfilled, algorithms 3.36 and 3.48, respectively, enable the fully automated calculation of an explicit PHS from a given bond graph. The two algorithms are the first to

[18]For the case of no feedthrough, this class of PHSs particularizes to the class of PHSs with nonlinear resistive structure introduced by van der Schaft [2017, Def. 6.1.4].

[19]On the other hand, the PHS obtained in Section B.4 does not have the favorable symmetry and definiteness properties of (2.12) which is why we do not consider the PHS from Section B.4 any further.

enable such a fully automated calculation and the main practical result of this chapter.

A PHS obtained from algorithms 3.36 and 3.48 has Property 3.8 and thus correctly reflects the source elements of the underlying bond graph. The state vector of the PHS consists of the states of the independent energy storages. This is in line with the literature, where the order of an ODE model of a bond graph is known to be equal to the number of independent storages [Borutzky, 2010, p. 119]. The Hamiltonian of the PHS is the sum of the storage functions of the independent storages. Hence, the properties of the energy storages directly translate from the bond graph into the explicit PHS.

The calculation laws in algorithms 3.36 and 3.48 reveal state-modulated resistors, transformers, or gyrators to result in an explicit PHS with state-dependent matrices. On the other hand, if all bond graph elements of the types R, TF, and GY are non-modulated, the matrices of the explicit PHS are constant. If, in addition, the storages obey quadratic storage functions (i.e., Assumption 3.38 holds), the resulting PHS is linear. For the feedthrough-case, algorithms 3.36 and 3.48 reveal the parameters of R-type elements to possibly appear in each of the matrices of the PHS—and not only in the dissipation matrix $\boldsymbol{R}(\boldsymbol{x})$. This is in line with previous findings in the literature from Donaire and Junco [2009][p. 145, Remark] and van der Schaft and Jeltsema [2014, Example 4.1].

As can be seen, algorithms 3.36 and 3.48 ensure important properties of the system to translate into the explicit PHS. On the other hand, the results show that the physical interpretability of explicit models has natural limits. Introducing a causality from inputs over states to outputs leads to inter-subsystem dependencies that go beyond the physical interconnection structure. Therefore, the matrices of an explicit model usually do not have a subsystem-wise block structure.

A practical limitation of algorithms 3.36 and 3.48 is given for the development of fully symbolic models. The algorithms require matrix inversions, see (3.38), (3.67g), and (3.90). The inversion of large symbolic matrices, however, may lead to large expressions. Thereby, one cannot state an absolute limit for the allowed matrix sizes. The practical invertibility of a symbolic matrix strongly depends on the specific structure of the matrix. If the inversion of a symbolic matrix is intractable, one can still approach with local models.

The development of numeric models is free of such limitations. In the numeric case, the maximal size of the involved matrices is determined by the memory of the applied computer. Thereby, the sizes of the matrices grow with $O(K^2)$ where K is the dimension of the bond graph. The practical limit, however, is given by the computation time for the matrix inversions. In the worst case, the computation time for a matrix inverse grows polynomially with $O(s^3)$ where s is the matrix order [Lyche, 2020, p. 65]. In both algorithms, the largest matrix to be inverted is given in (3.38). This matrix is of order KN. Hence, the upper bound for the computation time for algorithms 3.36 and 3.48 grows with $O(K^3N^3)$, i.e., cubically with the product of the dimension and the sum of elements in the bond graph. It is noteworthy, that in practice this worst case rarely occurs as the matrix in (3.38) is typically a sparse matrix.

In conclusion, for a large class of interconnected systems the methods from this chapter enable an efficient automated generation of explicit port-Hamiltonian models. The obtained models transparently reflect important properties of the underlying physical systems.

3.4 Summary and Contributions

Modeling an interconnected system as an explicit state-space system is a cumbersome task. The methods and algorithms from this chapter provide remedy as they enable an automated generation of explicit port-Hamiltonian models on the basis of a bond graph representation of the system. The main contributions of this chapter are:

(C1.1) necessary and sufficient conditions for the existence of an explicit port-Hamiltonian formulation of bond graphs (theorems 3.35 and 3.47);

(C1.2) fully automatable algorithms which allow to compute an explicit PHS based on a given bond graph (algorithms 3.36 and 3.48).

The methods and algorithms from this chapter are the first to allow for an automated generation of physical-based explicit state-space models for a large class of interconnected systems. Therewith, these contributions meet the research objective O1 from Section 1.3. Successive examples verified the theoretical findings and illustrated the algorithmic nature of the automated model generation.

Chapter **4**

Automated Observer Design

The previous chapter presented methods for an automated generation of port-Hamiltonian models for interconnected systems. The obtained models may serve as the starting point for numerical simulations as well as for the design of controllers and observers. This chapter focuses on the latter, i.e., the PHS-based design of observers. Thereby, we aim for an automatable observer design which exploits the port-Hamiltonian structure of the model.

First, the state of the art in observer design methods for PHSs is reviewed in Section 4.1. We will identify two research gaps hampering an automated observer design for a large class of PHSs. From the research gaps we deduce the objectives of the chapter. To reach these objectives, new PHS-based observers which allow for an automated design are derived in Section 4.2. Finally, in Section 4.3 and Section 4.4 we discuss the new developed methods in the light of the existing literature and summarize the contributions from this chapter, respectively.

4.1 Literature Review

The first notable work on the observer design for PHSs was conducted by Sira-Ramírez and Cruz-Hernández [2001]. The authors propose a passivity-based observer design for so-called *generalized Hamiltonian systems*, i.e., autonomous PHSs. Since then, several methods addressing the observer design for PHSs have been reported in the literature. These methods will be reviewed in the sequel. For the review, the approaches are classified into observer designs for linear PHSs and observer designs for nonlinear PHSs.

Linear PHSs are a special class of linear state-space systems. Hence, for the state reconstruction of such systems it is natural to approach with a standard Luenberger observer or Kalman filter, see, e.g., Khalil et al. [2012]. As argued in Section 1.2, the design of such an observer can be automated by applying well-known methods. Cardoso Ribeiro [2016] show that the Luenberger observer is also a viable option if the linear model arises from the structure-preserving discretization of an infinite-dimensional PHS.[1] Kotyczka and Wang [2015] design a compensator for linear PHSs based on a dual observer. However, the dual observer does not provide an explicit reconstruction of the system state (cf. Luenberger [1971]). Atitallah et al. [2015] address the combined input and state reconstruction for linear PHSs. The authors propose two observers: the first observer determines a reconstruction of the state; the second observer reconstructs the unknown input. As argued in Section 2.2, besides the state and input, the output of a PHS is in general also not fully available for measurement. This leads to an input-state-output reconstruction problem as was first tackled in a preliminary work of this thesis authored by Pfeifer et al. [2019a]. The authors propose an interval input-state-output estimator for linear PHSs. The approach from Pfeifer et al. [2019a] can be automated but involves an observer existence condition that is rather restrictive.

In the above, we considered linear observer methods. For nonlinear PHSs, there exist also several observer methods. For the nonlinear observer methods, one can distinguish between two kinds of nonlinearities, viz. (a) nonlinearities in the interconnection structure and (b) nonlinearities in the storages. The former are characterized by state-dependent matrices of the PHSs; the latter are characterized by possibly non-quadratic Hamiltonians.

Wang et al. [2005] were the first to address the design of observers for nonlinear PHSs. The authors develop adaptive and non-adaptive state observers for systems with nonlinear interconnection structures and nonlinear storages. However, the observers are only asymptotic when the system reaches a steady state. Venkatraman and van der Schaft [2010] present a passivity-based, globally exponentially stable observer for PHSs with nonlinear interconnection structure and nonlinear storages. The proposed observer design is delicate as it requires the solution of a set of algebraic equations and partial differential equations (PDEs). A closed-form solution for such a system of equations exists only in special cases and is difficult to determine. Hence, the observer design from Venkatraman and van der Schaft [2010] cannot be automated. Vincent et al. [2016] present two nonlinear, passivity-based observers for PHSs with nonlinear interconnection structure: a proportional observer and a proportional observer with integral action. An interconnection and damping assignment (IDA) passivity-

[1]Kotyczka et al. [2019] follow an approach similar to Cardoso Ribeiro [2016].

based control (PBC)-like observer design for PHSs with nonlinear storages has been proposed by Yaghmaei and Yazdanpanah [2019b].[2] As with the approach from Venkatraman and van der Schaft [2010], the observer design of Yaghmaei and Yazdanpanah [2019b] requires the solution of a set of PDEs which hinders its automation.

Another two notable publications in this field stem from Biedermann et al. [2018] and Biedermann and Meurer [2021]: the former present a passivity-based observer design for a class of state-affine systems; the latter propose a dissipativity- and IDA-PBC-based observer design for nonlinear systems that can be decomposed into a time varying state affine part, a nonlinear feedback part, and a perturbation term. Thereby, the observer designs from Biedermann et al. [2018] and Biedermann and Meurer [2021] can be also applied to a class of PHSs with nonlinear interconnection structure and linear storages.

Table 4.1 provides an overview of the existing observers for linear and nonlinear PHSs. In the first column, the publications are listed in the order they are mention above. The second column of Table 4.1 differentiates between linear and nonlinear interconnection structures in the underlying port-Hamiltonian model. Similar, the third column distinguishes between linear and nonlinear storages. The fourth column shows the variables to be reconstructed by the respective observer.

From Table 4.1 it can be seen that the method from Pfeifer et al. [2019a] is the only approach which enables a reconstruction of inputs, states, and outputs. However, as mentioned above, this approach involves an existence condition which is rather strict. Hence, the method from Pfeifer et al. [2019a] is applicable to only a small class of linear PHSs.

The second and third columns in Table 4.1 show that the observers from Wang et al. [2005] and Venkatraman and van der Schaft [2010] are the only two approaches which are applicable to PHSs with nonlinearities in both, the interconnection structure and storages. However, the observer of Wang et al. [2005] is in general not asymptotic and the observer design from Venkatraman and van der Schaft [2010] cannot be automated.

To summarize, there exist powerful observer methods for linear and nonlinear PHSs in the literature However, there are two research gaps which hamper the automated design of observers for interconnected systems. First, the only existing

[2]The observer from Yaghmaei and Yazdanpanah [2019b] allows for a separation principle as known from linear systems theory, see Yaghmaei and Yazdanpanah [2019a].

Table 4.1: Classification of existing observers for PHSs with respect to the interconnection structure, storages, and reconstructed variables

Publication	Intercon.	Storages	Reconstructions
Khalil et al. [2012]	linear	linear	states
Cardoso Ribeiro [2016]	linear	linear	states
Kotyczka and Wang [2015]	linear	linear	–
Atitallah et al. [2015]	linear	linear	states, inputs
Pfeifer et al. [2019a]	linear	linear	states, inputs, outputs
Wang et al. [2005]	nonlinear	nonlinear	states
Venkatraman and van der Schaft [2010]	nonlinear	nonlinear	states
Vincent et al. [2016]	nonlinear	linear	states
Yaghmaei and Yazdanpanah [2019b]	linear	nonlinear	states
Biedermann et al. [2018]	nonlinear	linear	states
Biedermann and Meurer [2021]	nonlinear	linear	states

observer for the reconstruction of inputs, states, and outputs of a linear PHS is the estimator from Pfeifer et al. [2019a]. Unfortunately, this approach is applicable to only a small class of systems. Second, as argued above, the available observer design schemes for PHSs with nonlinearities in both, the interconnection structure and storages, cannot be automated.

This chapter addresses these research gaps by developing automated observer design methods for linear and nonlinear PHSs. Specifically, the objectives of this chapter are

(i) to develop an automatable design for an observer that is able to reconstruct the inputs, states, and outputs in a large class of linear PHSs, and
(ii) to derive an observer with an automatable design scheme for PHSs with nonlinear interconnection structures and nonlinear storages.

The objectives (i) and (ii) focus on the design of centralized observers based on global model knowledge. For interconnected systems, however, the design of distributed observers based on local model information is also of high interest (cf. Chapter 2). Hence, a secondary objective of this chapter is

(iii) to investigate how the methods from (i) and (ii) can be applied for an automated design of distributed observers based on local models.

4.2 Main Results

The previous section stated the aims of this chapter. To reach these aims we will now derive observers and corresponding design schemes for different classes of PHSs. First, in Subsection 4.2.1 we consider an automatable design scheme for an observer that reconstructs the inputs, states, and outputs of a linear PHS. Next, Subsection 4.2.2 outlines the design of a globally exponentially convergent state-output observer for a class of nonlinear PHSs. The methods from subsections 4.2.1 and 4.2.2 are presented in the light of an automated design of centralized observers based on global model knowledge. Subsection 4.2.3 shows that these methods can also be used for an automated design of distributed observers based on local model knowledge. To this end, we first reconsider the methods from Subsections 4.2.1 for the design of a distributed observer for linear interconnected systems with unknown subsystem interactions. Afterwards, we particularize the methods from Subsection 4.2.2 for the design of a distributed state observer for nonlinear interconnected systems with known subsystem interactions. Figure 4.1 depicts an overview of the observers from this section and illustrates their relations.

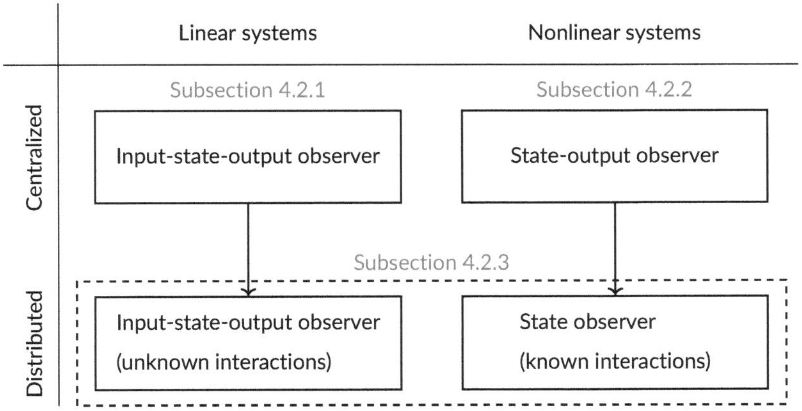

Figure 4.1: Overview of the observers developed in Section 4.2

4.2.1 Automated Design of an Input-State-Output Observer for Linear PHSs

Consider an interconnected system with linear interconnection structure, linear dissipation, and linear storages. In Chapter 3, it was shown that a large class of such systems can be described by linear input-state-output port-Hamiltonian models. For such models, we have seen that the entries of the inputs $\boldsymbol{u}$ and outputs $\boldsymbol{y}$ are determined by physical constraints at the system boundary. The corresponding variables may or may not be available for measurement. Thus, the individual entries of $\boldsymbol{u}$ and $\boldsymbol{y}$ may or may not be known during runtime. This is in contrast to classical observer theory, where the complete system input and output are assumed to be known or measured (cf. Luenberger [1971]). As a remedy, let us introduce a measurement vector $\boldsymbol{m}$ which contains the variables that are available for measurement. The aim is then to determine an asymptotic observer which provides reconstructions of $\boldsymbol{u}$, $\boldsymbol{x}$, and $\boldsymbol{y}$ based on knowledge of $\boldsymbol{m}$. Following the notion of a *state observer*, we denote such an observer as *input-state-output observer*. In the following, we formalize the problem of designing such an input-state-output observer in an automated manner.

Problem 4.1 (Automated input-state-output observer design)
Given a linear explicit PHS

$$\dot{\boldsymbol{x}} = (\boldsymbol{J}-\boldsymbol{R})\ \boldsymbol{Q}\boldsymbol{x}+(\boldsymbol{G}-\boldsymbol{P})\boldsymbol{u}, \quad \boldsymbol{x}|_{t=0} = \boldsymbol{x}_0, \tag{4.1a}$$
$$\boldsymbol{y} = (\boldsymbol{G}+\boldsymbol{P})^\top \boldsymbol{Q}\boldsymbol{x}+(\boldsymbol{M}+\boldsymbol{S})\boldsymbol{u}, \tag{4.1b}$$

with $\boldsymbol{u},\boldsymbol{y} \in \mathbb{R}^p$ and $\boldsymbol{x} \in \mathbb{R}^n$. For the storage matrix $\boldsymbol{Q} \in \mathbb{R}^{n\times n}$ we have $\boldsymbol{Q} = \boldsymbol{Q}^\top \succ 0$. The remaining matrices are constant matrices of appropriate sizes which satisfy the usual symmetry and definiteness properties of an explicit PHS (cf. Definition 2.23).
We assume that $\boldsymbol{u}$, $\boldsymbol{x}$, and $\boldsymbol{y}$ are (in general) non-measurable. Instead, let us consider measurements $\boldsymbol{m} \in \mathbb{R}^q$ of the form

$$\boldsymbol{m} = \boldsymbol{C}_u\boldsymbol{u}+\boldsymbol{C}_x\boldsymbol{x}+\boldsymbol{C}_y\boldsymbol{y}, \tag{4.2}$$

where $\boldsymbol{C}_u$, $\boldsymbol{C}_x$, and $\boldsymbol{C}_y$ are constant matrices of proper sizes.
The aim of this subsection is to give an answer to the following problem: How can we–in an automated manner–design an asymptotic observer that produces reconstructions of $\boldsymbol{u}$, $\boldsymbol{x}$, and $\boldsymbol{y}$ based on the model (4.1) and the measurements (4.2)?

Problem 4.1 requires to develop an observer that is able to reconstruct the inputs, states, and outputs of the system (4.1). The leading idea is to rewrite the system step by step until we obtain a representation which allows to apply existing methods from linear observer theory.

We now consider the first step of this procedure. The input-state-output reconstruction from Problem 4.1 involves three equations, viz. a dynamics equation (4.1a), an output equation (4.1b), and a measurement equation (4.2). The following lemma shows that Problem 4.1 can be reduced to a reconstruction problem which involves only two equations:

Lemma 4.2 (Output reconstruction based on $\hat{\boldsymbol{u}}$ and $\hat{\boldsymbol{x}}$)

Consider the situation in Problem 4.1. Let $\hat{\boldsymbol{u}}$ and $\hat{\boldsymbol{x}}$ be reconstructions of $\boldsymbol{u}$ and $\boldsymbol{x}$, respectively, with $\hat{\boldsymbol{u}} \to \boldsymbol{u}$ and $\hat{\boldsymbol{x}} \to \boldsymbol{x}$ for $t \to \infty$. Then, we can calculate an output reconstruction

$$\hat{\boldsymbol{y}} = (\boldsymbol{G}+\boldsymbol{P})^\top \boldsymbol{Q}\hat{\boldsymbol{x}} + (\boldsymbol{M}+\boldsymbol{S})\,\hat{\boldsymbol{u}}, \tag{4.3}$$

with $\hat{\boldsymbol{y}} \to \boldsymbol{y}$ for $t \to \infty$.

Proof:
Let $\hat{\boldsymbol{u}}$ and $\hat{\boldsymbol{x}}$ be reconstructions of $\boldsymbol{u}$ and $\boldsymbol{x}$ with $\hat{\boldsymbol{u}} \to \boldsymbol{u}$ and $\hat{\boldsymbol{x}} \to \boldsymbol{x}$ for $t \to \infty$, respectively. By substituting $\boldsymbol{u}$, $\boldsymbol{x}$, and $\boldsymbol{y}$ in (4.1b) with their respective reconstructions, we obtain (4.3). In particular, we have $\hat{\boldsymbol{y}} \to \boldsymbol{y}$ for $t \to \infty$. □

From Lemma 4.2 it follows that the input-state-output reconstruction problem can be reduced to an input-state reconstruction problem. To solve the input-state reconstruction problem, it is desirable to formulate the system only in terms of inputs and states, i.e., without outputs. In the next step, we again use the output equation (4.1b) to eliminate the outputs from the measurement equation (4.2).

Lemma 4.3 (State-space system in standard form)
Equations (4.1a) and (4.2) can be written as a state-space system in standard form

$$\dot{\boldsymbol{x}} = \boldsymbol{A}\boldsymbol{x} + \boldsymbol{B}\boldsymbol{u}, \quad \boldsymbol{x}|_{t=0} = \boldsymbol{x}_0, \tag{4.4a}$$
$$\boldsymbol{m} = \boldsymbol{C}\boldsymbol{x} + \boldsymbol{D}\boldsymbol{u}, \tag{4.4b}$$

where $\boldsymbol{A} = (\boldsymbol{J} - \boldsymbol{R})\boldsymbol{Q}$, $\boldsymbol{B} = (\boldsymbol{G} - \boldsymbol{P})$, $\boldsymbol{C} = \boldsymbol{C}_x + \boldsymbol{C}_y(\boldsymbol{G} + \boldsymbol{P})^\top \boldsymbol{Q}$, *and* $\boldsymbol{D} = \boldsymbol{C}_u + \boldsymbol{C}_y(\boldsymbol{M} + \boldsymbol{S})$.

Proof:
From (4.1a), we directly obtain the calculation rules for the matrices $\boldsymbol{A}$ and $\boldsymbol{B}$. Inserting (4.1b) into (4.2) yields the expressions for $\boldsymbol{C}$ and $\boldsymbol{D}$. □

Lemma 4.3 shows that a system with dynamics (4.1a) and measurements (4.2) can be formulated by means of a state-space system in standard form. Note that in (4.4) we have a clear differentiation in the description of the system's physics (i.e., (4.4a)) and the system's measurements (i.e., (4.4b)). Thereby, the input $\boldsymbol{u}$ of the system (4.4) is in general unknown. Instead, all measurements are specified by $\boldsymbol{m}$ in (4.4b). Nevertheless, $\boldsymbol{m}$ may contain measurements of an input, that is, a row of (4.4b) in which $\boldsymbol{D}$ contains exactly one "1" while the corresponding row of $\boldsymbol{C}$ contains exclusively zeros. For the upcoming considerations, it will be helpful (i) to split the input vector $\boldsymbol{u}$ into known inputs $\boldsymbol{u}_\mathrm{k}$ and unknown inputs $\boldsymbol{u}_\mathrm{u}$ and (ii) to split the vector $\boldsymbol{m}$ accordingly into the measured inputs $\boldsymbol{u}_\mathrm{k}$ and the remaining measurements $\tilde{\boldsymbol{m}}$. By doing so, the system (4.4) reads

$$\dot{\boldsymbol{x}} = \boldsymbol{A}\boldsymbol{x} + \boldsymbol{B}_\mathrm{k}\boldsymbol{u}_\mathrm{k} + \boldsymbol{B}_\mathrm{u}\boldsymbol{u}_\mathrm{u}, \quad \boldsymbol{x}|_{t=0} = \boldsymbol{x}_0, \tag{4.5a}$$
$$\tilde{\boldsymbol{m}} = \tilde{\boldsymbol{C}}\boldsymbol{x} + \tilde{\boldsymbol{D}}_\mathrm{k}\boldsymbol{u}_\mathrm{k} + \tilde{\boldsymbol{D}}_\mathrm{u}\boldsymbol{u}_\mathrm{u}, \tag{4.5b}$$

where $\boldsymbol{u}_\mathrm{k} \in \mathbb{R}^{p_\mathrm{k}}$, $\boldsymbol{u}_\mathrm{u} \in \mathbb{R}^{p_\mathrm{u}}$, $\boldsymbol{x} \in \mathbb{R}^n$, $\tilde{\boldsymbol{m}} \in \mathbb{R}^{\tilde{q}}$ with $p_\mathrm{k} + p_\mathrm{u} = p$ and $\tilde{q} = q - p_\mathrm{k}$. The matrices $\tilde{\boldsymbol{C}} \in \mathbb{R}^{\tilde{q}\times n}$ and $\left(\tilde{\boldsymbol{D}}_\mathrm{k}\ \tilde{\boldsymbol{D}}_\mathrm{u}\right) \in \mathbb{R}^{\tilde{q}\times p}$ can be obtained by removing those rows from the matrices $\boldsymbol{C}$ and $\boldsymbol{D}$, respectively, that correspond to $\boldsymbol{u}_\mathrm{k}$ in $\boldsymbol{m}$.

Equation (4.5) represents a linear system with known and unknown inputs. For such a system, it would be natural to apply an unknown-input observer. However, such an approach is significantly hampered by the fact that the measurement vector $\tilde{\boldsymbol{m}}$ directly depends on the unknown inputs $\boldsymbol{u}_\mathrm{u}$. Hence, it is desirable to find a formulation of (4.5) without a feedthrough of $\boldsymbol{u}_\mathrm{u}$. The next step of our procedure presents such a formulation.

Without loss of generality we may assume that $\mathrm{rank}(\boldsymbol{B}_\mathrm{u}) = p_\mathrm{u} (\leq n)$ and $\mathrm{rank}(\tilde{\boldsymbol{C}}) = \tilde{q} (\leq n)$. If either the first or the second statement is violated, we introduce new inputs or outputs, respectively, for which these conditions are fulfilled [Singer, 2019, p. 19]. In order to eliminate the feedthrough from the system, we make the assumption that in the system (4.5) the number of measurements is greater than the number of unknown inputs:

Assumption 4.4 (Number of measurements and unknown inputs)

We have $\tilde{q} > p_\mathrm{u}$.

In most practical systems, Assumption 4.4 can be satisfied by providing a sufficient number of measurements. Based on Assumption 4.4, we can now eliminate the feedthrough from the system.

Lemma 4.5 (Feedthrough elimination)

Consider the system with feedthrough (4.5). Let Assumption 4.4 hold. Then, a formulation of (4.5) without feedthrough is given by

$$\dot{\boldsymbol{x}} = \boldsymbol{A}\boldsymbol{x} + \boldsymbol{B}_\mathrm{k}\boldsymbol{u}_\mathrm{k} + \boldsymbol{B}_\mathrm{u}\boldsymbol{u}_\mathrm{u}, \quad \boldsymbol{x}\big|_{t=0} = \boldsymbol{x}_0, \tag{4.6a}$$

$$\bar{\boldsymbol{m}} = \bar{\boldsymbol{C}}\boldsymbol{x}. \tag{4.6b}$$

The reduced measurement output $\bar{\boldsymbol{m}} \in \mathbb{R}^{\tilde{q}-r}$ is given by $\bar{\boldsymbol{m}} = \boldsymbol{U}_2^\top \left(\tilde{\boldsymbol{m}} - \tilde{\boldsymbol{D}}_\mathrm{k}\boldsymbol{u}_\mathrm{k}\right)$ where $r = \mathrm{rank}(\tilde{\boldsymbol{D}}_\mathrm{u})$. The matrix $\bar{\boldsymbol{C}} \in \mathbb{R}^{\tilde{q}-r\times n}$ has full column rank and can be calculated as $\bar{\boldsymbol{C}} = \boldsymbol{U}_2^\top\tilde{\boldsymbol{C}}$. The matrix $\boldsymbol{U}_2$ is obtained from a singular value decomposition of $\boldsymbol{D}_\mathrm{u}$, i.e.,

$$\boldsymbol{D}_\mathrm{u} = \begin{pmatrix}\boldsymbol{U}_1 & \boldsymbol{U}_2\end{pmatrix}\boldsymbol{\Sigma}\boldsymbol{V}^\top, \tag{4.7}$$

with $\boldsymbol{U}_1 \in \mathbb{R}^{\tilde{q}\times r}$, $\boldsymbol{U}_2 \in \mathbb{R}^{\tilde{q}\times\tilde{q}-r}$, $\boldsymbol{\Sigma} \in \mathbb{R}^{\tilde{q}\times p_\mathrm{u}}$, and $\boldsymbol{V} \in \mathbb{R}^{p_\mathrm{u}\times p_\mathrm{u}}$. The matrix $\boldsymbol{\Sigma}$ contains the singular values of $\boldsymbol{D}_\mathrm{u}$ on its diagonal.

Proof:
We first show that the construction law of the vector $\bar{\boldsymbol{m}}$ indeed leads to an elimination of the feedthrough. Afterwards, we proof that the matrix $\bar{\boldsymbol{C}}$ has full column rank.

Let us first introduce a new measurement output without feedthrough of the known inputs:

$$\boldsymbol{m}^* := \tilde{\boldsymbol{m}} - \tilde{\boldsymbol{D}}_{\mathrm{k}}\boldsymbol{u}_{\mathrm{k}} = \tilde{\boldsymbol{C}}\boldsymbol{x} + \tilde{\boldsymbol{D}}_{\mathrm{u}}\boldsymbol{u}_{\mathrm{u}}. \tag{4.8}$$

Note that the vector $\boldsymbol{m}^*$ can be easily computed online during runtime.

Now we eliminate the unknown inputs by applying the idea from Irle [2016, p. 106]. Let Assumption 4.4 hold. Consider the singular value decomposition (4.7). For the case $\tilde{q} \neq 0$ and $p_{\mathrm{u}} \neq 0$, Assumption 4.4 implies $\boldsymbol{U}_2$ to exist. By multiplying $\boldsymbol{m}^*$ from the left with $\boldsymbol{U}_2^\top$ and by the properties of orthogonal matrices we obtain

$$\underbrace{\boldsymbol{U}_2^\top \boldsymbol{m}^*}_{=:\bar{\boldsymbol{m}}} = \underbrace{\boldsymbol{U}_2^\top \tilde{\boldsymbol{C}}}_{=:\bar{\boldsymbol{C}}}\boldsymbol{x} + \underbrace{\begin{pmatrix} \boldsymbol{0} & \boldsymbol{U}_2^\top \boldsymbol{U}_2 \end{pmatrix} \begin{pmatrix} \operatorname{diag}(\sigma_i) & \boldsymbol{0} \\ \boldsymbol{0} & \boldsymbol{0} \end{pmatrix} \boldsymbol{V}^\top \boldsymbol{u}_{\mathrm{u}}}_{=\boldsymbol{0}}, \tag{4.9}$$

where σ_i are the singular values of $\tilde{\boldsymbol{D}}_{\mathrm{u}}$ for $i = 1, \ldots, r$.
Now for the rank of $\bar{\boldsymbol{C}}$. From the Sylvester rank inequality we obtain

$$\operatorname{rank}\left(\boldsymbol{U}_2^\top \tilde{\boldsymbol{C}}\right) \geq \underbrace{\operatorname{rank}(\boldsymbol{U}_2^\top)}_{\tilde{q}-r} + \underbrace{\operatorname{rank}(\tilde{\boldsymbol{C}})}_{\tilde{q}} - \tilde{q} = \tilde{q} - r. \tag{4.10}$$

On the other hand, we have

$$\operatorname{rank}\left(\boldsymbol{U}_2^\top \tilde{\boldsymbol{C}}\right) \leq \min\{\operatorname{rank}(\boldsymbol{U}_2^\top), \operatorname{rank}(\tilde{\boldsymbol{C}})\} = \tilde{q} - r. \tag{4.11}$$

Combining (4.10) and (4.11) then yields $\operatorname{rank}\left(\bar{\boldsymbol{C}}\right) = \tilde{q} - r$. □

Remark 4.6 (Case $\tilde{\boldsymbol{D}}_{\mathrm{u}} = \boldsymbol{0}$). In (4.5), suppose $\tilde{\boldsymbol{D}}_{\mathrm{u}} = \boldsymbol{0}$, i.e., there is no feedthrough of the unknown-input to eliminate. This case obviates the singular value decomposition and therewith the need for Assumption 4.4.

The previous lemmas constitute a stepwise procedure to formulate a PHS (4.1) with measurements (4.2) as linear unknown-input system without feedthrough (4.6). In the literature, there exists a number of state observers for such an unknown-input system. The thesis of Singer [2019] provides a systematic comparison of these observers with respect to the automatability of their designs. An approach that is particularly easy to automate is the unknown-input observer design from Darouach et al. [1994]. The following assumption represents a necessary and sufficient condition for the existence of this observer (cf. Darouach et al. [1994, Theorem 1] and Singer [2019, Remark 3.2]):

Assumption 4.7 (Strong* detectability)
The system (4.6) is strong detectable, i.e.,* $\lim_{t\to\infty} \bar{\boldsymbol{m}} \to \boldsymbol{0}$ *implies* $\lim_{t\to\infty} \boldsymbol{x} \to \boldsymbol{0}$.

For a brief introduction to the concept of strong* detectability, the reader is referred to Appendix C.1. The existence condition from Assumption 4.7 now allows to state the main theorem of this subsection. In this theorem, the state observer from Darouach et al. [1994] is extended by an input and an output reconstruction.

Theorem 4.8 (Input-state-output observer)
Consider a PHS (4.1) with measurements (4.2). Let Assumption 4.4 hold which permits to formulate an unknown-input system (4.6). For the system (4.6), let Assumption 4.7 hold. Then, there exist matrices $\boldsymbol{N} \in \mathbb{R}^{n\times n}$, $\boldsymbol{L} \in \mathbb{R}^{n\times(\tilde{q}-r)}$, $\boldsymbol{F} \in \mathbb{R}^{n\times p_\mathrm{k}}$, *and* $\boldsymbol{E} \in \mathbb{R}^{n\times(\tilde{q}-r)}$ *such that the system*

$$\dot{\boldsymbol{z}} = \boldsymbol{N}\boldsymbol{z} + \boldsymbol{L}\bar{\boldsymbol{m}} + \boldsymbol{F}\boldsymbol{u}_\mathrm{k}, \quad \boldsymbol{z}|_{t=0} = \boldsymbol{z}_0, \tag{4.12a}$$

$$\hat{\boldsymbol{x}} = \boldsymbol{z} - \boldsymbol{E}\bar{\boldsymbol{m}}, \tag{4.12b}$$

$$\hat{\boldsymbol{u}}_\mathrm{u} = \left(\bar{\boldsymbol{C}}\boldsymbol{B}_\mathrm{u}\right)^+ \left(\dot{\bar{\boldsymbol{m}}} - \bar{\boldsymbol{C}}\boldsymbol{A}\hat{\boldsymbol{x}} - \bar{\boldsymbol{C}}\boldsymbol{B}_\mathrm{k}\boldsymbol{u}_\mathrm{k}\right), \tag{4.12c}$$

$$\hat{\boldsymbol{y}} = (\boldsymbol{G}+\boldsymbol{P})^\top \boldsymbol{Q}\hat{\boldsymbol{x}} + (\boldsymbol{M}+\boldsymbol{S})\left(\boldsymbol{u}_\mathrm{k}^\top \ \hat{\boldsymbol{u}}_\mathrm{u}^\top\right)^\top, \tag{4.12d}$$

is an asymptotic input-state-output observer for the PHS (4.1) based on measurements (4.2). In (4.12), the vector $\boldsymbol{z} \in \mathbb{R}^n$ *is the observer state and the term* $\left(\bar{\boldsymbol{C}}\boldsymbol{B}_\mathrm{u}\right)^+$ *is the Moore-Penrose inverse of* $\bar{\boldsymbol{C}}\boldsymbol{B}_\mathrm{u}$.

Proof:
We first prove that (4.12a) and (4.12b) yield an asymptotic reconstruction of the system state—independently of the unknown inputs. To this end, we follow the

approach from Darouach et al. [1994]. Afterwards, we show that (4.12c) and (4.12d) yield an asymptotic reconstruction of the unknown inputs and outputs, respectively.

Let Assumption 4.7 hold. Consider the reconstruction error $\boldsymbol{\varepsilon} = \hat{\boldsymbol{x}} - \boldsymbol{x}$. With (4.6) and (4.12), the error dynamics read

$$\dot{\boldsymbol{\varepsilon}} = \boldsymbol{N}\boldsymbol{\varepsilon} + \left(\boldsymbol{N}\boldsymbol{K} + \boldsymbol{L}\bar{\boldsymbol{C}} - \boldsymbol{K}\boldsymbol{A}\right)\boldsymbol{x} + (\boldsymbol{F} - \boldsymbol{K}\boldsymbol{B}_{\mathrm{k}})\,\boldsymbol{u}_{\mathrm{k}} - \boldsymbol{K}\boldsymbol{B}_{\mathrm{u}}\boldsymbol{u}_{\mathrm{u}}, \tag{4.13}$$

where $\boldsymbol{K} = \boldsymbol{I} + \boldsymbol{E}\bar{\boldsymbol{C}}$. Suppose we have

$$\boldsymbol{0} = \boldsymbol{N}\boldsymbol{K} + \boldsymbol{L}\bar{\boldsymbol{C}} - \boldsymbol{K}\boldsymbol{A}, \tag{4.14a}$$
$$\boldsymbol{0} = \boldsymbol{F} - \boldsymbol{K}\boldsymbol{B}_{\mathrm{k}}, \tag{4.14b}$$
$$\boldsymbol{0} = \boldsymbol{K}\boldsymbol{B}_{\mathrm{u}}. \tag{4.14c}$$

If the conditions in (4.14) hold, the error dynamics read $\dot{\boldsymbol{\varepsilon}} = \boldsymbol{N}\boldsymbol{\varepsilon}$. If, in addition, $\boldsymbol{N}$ is a Hurwitz matrix, we have $\boldsymbol{\varepsilon} \to \boldsymbol{0}$ for $t \to \infty$.

Next, we show that we can always find matrices $\boldsymbol{N}$, $\boldsymbol{L}$, $\boldsymbol{F}$, and $\boldsymbol{E}$ such that (4.14) is fulfilled and $\boldsymbol{N}$ is a Hurwitz matrix.

Assumption 4.7 implies $\operatorname{rank}\left(\bar{\boldsymbol{C}}\boldsymbol{B}_{\mathrm{u}}\right) = \operatorname{rank}\left(\boldsymbol{B}_{\mathrm{u}}\right) = p_{\mathrm{u}}$ [Hautus, 1983, Theorem 1.6]. Hence, from (4.14c) we obtain the following solution set for $\boldsymbol{E}$:

$$\boldsymbol{E} = -\boldsymbol{B}_{\mathrm{u}}\left(\bar{\boldsymbol{C}}\boldsymbol{B}_{\mathrm{u}}\right)^{+} + \boldsymbol{Y}\left(\boldsymbol{I} - \left(\bar{\boldsymbol{C}}\boldsymbol{B}_{\mathrm{u}}\right)\left(\bar{\boldsymbol{C}}\boldsymbol{B}_{\mathrm{u}}\right)^{+}\right), \tag{4.15}$$

where

$$\left(\bar{\boldsymbol{C}}\boldsymbol{B}_{\mathrm{u}}\right)^{+} = \left(\left(\bar{\boldsymbol{C}}\boldsymbol{B}_{\mathrm{u}}\right)^{\top}\left(\bar{\boldsymbol{C}}\boldsymbol{B}_{\mathrm{u}}\right)\right)^{-1}\left(\bar{\boldsymbol{C}}\boldsymbol{B}_{\mathrm{u}}\right)^{\top} \tag{4.16}$$

and $\boldsymbol{Y} \in \mathbb{R}^{n \times p_{\mathrm{u}}}$ is an arbitrary matrix [Rao and Mitra, 1972]. Based on a particular solution for $\boldsymbol{E}$ we can then calculate matrix $\boldsymbol{F}$ from (4.14b):

$$\boldsymbol{F} = \left(\boldsymbol{I} + \boldsymbol{E}\bar{\boldsymbol{C}}\right)\boldsymbol{B}_{\mathrm{k}}. \tag{4.17}$$

For the determination of the matrix $\boldsymbol{N}$ we write (4.14a) as

$$\boldsymbol{N} = \boldsymbol{KA} - \boldsymbol{Z}\bar{\boldsymbol{C}}, \tag{4.18}$$

where $\boldsymbol{Z} = (\boldsymbol{L} + \boldsymbol{NE})$. From Darouach et al. [1994, Theorem 2] and Hautus [1983, Theorem 1.5] it follows that Assumption 4.7 implies the pair $(\boldsymbol{KA}, \bar{\boldsymbol{C}})$ to be detectable. Thus, there always exists a matrix $\boldsymbol{Z}$ such that $\boldsymbol{N}$ is Hurwitz. Substituting (4.18) into $\boldsymbol{Z} = (\boldsymbol{L} + \boldsymbol{NE})$ yields the following expression for $\boldsymbol{L}$:

$$\boldsymbol{L} = \boldsymbol{Z}\left(\boldsymbol{I} + \bar{\boldsymbol{C}}\boldsymbol{E}\right) - \boldsymbol{KAE}, \tag{4.19}$$

which then satisfies (4.14a). Hence, we can always find matrices $(\boldsymbol{N}, \boldsymbol{L}, \boldsymbol{F}, \boldsymbol{E})$ such that the conditions in (4.14) are fulfilled and such that $\boldsymbol{N}$ is a Hurwitz matrix.

Now for the reconstruction of the unknown inputs. We apply the idea of Ding [2008, p. 143]. Deriving (4.6b) with respect to time gives

$$\dot{\boldsymbol{m}} = \bar{\boldsymbol{C}}\dot{\boldsymbol{x}} = \bar{\boldsymbol{C}}\boldsymbol{A}\boldsymbol{x} + \bar{\boldsymbol{C}}\boldsymbol{B}_{\mathrm{k}}\boldsymbol{u}_{\mathrm{k}} + \bar{\boldsymbol{C}}\boldsymbol{B}_{\mathrm{u}}\boldsymbol{u}_{\mathrm{u}}. \tag{4.20}$$

With $\left(\bar{\boldsymbol{C}}\boldsymbol{B}_{\mathrm{u}}\right)^{+}$ we may solve (4.20) for $\boldsymbol{u}_{\mathrm{u}}$:

$$\boldsymbol{u}_{\mathrm{u}} = \left(\bar{\boldsymbol{C}}\boldsymbol{B}_{\mathrm{u}}\right)^{+}\left(\dot{\boldsymbol{m}} - \bar{\boldsymbol{C}}\boldsymbol{A}\boldsymbol{x} - \bar{\boldsymbol{C}}\boldsymbol{B}_{\mathrm{k}}\boldsymbol{u}_{\mathrm{k}}\right). \tag{4.21}$$

In (4.21), we substitute $\boldsymbol{u}_{\mathrm{u}}$ with $\hat{\boldsymbol{u}}_{\mathrm{u}}$ and $\boldsymbol{x}$ with $\hat{\boldsymbol{x}}$ and obtain (4.12c). By comparing (4.21) and (4.12c), we may deduce $\hat{\boldsymbol{u}}_{\mathrm{u}} \to \boldsymbol{u}_{\mathrm{u}}$ from $\hat{\boldsymbol{x}} \to \boldsymbol{x}$ for $t \to \infty$.

Finally, Lemma 4.2 proves (4.12d) to yield an output reconstruction $\hat{\boldsymbol{y}}$ with $\hat{\boldsymbol{y}} \to \boldsymbol{y}$ for $t \to \infty$. □

In order to automate the design of the observer from Theorem 4.8, a computationally evaluable formulation of Assumption 4.7 is required. The subsequent proposition provides such a formulation.

Proposition 4.9 (Evaluation of Assumption 4.7)

Assumption 4.7 is satisfied if and only if the following two conditions hold:

i. $\operatorname{rank}\left(\bar{\boldsymbol{C}}\boldsymbol{B}_u\right) = \operatorname{rank}\left(\boldsymbol{B}_u\right) = p_u,$ (4.22a)

ii. $\operatorname{rank}\begin{pmatrix} s\boldsymbol{I}-\boldsymbol{A} & -\boldsymbol{B}_u \\ \bar{\boldsymbol{C}} & \boldsymbol{0} \end{pmatrix} = n + p_u, \quad \forall s \in \operatorname{Spec}(\boldsymbol{A}) : Re(s) \geq 0.$ (4.22b)

Proof:

The proof follows directly from Darouach et al. [1994, Theorem 2] and Hautus [1983, Theorem 1.5] by noting that, in condition *ii.*, "$\forall s \in \mathbb{C}$" is equivalent to "$\forall s \in \operatorname{Spec}(\boldsymbol{A})$" (cf. Hautus [1969]). □

Algorithm 4.10 now summarizes the above insights in an automatable design scheme which determines the free observer parameters of the input-state-output observer from Theorem 4.8, i.e., the matrices $\boldsymbol{N}, \boldsymbol{L}, \boldsymbol{F}$, and $\boldsymbol{E}$.

In line 20 of Algorithm 4.10, the choice of the eigenvalues of the matrix $\boldsymbol{N}$ determines the dynamics of the state reconstruction error (cf. (4.13)). Appendix C.2 provides a straightforward method for the automated placement of the eigenvalues of $\boldsymbol{N}$. Based on the eigenvalues chosen for $\boldsymbol{N}$, the determination of the matrix $\boldsymbol{Z}$ in code line 21 can be accomplished by the standard pole placement techniques presented in Section 1.2.

The block diagram in Figure 4.2 depicts the structure of the input-state-output observer. As can be seen, the observer consists of two stages, viz. a pre-processing of the measurements and the actual observer (4.12).

The input-state-output observer in Figure 4.2 yields asymptotic reconstructions $\hat{\boldsymbol{u}}$, $\hat{\boldsymbol{x}}$, and $\hat{\boldsymbol{y}}$ of the inputs, states, and outputs of the PHS. Theorem 4.8 gives a sufficient condition for the existence of such an input-state-output observer; Algorithm 4.10 provides a corresponding design scheme which can be fully automated. Hence, these methods solve Problem 4.1.

Algorithm 4.10 Automated design of an input-state-output observer

Input: PHS (4.1) with measurements (4.2)
Calculate matrices $\boldsymbol{A}$, $\boldsymbol{B}$, $\boldsymbol{C}$, and $\boldsymbol{D}$ according to Lemma 4.3
Split $\boldsymbol{u}$ into $(\boldsymbol{u}_\mathrm{k}, \boldsymbol{u}_\mathrm{u})$ and $\boldsymbol{m}$ into $(\boldsymbol{u}_\mathrm{k}, \tilde{\boldsymbol{m}})$ (possibly after permutations)
Calculate $\boldsymbol{B}_\mathrm{k}$, $\boldsymbol{B}_\mathrm{u}$, $\tilde{\boldsymbol{D}}_\mathrm{k}$, and $\tilde{\boldsymbol{D}}_\mathrm{u}$
if $\tilde{\boldsymbol{D}}_\mathrm{u} = 0$ **then**
    $\boldsymbol{U}_2 \leftarrow \boldsymbol{I}$
else
    if Assumption 4.4 is met **then**
        $[(\boldsymbol{U}_1, \boldsymbol{U}_2), \boldsymbol{\Sigma}, \boldsymbol{V}] \leftarrow \mathrm{svd}(\tilde{\boldsymbol{D}}_\mathrm{u})$
    else
        print "Feedthrough could not be eliminated. Algorithm terminates!"
        terminate
    end if
end if
$\bar{\boldsymbol{C}} \leftarrow \boldsymbol{U}_2^\top \boldsymbol{C}$
if (4.22a) or (4.22b) (i.e., Assumption 4.7) is violated **then**
    print "Observer existence condition is violated. Algorithm terminates!"
    terminate
end if
Calculate $\boldsymbol{E}$ and $\boldsymbol{F}$ from (4.15) and (4.17), respectively
Specify the eigenvalues of $\boldsymbol{N}$ that correspond to the observable subsystem of $(\boldsymbol{KA}, \bar{\boldsymbol{C}})$
Calculate $\boldsymbol{Z}$ from (4.18) by pole placement techniques
Calculate $\boldsymbol{N}$ and $\boldsymbol{L}$ from (4.18) and (4.19), respectively
return $(\boldsymbol{N}, \boldsymbol{L}, \boldsymbol{F}, \boldsymbol{E})$

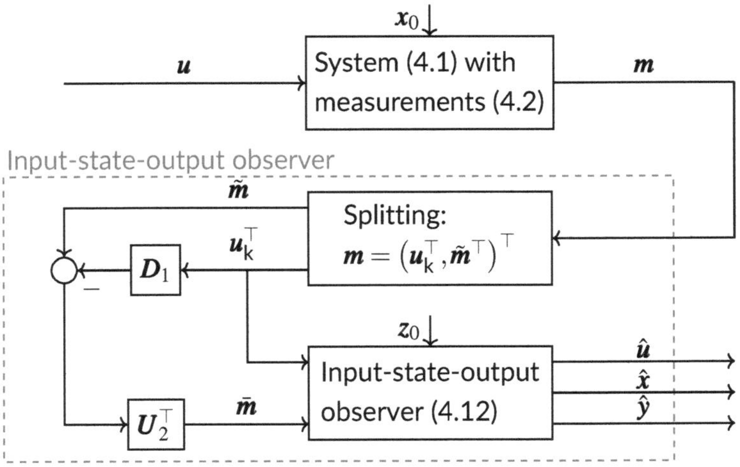

Figure 4.2: Block diagram of the system with the input-state-output observer

Example 4.11 illustrates the results from above.

Example 4.11:

Consider the following linear PHS:

$$\begin{pmatrix} \dot{x}_1 \\ \dot{x}_2 \end{pmatrix} = \begin{pmatrix} 0 & -1 \\ 1 & -d \end{pmatrix} \frac{\partial H}{\partial \boldsymbol{x}} + \begin{pmatrix} 1 \\ d \end{pmatrix} u, \tag{4.23a}$$

$$y = \begin{pmatrix} 1 & -d \end{pmatrix} \frac{\partial H}{\partial \boldsymbol{x}} + du, \tag{4.23b}$$

with $d > 0$ and Hamiltonian $H(\boldsymbol{x}) = \boldsymbol{x}^\top \boldsymbol{x}$. In the system, u, $\boldsymbol{x}$, and y are unknown. Instead, we consider $q = 2$ measurements

$$\boldsymbol{m} = \underbrace{\begin{pmatrix} -d \\ 0 \end{pmatrix}}_{=\boldsymbol{C}_u} u + \underbrace{\begin{pmatrix} 0 & 2d \\ 0 & 2 \end{pmatrix}}_{=\boldsymbol{C}_x} \boldsymbol{x} + \underbrace{\begin{pmatrix} 1 \\ 0 \end{pmatrix}}_{=\boldsymbol{C}_y} y. \tag{4.24}$$

With the formulas from Lemma 4.3, we can write (4.23a) and (4.24) as the following state-space system:

$$\dot{\boldsymbol{x}} = \begin{pmatrix} 0 & -2 \\ 2 & -2d \end{pmatrix} \boldsymbol{x} + \begin{pmatrix} 1 \\ d \end{pmatrix} u, \tag{4.25a}$$

$$\boldsymbol{m} = \begin{pmatrix} 2 & 0 \\ 0 & 2 \end{pmatrix} \boldsymbol{x}, \tag{4.25b}$$

where u is an unknown input. As can be seen, in (4.25) we have no feed-through which is why Assumption 4.4 is irrelevant in this example (cf. Remark 4.6). Moreover, the conditions in (4.22) are satisfied. Hence, Assumption 4.7 is met and we can design an input-state-output observer. From (4.15) and (4.17) we obtain

$$\boldsymbol{E} = -\frac{1}{2} \begin{pmatrix} d^{-1} & 0 \\ 1 & 0 \end{pmatrix} \tag{4.26}$$

and $\boldsymbol{F} \in \mathbb{R}^{2\times 0}$, respectively. The eigenvalues of the matrix $\boldsymbol{N}$ are placed at $\lambda_{min} = -1$ and $\lambda_{min} = -2$ where $\lambda_{min} = \min\{Re(\lambda_i)\}$ with λ_i the eigenvalues of the matrix $\boldsymbol{A}$ from (4.25) ($i = 1,2$). Via pole placement, we obtain

$$\boldsymbol{Z} = \frac{1}{2}\begin{pmatrix} 0 & Re\left(\sqrt{d^2-4}\right)+d-\frac{2}{d}+2 \\ Re\left(\sqrt{d^2-4}\right)+d+1 & 0 \end{pmatrix}. \tag{4.27}$$

Based on (4.27), we may now calculate $\boldsymbol{N}$ and $\boldsymbol{L}$ with (4.18) and (4.19) as

$$\boldsymbol{N} = \begin{pmatrix} -Re\left(\sqrt{d^2-4}\right)-d-2 & 0 \\ 0 & -Re\left(\sqrt{d^2-4}\right)-d-1 \end{pmatrix} \tag{4.28a}$$

and

$$\boldsymbol{L} = \begin{pmatrix} -\frac{Re\left(\sqrt{d^2-4}\right)+d+2}{2d} & \frac{1}{2}\left(Re\left(\sqrt{d^2-4}\right)+d-\frac{2}{d}+2\right) \\ 0 & 0 \end{pmatrix}, \tag{4.28b}$$

respectively. This concludes the observer design.

The obtained input-state-output observer are illustrated by means of the results obtained from numerical simulations. The parameter d is chosen to $d = 1$. The initial values of the system and the observer are $\boldsymbol{x}_0 = (0\ \ 0)^\top$ and $\hat{\boldsymbol{z}}_0 = (1\ \ 1)^\top$ $(= \hat{\boldsymbol{x}}_0)$, respectively. The unknown input is a unit sawtooth signal with an angular frequency of 1 s^{-1}.

Figure 4.3 depicts the inputs, states, and outputs of the system (4.23) (solid, blue) together with the corresponding reconstructions from the input-state-output observer (dashed, red).

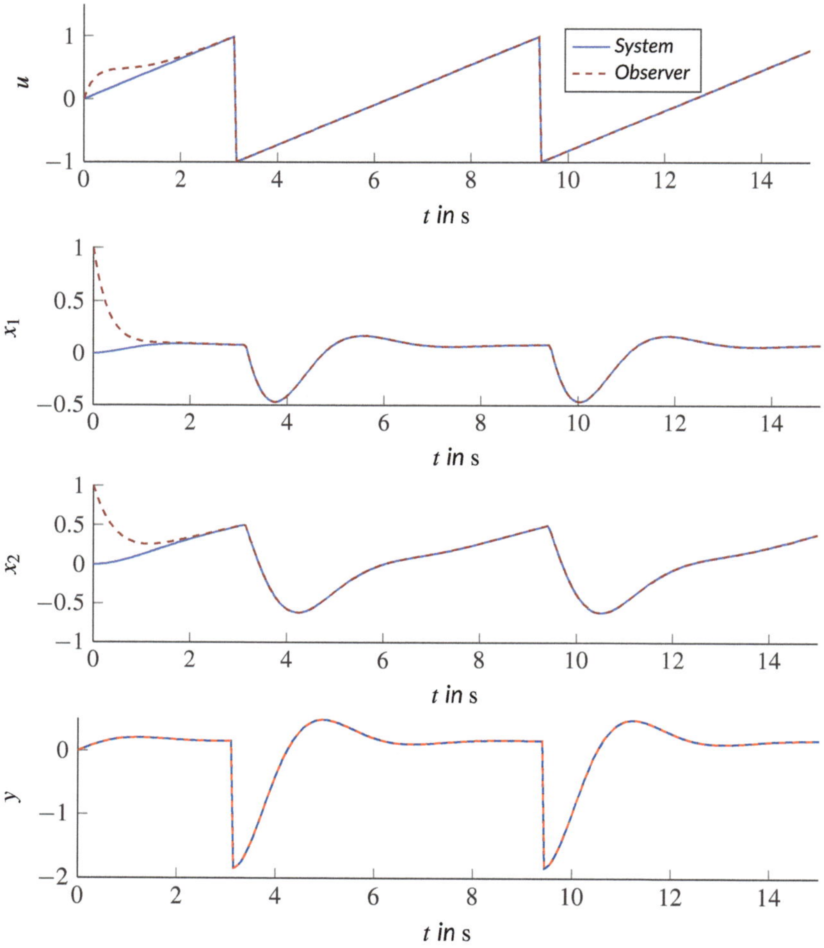

Figure 4.3: Inputs, states, and outputs of the PHS (4.23) (solid, blue) and corresponding observer reconstructions (dashed, red)

As can be seen, after three seconds the reconstructions of the input and states lie on top of the true system variables. Moreover, we obtain an output reconstruction which is identical to the true output. Unfortunately, this identity is coincidence and cannot be regarded as a general property of the observer.

The previous example illustrates the algorithmic design of a symbolic input-state-output observer for a PHS (4.1). The obtained observer yields asymptotic reconstructions of the unknown system variables which confirms the theoretical results from this subsection.

The present subsection was devoted to the development of an observer for linear PHSs. In the next subsection, we turn our attention to the automated observer design for nonlinear PHSs.

4.2.2 Automated Observer Design for a Class of Nonlinear PHSs

Consider an interconnected system with nonlinearities in the interconnection structure and storages. The former kind of nonlinearity appears, for example, in the transformations between different reference frames in electric and mechanic systems as, e.g., the *dq*-transformation or generalized coordinate transformations, respectively (cf. Karnopp [1969]); the latter kind is present, for instance, in systems involving the potential energy of a mass[3] (cf. van der Schaft [2009, p. 66]) or systems exhibiting effects from relativistic mechanics (cf. Wellstead [1979, p. 27]). Systems with nonlinear interconnection structure and nonlinear storages lead to explicit port-Hamiltonian models with state-dependent matrices and possibly non-quadratic Hamiltonians (cf. Section 3.3). For many of such PHSs, the state and output of the system are unknown and have to be reconstructed. However, in contrast to Subsection 4.2.1, we now consider the situation where at least the input of the system is known. Besides the system input, we are given measurements that may be nonlinear in the states. The aim of this section is then to find an automatable design scheme for an asymptotic observer which reconstructs the states and outputs of the system. This is formalized in the following problem:

[3]Recall that the potential energy $V(x)$ of a mass m is given by $V(x) = mgx$, where g is the gravitional constant. This storage function is non-quadratic in the state and the corresponding storage element therewith nonlinear.

Problem 4.12 (Automated observer design for nonlinear PHSs)
Consider an explicit PHS of the form

$$\frac{\mathrm{d}}{\mathrm{d}t}\begin{pmatrix}\boldsymbol{x}'\\ \boldsymbol{x}''\end{pmatrix} = \left(\boldsymbol{J}(\boldsymbol{x}') - \boldsymbol{R}(\boldsymbol{x}')\right)\frac{\partial H}{\partial \boldsymbol{x}}(\boldsymbol{x}) + \boldsymbol{G}(\boldsymbol{x}')\boldsymbol{u}, \quad \boldsymbol{x}|_{t=0} = \boldsymbol{x}_0, \tag{4.29a}$$

$$\boldsymbol{y} = \boldsymbol{G}^\top(\boldsymbol{x}')\frac{\partial H}{\partial \boldsymbol{x}}(\boldsymbol{x}), \tag{4.29b}$$

with $\boldsymbol{x}' \in \mathbb{X}' \subset \mathbb{R}^{n_1}$, $\boldsymbol{x}'' \in \mathbb{X}'' \subset \mathbb{R}^{n-n_1}$, $\boldsymbol{u} \in \mathbb{U} \subset \mathbb{R}^p$, and $\boldsymbol{y} \in \mathbb{Y} \subset \mathbb{R}^p$, where $\mathbb{X}'$ and $\mathbb{X}''$ are closed and bounded and therewith compact. The overall state vector is defined as $\boldsymbol{x} := (\boldsymbol{x}'^\top\ \boldsymbol{x}''^\top)^\top \in \mathbb{X} = \mathbb{X}' \times \mathbb{X}''$, where $\mathbb{X}$ is then also compact. The matrices in (4.29) are of proper sizes, continuously differentiable in $\boldsymbol{x}'$, and have the usual symmetry and definiteness properties from Definition 2.23. Let the Hamiltonian of (4.29) be of the form

$$H(\boldsymbol{x}) = \frac{1}{2}\left(\boldsymbol{x}'^\top\ \ \boldsymbol{x}''^\top\right)\begin{pmatrix}\boldsymbol{Q}' & \boldsymbol{0}\\ \boldsymbol{0} & \boldsymbol{Q}''\end{pmatrix}\begin{pmatrix}\boldsymbol{x}'\\ \boldsymbol{x}''\end{pmatrix} + \boldsymbol{N}(\boldsymbol{x}'), \tag{4.30}$$

where $\boldsymbol{Q} := \mathrm{blkdiag}\left(\boldsymbol{Q}', \boldsymbol{Q}''\right) = \boldsymbol{Q}^\top \succ 0$ and $\boldsymbol{N} : \mathbb{X}' \to \mathbb{R}$, $\boldsymbol{x}' \mapsto \boldsymbol{N}(\boldsymbol{x}')$. The function $\boldsymbol{N}$ may be any function that is positive semi-definite and twice continuously differentiable in $\boldsymbol{x}'$.
Suppose $\boldsymbol{u}$ is known but $\boldsymbol{x}$ and $\boldsymbol{y}$ are unknown. Moreover, assume measurements $\boldsymbol{m} \in \mathbb{R}^q$ with $q \geq n_1$ of the form $\boldsymbol{m} = \boldsymbol{C}(\boldsymbol{x}')\boldsymbol{Q}\boldsymbol{x}$ where $\boldsymbol{C}(\boldsymbol{x}')$ is continuous in $\boldsymbol{x}'$:

$$\begin{pmatrix}\boldsymbol{m}_1\\ \boldsymbol{m}_2\end{pmatrix} = \begin{pmatrix}\boldsymbol{Q}'^{-1} & \boldsymbol{0}\\ \boldsymbol{C}'(\boldsymbol{x}') & \boldsymbol{C}''(\boldsymbol{x}')\end{pmatrix}\begin{pmatrix}\boldsymbol{Q}' & \boldsymbol{0}\\ \boldsymbol{0} & \boldsymbol{Q}''\end{pmatrix}\begin{pmatrix}\boldsymbol{x}'\\ \boldsymbol{x}''\end{pmatrix}. \tag{4.31}$$

Note that we have $\boldsymbol{m}_1 = \boldsymbol{x}'$, i.e., $\boldsymbol{x}'$ is the measured part of the state vector $\boldsymbol{x}$.
What is an asymptotic observer for (4.29) that produces reconstructions of $\boldsymbol{x}$ and $\boldsymbol{y}$ based on knowledge on $\boldsymbol{m}$? How can we design such an observer in an automated manner?

Remark 4.13 (Class of systems). At first glance, the class of systems addressed in Problem 4.12 may seem rather restrictive. However, as Venkatraman and van der Schaft [2010] point out, this class covers a considerable number of physical examples such as mechanical and electromechanical PHSs, see, e.g., Yaghmaei and Yazdanpanah [2019b, Eq. (23) and (27)]. Moreover, note that the measurement equation can also be written as

$$\begin{pmatrix} \boldsymbol{m}_1 \\ \boldsymbol{m}_2 \end{pmatrix} = \begin{pmatrix} \boldsymbol{I} & \boldsymbol{0} \\ \tilde{\boldsymbol{C}}'(\boldsymbol{x}') & \tilde{\boldsymbol{C}}''(\boldsymbol{x}') \end{pmatrix} \begin{pmatrix} \boldsymbol{x}' \\ \boldsymbol{x}'' \end{pmatrix}.^{4} \tag{4.32}$$

In (4.32), we have $\boldsymbol{m}_1 = \boldsymbol{x}'$, $\boldsymbol{m}_2 = \tilde{\boldsymbol{C}}(\boldsymbol{x}')\boldsymbol{x}$ which reveals the generality of this formulation.

Remark 4.14 (System feedthrough). In Problem 4.12, we consider a PHS without feedthrough. This is in contrast to Problem 4.1 from Subsection 4.2.1. However, due to the known inputs there is no loss of generality in neglecting the feedthrough (cf. Ludyk [1995, p. 7]). Throughout this section the omission of the feedthrough will allow for a compact notation.

Problem 4.12 contains a state-output reconstruction problem which involves three equations, viz. a dynamics equation (4.29a), an output equation (4.29b), and a measurement equation (4.31). Note that the measurement equation may be nonlinear in the states. Similar to Lemma 4.2, the following lemma shows that the state-output reconstruction problem from Problem 4.12 can be easily reduced to a state reconstruction problem which involves only two equations. Thereby, we consider an exponential convergence of the reconstructions.

Lemma 4.15 (Output reconstruction based on $\hat{\boldsymbol{x}}$)

Consider the situation in Problem 4.12. Let $\hat{\boldsymbol{x}}$ be a reconstruction of $\boldsymbol{x}$ with $||\boldsymbol{x} - \hat{\boldsymbol{x}}|| \leq k_1 e^{-k_2 t}$ for $t \geq 0$ and some positive constants $k_1, k_2 \in \mathbb{R}_{>0}$. Then, we can calculate an output reconstruction

$$\hat{\boldsymbol{y}} = \boldsymbol{G}^{\top}(\hat{\boldsymbol{x}}') \frac{\partial H}{\partial \boldsymbol{x}}(\hat{\boldsymbol{x}}), \tag{4.33}$$

with $||\boldsymbol{y} - \hat{\boldsymbol{y}}|| \leq k_3 e^{-k_2 t}$ for all $t \geq 0$ and some positive constant $k_3 \in \mathbb{R}_{>0}$.

[4] To bring (4.32) to the form (4.31), we write $\boldsymbol{m} = \tilde{\boldsymbol{C}}(\boldsymbol{x}')\boldsymbol{Q}^{-1}\boldsymbol{Q}\boldsymbol{x} = \boldsymbol{C}(\boldsymbol{x}')\boldsymbol{Q}\boldsymbol{x}$ with $\boldsymbol{C}(\boldsymbol{x}') = \tilde{\boldsymbol{C}}(\boldsymbol{x}')\boldsymbol{Q}^{-1}$.

Proof:
Because $\mathbb{X}'$ and $\mathbb{X}$ are compact, $\boldsymbol{G}^\top$ and $\frac{\partial H}{\partial \boldsymbol{x}}$ are bounded in $\boldsymbol{x}'$ and $\boldsymbol{x}$, respectively, i.e., there exist constants $k_G, k_H \in \mathbb{R}_{>0}$ such that $||\boldsymbol{G}^\top(\boldsymbol{x}')|| < k_G$ and $||\frac{\partial H}{\partial \boldsymbol{x}}(\boldsymbol{x})|| < k_H$ for all $\boldsymbol{x}' \in \mathbb{X}'$ and $\boldsymbol{x} \in \mathbb{X}$. Moreover, since $\boldsymbol{G}^\top$ is continuously differentiable and $\mathbb{X}'$ is compact, $\boldsymbol{G}^\top$ is Lipschitz continuous on $\mathbb{X}'$ with constant $L_G = \sup_{\boldsymbol{x}' \in \mathbb{X}'} ||\frac{\partial G}{\partial \boldsymbol{x}'}(\boldsymbol{x}')||$ that is $||\boldsymbol{G}^\top(\boldsymbol{x}_1') - \boldsymbol{G}^\top(\boldsymbol{x}_2')|| \leq L_G ||\boldsymbol{x}_1' - \boldsymbol{x}_2'||$ for all $\boldsymbol{x}_1', \boldsymbol{x}_2' \in \mathbb{X}'$. Likewise $\frac{\partial H}{\partial \boldsymbol{x}}$ is Lipschitz continuous with a constant L_H on $\mathbb{X}$.

We now can conclude

$$\begin{aligned}
||\boldsymbol{y} - \hat{\boldsymbol{y}}|| &= ||\boldsymbol{G}^\top(\boldsymbol{x}')\frac{\partial H}{\partial \boldsymbol{x}}(\boldsymbol{x}) - \boldsymbol{G}^\top(\hat{\boldsymbol{x}}')\frac{\partial H}{\partial \boldsymbol{x}}(\hat{\boldsymbol{x}})|| \\
&\leq ||\boldsymbol{G}^\top(\boldsymbol{x}')\frac{\partial H}{\partial \boldsymbol{x}}(\boldsymbol{x}) - \boldsymbol{G}^\top(\boldsymbol{x}')\frac{\partial H}{\partial \boldsymbol{x}}(\hat{\boldsymbol{x}})|| + ||\boldsymbol{G}^\top(\boldsymbol{x}')\frac{\partial H}{\partial \boldsymbol{x}}(\hat{\boldsymbol{x}}) - \boldsymbol{G}^\top(\hat{\boldsymbol{x}}')\frac{\partial H}{\partial \boldsymbol{x}}(\hat{\boldsymbol{x}})|| \\
&\leq ||\boldsymbol{G}^\top(\boldsymbol{x}')|| \, ||\frac{\partial H}{\partial \boldsymbol{x}}(\boldsymbol{x}) - \frac{\partial H}{\partial \boldsymbol{x}}(\hat{\boldsymbol{x}})|| + ||\boldsymbol{G}^\top(\boldsymbol{x}') - \boldsymbol{G}^\top(\hat{\boldsymbol{x}}')|| \, ||\frac{\partial H}{\partial \boldsymbol{x}}(\hat{\boldsymbol{x}})|| \\
&\leq k_G L_H ||\boldsymbol{x} - \hat{\boldsymbol{x}}|| + L_G ||\boldsymbol{x}' - \hat{\boldsymbol{x}}'|| k_H \\
&\leq (k_G L_H + L_G k_H) k_1 e^{-k_2 t},
\end{aligned} \tag{4.34}$$

where in the last step we used $||\boldsymbol{x}' - \hat{\boldsymbol{x}}'|| \leq ||\boldsymbol{x} - \hat{\boldsymbol{x}}||$ and $||\boldsymbol{x} - \hat{\boldsymbol{x}}|| \leq k_1 e^{-k_2 t}$. □

Lemma 4.15 shows that an exponentially convergent reconstruction of the output can always be obtained from an exponentially convergent reconstruction of the state. Hence, Problem 4.12 can be formulated as an ordinary state reconstruction problem that involves two equations, viz. (4.29a) and (4.31). This motivates to approach with a Luenberger-like observer consisting of an internal model of the system dynamics and a measurement error injection term. This is the approach we follow in the subsequent lemma.

Lemma 4.16 (Asymptotic state observer)

Consider a system with dynamics (4.29a) and measurements (4.31). Suppose there exists a matrix $\boldsymbol{L} \in \mathbb{R}^{n \times q}$ depending continuously on $\boldsymbol{x}'$ such that

$$\boldsymbol{R}(\boldsymbol{x}') + \frac{1}{2}\boldsymbol{L}(\boldsymbol{x}')\boldsymbol{C}(\boldsymbol{x}') + \frac{1}{2}\boldsymbol{C}^\top(\boldsymbol{x}')\boldsymbol{L}^\top(\boldsymbol{x}') \succ 0, \tag{4.35}$$

for all $x' \in \mathbb{X}'$. Then, there exists a globally exponentially convergent state observer of the form

$$\dot{\hat{x}} = \left(J(x') - R(x')\right) \frac{\partial H}{\partial x}(\hat{x}) + G(x')u + L(x')\left(m - C(x')Q\hat{x}\right), \tag{4.36}$$

with initial value $\hat{x}|_{t=0} = \hat{x}_0$. The vectors $\hat{x}' \in \mathbb{X}'$ and $\hat{x}'' \in \mathbb{X}''$ of the splitting $\hat{x} = \left(\hat{x}'^\top \; \hat{x}''^\top\right)^\top$ are mimicking the splitting of $x = (x'^\top \; x''^\top)^\top$.

Proof:

Let us define the reconstruction error as $\varepsilon := x - \hat{x}$. With (4.29a), (4.30), (4.31), and (4.36), the error dynamics can be expressed as

$$\dot{\varepsilon} = \left(J(x') - R(x') - L(x')C(x')\right) Q\varepsilon, \tag{4.37}$$

with initial value $\varepsilon_0 = x_0 - \hat{x}_0$. Obviously, $\varepsilon \equiv 0$ is an equilibrium of (4.37). Next, we analyze the stability of this equilibrium by using Lyapunov's direct method. Consider the Lyapunov candidate

$$V(\varepsilon) = \frac{1}{2}\varepsilon^\top Q\varepsilon. \tag{4.38}$$

As shown in Proposition C.4 from Appendix C.3, for a system and a Lyapunov candidate of the form (4.37) and (4.38), respectively, we obtain

$$\dot{V}(\varepsilon) = -\varepsilon^\top Q \underbrace{R(x') + \frac{1}{2}L(x')C(x') + \frac{1}{2}C^\top(x')L^\top(x')}_{=:\Gamma} Q\varepsilon. \tag{4.39}$$

It is noteworthy that (4.39) is independent of the matrix $J(x')$. Now let (4.35) hold. We then have $\Gamma = \Gamma^\top \succ 0$ which is equivalent to $Q\Gamma Q = (Q\Gamma Q)^\top \succ 0$. From this follows that $\dot{V}(\varepsilon)$ is negative-definite and thus $\varepsilon \equiv 0$ an asymptotically stable equilibrium of (4.37). Moreover, as shown in Proposition C.5 from Appendix C.3, the positive definiteness of Q and $Q\Gamma Q$ implies the existence of positive constants $k_1, k_2, k_3 \in \mathbb{R}_{>0}$ such that

$$k_1 \|\boldsymbol{\varepsilon}\|^2 \leq V(\boldsymbol{\varepsilon}) \leq k_2 \|\boldsymbol{\varepsilon}\|^2 \tag{4.40a}$$

and

$$\dot{V}(\boldsymbol{\varepsilon}) \leq -k_3 \|\boldsymbol{\varepsilon}\|^2 \tag{4.40b}$$

hold for all $\boldsymbol{x} \in \mathbb{X}$. Hence, $\boldsymbol{\varepsilon} \equiv \mathbf{0}$ is a globally exponentially stable equilibrium of (4.37) [Khalil, 2002, Theorem 4.10]. This implies (4.36) to be a globally exponentially convergent observer for the system consisting of (4.29) and (4.31). □

Equation (4.35) is a sufficient condition for the existence of an asymptotic observer of the form (4.36). Thus, the observer design problem is to find a matrix $\boldsymbol{L}(\boldsymbol{x}')$ such that (4.35) is fulfilled. For the case where $\boldsymbol{R}(\boldsymbol{x}') = \boldsymbol{R} = \text{const.}$, $\boldsymbol{C}(\boldsymbol{x}') = \boldsymbol{C} = \text{const.}$, we can approach with a constant matrix $\boldsymbol{L}(\boldsymbol{x}') = \boldsymbol{L}$. In this case, (4.35) represents a LMI. There exist powerful numerical methods to solve such an LMI, see, e.g., Boyd et al. [1994]. On the other hand, such an LMI-based design approach suffers from two shortcomings: (a) it is restricted to the case $\boldsymbol{R}(\boldsymbol{x}') = \boldsymbol{R} = \text{const.}$, $\boldsymbol{C}(\boldsymbol{x}') = \boldsymbol{C} = \text{const.}$; (b) in general it is limited to a numeric observer design, i.e., it disallows for a symbolic observer design. Hence, in the sequel, we consider an alternative approach for finding a matrix $\boldsymbol{L}(\boldsymbol{x}')$ such that (4.35) is satisfied.

Recall that $\boldsymbol{R}(\boldsymbol{x}') \succeq 0$ for all $\boldsymbol{x}' \in \mathbb{X}'$. For (4.35) to hold, we need to find a matrix $\boldsymbol{L}(\boldsymbol{x}')$ which moves the zero eigenvalues of $-\boldsymbol{R}(\boldsymbol{x}')$ to the left. The following lemma proposes a choice of $\boldsymbol{L}(\boldsymbol{x}')$ which has the best chances to accomplish this:

Lemma 4.17 (Observer design)
Consider two matrices $\boldsymbol{R}(\boldsymbol{s}) \in \mathbb{R}^{n\times n}$ and $\boldsymbol{C}(\boldsymbol{s}) \in \mathbb{R}^{q\times n}$ depending on some parameter $\boldsymbol{s} \in \mathbb{S}$. Let $\boldsymbol{R}(\boldsymbol{s}) = \boldsymbol{R}^\top(\boldsymbol{s}) \succeq 0$ for all $\boldsymbol{s} \in \mathbb{S}$. There exists a matrix $\boldsymbol{L}(\boldsymbol{s}) \in \mathbb{R}^{n\times q}$ which satisfies

$$\boldsymbol{R}(\boldsymbol{s}) + \frac{1}{2}\boldsymbol{L}(\boldsymbol{s})\boldsymbol{C}(\boldsymbol{s}) + \frac{1}{2}\boldsymbol{C}^\top(\boldsymbol{s})\boldsymbol{L}^\top(\boldsymbol{s}) \succ 0, \quad \forall \boldsymbol{s} \in \mathbb{S}, \tag{4.41}$$

if and only if (4.41) is satisfied for $\boldsymbol{L}(\boldsymbol{s}) = \boldsymbol{C}^\top(\boldsymbol{s})$.

Proof:
We show that the following two statements are equivalent:

$$(i) \quad \forall \boldsymbol{s} \in \mathbb{S}\colon \exists \boldsymbol{L}(\boldsymbol{s}) \in \mathbb{R}^{n\times q} \text{ s.t. } \boldsymbol{R}(\boldsymbol{s}) + \frac{1}{2}\boldsymbol{L}(\boldsymbol{s})\boldsymbol{C}(\boldsymbol{s}) + \frac{1}{2}\boldsymbol{C}^\top(\boldsymbol{s})\boldsymbol{L}^\top(\boldsymbol{s}) \succ 0, \tag{4.42a}$$

$$(ii) \quad \forall \boldsymbol{s} \in \mathbb{S}\colon \boldsymbol{R}(\boldsymbol{s}) + \boldsymbol{C}^\top(\boldsymbol{s})\boldsymbol{C}(\boldsymbol{s}) \succ 0. \tag{4.42b}$$

By setting $\boldsymbol{L}(\boldsymbol{s}) = \boldsymbol{C}^\top(\boldsymbol{s})$ it is easy to see that that (ii) implies (i). We now show that (i) also implies (ii). To this end, we show the contraposition, i.e., that if $\boldsymbol{R}(\boldsymbol{s}) + \boldsymbol{C}^\top(\boldsymbol{s})\boldsymbol{C}(\boldsymbol{s})$ is not positive-definite, then the matrix in (i) is not positive-definite for all $\boldsymbol{L}(\boldsymbol{s})$.
Let (ii) be violated. The matrix $\boldsymbol{R}(\boldsymbol{s}) + \boldsymbol{C}^\top(\boldsymbol{s})\boldsymbol{C}(\boldsymbol{s})$ is positive semi-definite, i.e.,

$$\boldsymbol{R}(\boldsymbol{s}) + \boldsymbol{C}^\top(\boldsymbol{s})\boldsymbol{C}(\boldsymbol{s}) \succeq 0, \quad \forall \boldsymbol{s} \in \mathbb{S}, \tag{4.43}$$

as $\boldsymbol{R}(\boldsymbol{s}) \succeq 0$ and

$$\boldsymbol{v}^\top\boldsymbol{C}^\top(\boldsymbol{s})\boldsymbol{C}(\boldsymbol{s})\boldsymbol{v} = \|\boldsymbol{C}(\boldsymbol{s})\boldsymbol{v}\| \geq 0, \quad \forall \boldsymbol{s} \in \mathbb{S}, \forall \boldsymbol{v} \in \mathbb{R}^n, \tag{4.44}$$

i.e., $\boldsymbol{C}^\top(\boldsymbol{s})\boldsymbol{C}(\boldsymbol{s}) \succeq 0$ for all $\boldsymbol{s} \in \mathbb{S}$. From (4.43) and the negation of (*ii*) follows, that there exists a non-zero vector $\boldsymbol{v} \in \mathbb{R}^n$ and a value $\boldsymbol{s}_0 \in \mathbb{S}$ such that

$$\boldsymbol{v}^\top \left(\boldsymbol{R}(\boldsymbol{s}_0) + \boldsymbol{C}^\top(\boldsymbol{s}_0)\boldsymbol{C}(\boldsymbol{s}_0)\right) \boldsymbol{v} = 0. \tag{4.45}$$

For this $\boldsymbol{v}$ and $\boldsymbol{s}_0$ we have

$$\begin{aligned} & \boldsymbol{v}^\top \boldsymbol{R}(\boldsymbol{s}_0)\boldsymbol{v} + \boldsymbol{v}^\top \boldsymbol{C}^\top(\boldsymbol{s}_0)\boldsymbol{C}(\boldsymbol{s}_0)\boldsymbol{v} = 0, \\ \Leftrightarrow \quad & \boldsymbol{v}^\top \boldsymbol{R}(\boldsymbol{s}_0)\boldsymbol{v} = 0 \quad \wedge \quad \boldsymbol{v}^\top \boldsymbol{C}^\top(\boldsymbol{s}_0)\boldsymbol{C}(\boldsymbol{s}_0)\boldsymbol{v} = 0, \\ \Leftrightarrow \quad & \boldsymbol{v}^\top \boldsymbol{R}(\boldsymbol{s}_0)\boldsymbol{v} = 0 \quad \wedge \quad \boldsymbol{v} \in \ker\left(\boldsymbol{C}(\boldsymbol{s}_0)\right). \end{aligned} \tag{4.46}$$

For the left hand side of (4.42a) we obtain

$$\underbrace{\boldsymbol{v}^\top \boldsymbol{R}(\boldsymbol{s}_0)\boldsymbol{v}}_{=0} + \frac{1}{2}\boldsymbol{v}^\top \boldsymbol{L}(\boldsymbol{s}_0)\underbrace{\boldsymbol{C}(\boldsymbol{s}_0)\boldsymbol{v}}_{=\boldsymbol{0}} + \frac{1}{2}\underbrace{\boldsymbol{v}^\top \boldsymbol{C}^\top(\boldsymbol{s}_0)}_{=\boldsymbol{0}}\boldsymbol{L}^\top(\boldsymbol{s}_0)\boldsymbol{v} = 0. \tag{4.47}$$

Hence, for $\boldsymbol{s}_0 \in \mathbb{S}$ and for all $\boldsymbol{L}(\boldsymbol{s})$ the matrix

$$\boldsymbol{R}(\boldsymbol{s}_0) + \frac{1}{2}\boldsymbol{L}(\boldsymbol{s}_0)\boldsymbol{C}(\boldsymbol{s}_0) + \frac{1}{2}\boldsymbol{C}^\top(\boldsymbol{s}_0)\boldsymbol{L}^\top(\boldsymbol{s}_0) \tag{4.48}$$

is not positive-definite. This is the contraposition of statement (*i*). □

Remark 4.18 (Scaling of the observer gain). Lemma 4.17 holds also for an observer gain $\boldsymbol{L}(\boldsymbol{s}) = \alpha \boldsymbol{C}^\top(\boldsymbol{s})$ where $\alpha \in \mathbb{R}_{>0}$. The parameter α can be used to increase ($\alpha > 1$) or decrease ($\alpha < 1$) the convergence rate of those observer states that are influenced by the error injection.

Lemma 4.16 presents a state observer for the PHS (4.29). Lemma 4.17 provides a simple design for such an observer. From Lemma 4.15 we know, that a state observer can be easily extended to a state-output observer. In the following theorem, we summarize these insights to formulate a globally exponentially convergent state-output observer for the PHS (4.29):

Theorem 4.19 (State-output observer)

Consider a nonlinear PHS (4.29) with Hamiltonian (4.30) and measurements (4.31). Let

$$\boldsymbol{R}(\boldsymbol{x}') + \boldsymbol{C}^\top(\boldsymbol{x}')\boldsymbol{C}(\boldsymbol{x}') \succ 0, \quad \forall \boldsymbol{x}' \in \mathbb{X}'. \tag{4.49}$$

hold. A globally exponentially convergent state-output observer for the system is given by

$$\dot{\hat{\boldsymbol{x}}} = \left(\boldsymbol{J}(\boldsymbol{x}') - \boldsymbol{R}(\boldsymbol{x}')\right) \frac{\partial H}{\partial \boldsymbol{x}}(\hat{\boldsymbol{x}}) + \boldsymbol{G}(\boldsymbol{x}')\boldsymbol{u} + \boldsymbol{C}^\top(\boldsymbol{x}')\left(\boldsymbol{m} - \boldsymbol{C}(\boldsymbol{x}')\boldsymbol{Q}\hat{\boldsymbol{x}}\right), \tag{4.50a}$$

$$\hat{\boldsymbol{y}} = \boldsymbol{G}^\top(\boldsymbol{x}') \frac{\partial H}{\partial \boldsymbol{x}}(\hat{\boldsymbol{x}}), \tag{4.50b}$$

with initial value $\hat{\boldsymbol{x}}|_{t=0} = \hat{\boldsymbol{x}}_0$.

Proof:
The proof follows directly from Lemma 4.15, Lemma 4.16, and Lemma 4.17. In the latter we substitute $\boldsymbol{s} \in \mathbb{S}$ with $\boldsymbol{x}' \in \mathbb{X}'$. □

It is important to note that the observer from Theorem 4.19 is directly obtained from the system model. In particular, there are no free observer parameters which is why its design is inherently automatable. Hence, Theorem 4.19 solves Problem 4.12.

In the following, the nonlinear observer from Theorem 4.19 is illustrated. To this end, we resume with the example that was used to demonstrate the methods from Chapter 3. This illustrates the consistency between the methods from chapters 3 and 4.

Example 4.20:

Consider the PHS from Example 3.34 for $K = 1$:

$$\begin{pmatrix} \dot{x}_1 \\ \dot{x}_2 \\ \dot{x}_3 \end{pmatrix} = \left(\begin{pmatrix} 0 & 0 & 0 \\ 0 & 0 & -1 \\ 0 & 1 & 0 \end{pmatrix} - \begin{pmatrix} d & de^{\kappa x_1} & 0 \\ de^{\kappa x_1} & de^{2\kappa x_1} & 0 \\ 0 & 0 & 0 \end{pmatrix} \right) \frac{\partial H}{\partial \boldsymbol{x}} + \begin{pmatrix} 0 & d \\ 1 & de^{\kappa x_1} \\ 0 & 0 \end{pmatrix} \begin{pmatrix} u_1 \\ u_2 \end{pmatrix}, \tag{4.51a}$$

$$\begin{pmatrix} y_1 \\ y_2 \end{pmatrix} = \begin{pmatrix} 0 & 1 & 0 \\ -d & -de^{\kappa x_1} & 0 \end{pmatrix} \frac{\partial H}{\partial \boldsymbol{x}} + \begin{pmatrix} 0 & 0 \\ 0 & d \end{pmatrix} \begin{pmatrix} u_1 \\ u_2 \end{pmatrix}, \tag{4.51b}$$

with $d > 0$ and the non-quadratic Hamiltonian

$$H(\boldsymbol{x}) = \frac{1}{2} \boldsymbol{x}^\top \begin{pmatrix} q_1 & 0 & 0 \\ 0 & q_2 & 0 \\ 0 & 0 & q_3 \end{pmatrix} \boldsymbol{x} + \frac{1}{4} x_1^4, \tag{4.52}$$

where $q_1, q_2, q_3 > 0$. For the system, consider two measurements $m_1 = x_1$ and $m_2 = e^{\kappa x_1} x_3$. The corresponding measurement equation reads:

$$\boldsymbol{m} = \underbrace{\begin{pmatrix} q_1^{-1} & 0 & 0 \\ 0 & 0 & q_3^{-1} e^{\kappa x_1} \end{pmatrix}}_{= \boldsymbol{C}(x_1)} \begin{pmatrix} q_1 & 0 & 0 \\ 0 & q_2 & 0 \\ 0 & 0 & q_3 \end{pmatrix} \begin{pmatrix} x_1 \\ x_2 \\ x_3 \end{pmatrix}. \tag{4.53}$$

Following the notation from Problem 4.12, we have $\boldsymbol{x}' = x_1$ and $\boldsymbol{x}'' = (x_2\ x_3)^\top$.

Now for the observer. We have

$$\boldsymbol{R}(x_1) + \boldsymbol{C}^\top(x_1) \boldsymbol{C}(x_1) = \begin{pmatrix} d + q_1^{-2} & de^{\kappa x_1} & 0 \\ de^{\kappa x_1} & de^{2\kappa x_1} & 0 \\ 0 & 0 & q_3^{-2} e^{2\kappa x_1} \end{pmatrix} \succ 0, \tag{4.54}$$

for all $\boldsymbol{x} \in \mathbb{X}$. Thus, the observer existence condition (4.49) is satisfied and an asymptotic state-output observer is given by (4.50).

Now, the results obtained from numerical simulation of the system (4.51) and the observer (4.50) are illustrated. The system parameters are chosen to $d = 1$, $q_1 = \frac{1}{2}$, $q_2 = \frac{1}{3}$, $q_3 = \frac{1}{4}$, and $\kappa = 0.1$. The initial values of the system and the observer are given by $\boldsymbol{x}_0 = (0\ 0\ 0)^\top$ and $\hat{\boldsymbol{x}}_0 = (1\ 1\ 1)^\top$, respectively. The input signals are specified to $u_1 = \sigma(t - 10\,\mathrm{s})$ and $u_2 = \sin(0.1\,\mathrm{s}^{-1} t)$ where $\sigma(\cdot)$ is the unit step function.

Figure 4.4 depicts the states x_i (solid, blue) and the reconstructions $\hat{x}_i$ (dashed, red) for $i = 1, 2, 3$.

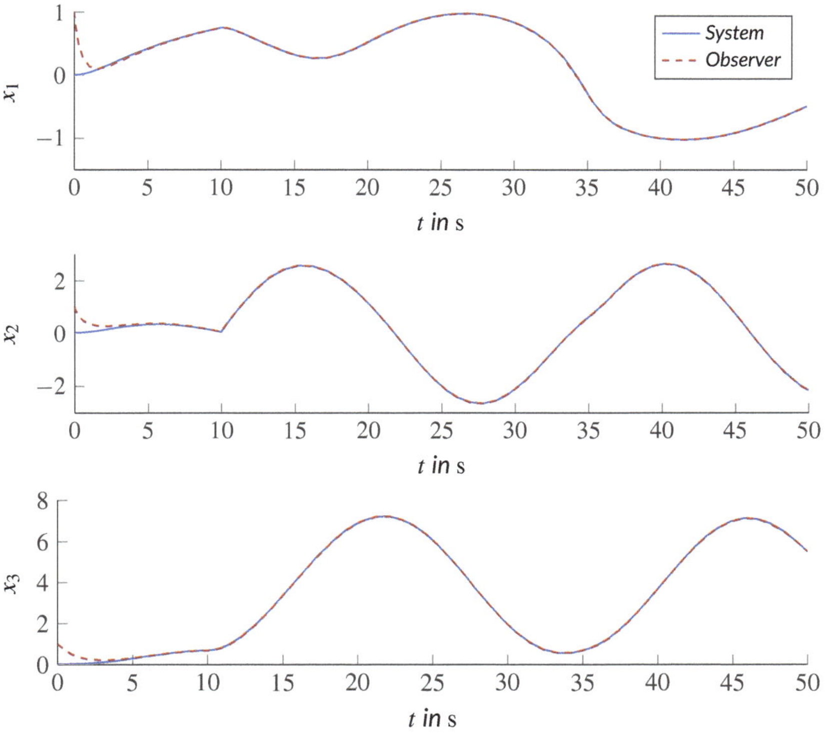

Figure 4.4: States of the system (4.51) (solid, blue) and corresponding reconstructions from the observer (4.50) (dashed, red)

As can be seen, the state reconstructions reach the true states in less than ten seconds. The reconstructions of the system output are given in the following figure:

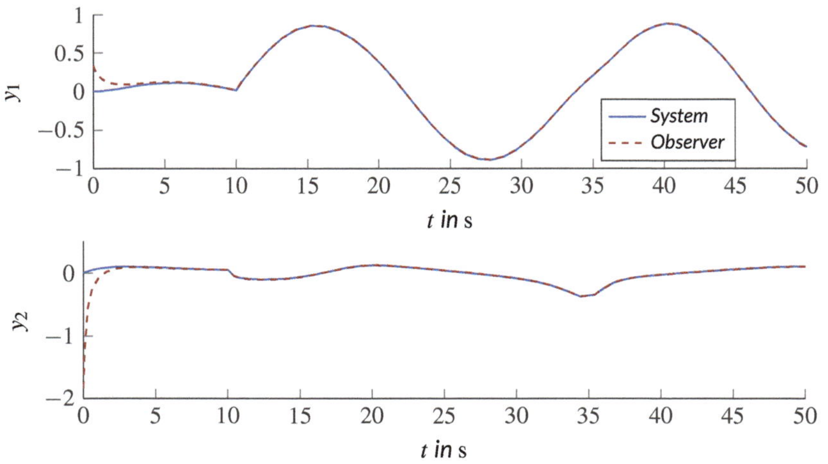

Figure 4.5: Outputs of the system (4.51) (solid, blue) and corresponding reconstructions from the observer (4.50) (dashed, red)

As can be seen, the reconstructed outputs also converge to the true outputs.

In the following two corollaries, we analyze the results obtained so far more in detail. First, we consider the case of linear measurements, i.e., the case where in (4.31) we have $\boldsymbol{C}(\boldsymbol{x}') = \boldsymbol{C} = \text{const}$.

Corollary 4.21 (Linear Measurements)

Given a system with dynamics (4.29a) and measurements (4.31) where the measurement matrix is a constant matrix $\boldsymbol{C}(\boldsymbol{x}') = \boldsymbol{C}$. The existence condition (4.35) for an observer of the form (4.36) is satisfied if and only if it is satisfied for the constant matrix $\boldsymbol{L} = \boldsymbol{C}^\top$.

Proof:

The claim follows from Lemma 4.16 and Lemma 4.17 under $\boldsymbol{C}(\boldsymbol{x}') = \boldsymbol{C}$. □

The main point from Corollary 4.21 is as follows. Despite the fact that the matrix $\boldsymbol{R}(\boldsymbol{x}')$ is parametrized over $\boldsymbol{x}'$, a constant observer gain $\boldsymbol{L}$ is sufficient to evaluate if the existence condition (4.35) is solvable or not. In other words, for $\boldsymbol{C}(\boldsymbol{x}') = \boldsymbol{C} = \text{const.}$, there is no benefit in approaching with a parametrized observer gain $\boldsymbol{L}(\boldsymbol{x}')$.[5] In this context, Corollary 4.21 reflects the idea behind Lemma 4.17. Loosely speaking, if the output error injection allows to access those parts of $-\boldsymbol{R}(\boldsymbol{x}')$ which corresponds to zero eigenvalues, we can shift them to the left. In the case of linear measurements, a constant observer gain which is independent of $\boldsymbol{x}'$ is sufficient towards this endeavor. On the other hand, if $\boldsymbol{R}(\boldsymbol{x}')$ is already positive-definite, the observer (4.36) is asymptotic without any error injection. This is addressed in the last corollary of this subsection.

Corollary 4.22 (Strictly passive systems)

Consider a strictly passive PHS (4.29) with measurements (4.31), i.e., the case where $\boldsymbol{R}(\boldsymbol{x}') \succ 0$ for all $\boldsymbol{x}' \in \mathbb{X}'$. A globally exponentially convergent state observer for the system is given by (4.36) with $\boldsymbol{L} = \boldsymbol{0}$.

Proof:

The statement follows from Lemma 4.16 under $\boldsymbol{R}(\boldsymbol{x}') \succ 0$ for all $\boldsymbol{x}' \in \mathbb{X}'$. □

Corollaries 4.21 and 4.22 conclude the observer design methods from this subsection. Thereby, like the previous Subsection 4.2.1, the subsection at hand addressed the design of a centralized observer based on global model knowledge. However, as argued in Chapter 2, the design of distributed observers based on local model knowledge is also of interest in the context of interconnected systems. This is the topic to be addressed in the next subsection.

4.2.3 Automated Design of Distributed Observers

In this subsection, we investigate how the methods from subsections 4.2.1 and 4.2.2 can be applied for an automated design of distributed observers based on local model knowledge.[6] First, the system setup is briefly outlined. Afterwards, two approaches for an automated design of distributed observers are presented. The first approach is based on the methods from Subsection 4.2.1; the second approach is based on the methods from Subsection 4.2.2.

Consider an interconnected system Σ described by an open graph $\mathscr{G} = (\mathbb{V}, \mathbb{B})$ where $\mathbb{V} = \mathbb{V}_{\mathrm{I}} \cup \mathbb{V}_{\mathrm{B}}$, $\mathbb{B} = \mathbb{B}_{\mathrm{I}} \cup \mathbb{B}_{\mathrm{B}}$ (cf. Term 2.3). Let the set of inner vertices

[5] This statement is limited to the property of an observer of being asymptotic.

[6] As outlined in Chapter 2, the here applied notion of a distributed observer follows that of Castanedo [2013] and Kupper [2019].

$\mathbb{V}_\mathrm{I}$ represent $N = |\mathbb{V}_\mathrm{I}|$ subsystems Σ_i ($i \in \mathbb{V}_\mathrm{I}$). The N subsystems are coupled through $M = |\mathbb{B}_\mathrm{I}|$ inner edges that represent power-conserving interconnections. Moreover, the subsystems Σ_i may interact with the environment of Σ via P boundary vertices and corresponding boundary edges ($P = |\mathbb{V}_\mathrm{B}| = |\mathbb{B}_\mathrm{B}|$). Each edge $j \in \mathbb{B}$ carries an effort $\boldsymbol{e}_j \in \mathbb{R}^{K_j}$ and a flow $\boldsymbol{f}_j \in \mathbb{R}^{K_j}$ where $K_j \in \mathbb{N}_{\geq 1}$ for all $j \in \mathbb{B}$.

Figure 4.6 depicts the situation for an example system Σ with $\mathbb{V}_\mathrm{I} = \{1,2,3\}$, $\mathbb{V}_\mathrm{B} = \{B_1, B_2\}$, $\mathbb{B}_\mathrm{I} = \{(1,2),(2,3)\}$, $B_\mathrm{B} = \{(B_1,1),(B_2,2)\}$, i.e., $N = 3$, $M = 2$, and $P = 2$.

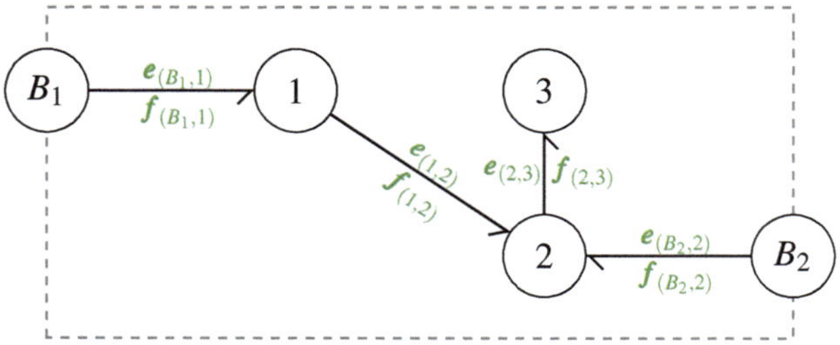

Figure 4.6: Exemplary interconnected system

In the following, we consider the situation where both—the observers and the model knowledge—are distributed according to the subsystems Σ_i of Σ. The first approach for an automated design of a distributed observer is based on the methods from Subsection 4.2.1.

Design of a Distributed Observer with the Methods from Subsection 4.2.1

Given an interconnected system Σ consisting of N subsystems as described above. Suppose that Σ is a linear system. Moreover, let us consider the case where the interactions between the subsystems are completely unknown.

Assumption 4.23 (Unknown interaction)
For each $j \in \mathbb{B}$, the effort-flow pair $(\boldsymbol{e}_j, \boldsymbol{f}_j)$ is not available for measurement.

From Assumption 4.23 we deduce that the input and output of each subsystem model is completely unknown. Moreover, subsystem states may also be unknown. Hence, for each subsystem the input, state, and output have to be reconstructed. To this end, we consider measurements that are linear combinations of the respective inputs, states, and outputs.

From a subsystem perspective, this setup is a special case of Problem 4.1 from Subsection 4.2.1. Accordingly, the following corollary reconsiders the methods from Subsection 4.2.1 in a subsystem-wise manner to design a distributed input-state-output observer:

Corollary 4.24 (Distributed observer for system with unknown interactions)

Consider an interconnected system Σ described by a graph $\mathscr{G} = (\mathbb{V}, \mathbb{B})$. Let the subsystems Σ_i be described by linear explicit PHSs of the form

$$\dot{\boldsymbol{x}}_i = (\boldsymbol{J}_i - \boldsymbol{R}_i)\ \boldsymbol{Q}_i \boldsymbol{x}_i + (\boldsymbol{G}_i - \boldsymbol{P}_i)\boldsymbol{u}_i, \quad \boldsymbol{x}_i|_{t=0} = \boldsymbol{x}_{i,0}, \tag{4.55a}$$

$$\boldsymbol{y}_i = (\boldsymbol{G}_i + \boldsymbol{P}_i)^\top \boldsymbol{Q}_i \boldsymbol{x}_i + (\boldsymbol{M}_i + \boldsymbol{S}_i)\boldsymbol{u}_i, \tag{4.55b}$$

with $\boldsymbol{u}_i, \boldsymbol{y}_i \in \mathbb{R}^{p_i}$ and $\boldsymbol{x}_i \in \mathbb{R}^{n_i}$ for $i \in \mathbb{V}_{\mathrm{I}}$. In (4.55), we have $\boldsymbol{Q}_i = \boldsymbol{Q}_i^\top \succ 0$; the remaining matrices satisfy the usual symmetry and definiteness conditions of such a PHS (cf. Definition 2.23). For each $i \in \mathbb{V}_{\mathrm{I}}$, consider $q_i > p_i$ linear independent measurements $\boldsymbol{m}_i \in \mathbb{R}^{q_i}$ of the form

$$\boldsymbol{m}_i = \boldsymbol{C}_{i,u}\boldsymbol{u}_i + \boldsymbol{C}_{i,x}\boldsymbol{x}_i + \boldsymbol{C}_{i,y}\boldsymbol{y}_i. \tag{4.56}$$

Then, for each $i \in \mathbb{V}_{\mathrm{I}}$ we can formulate an unknown-input system

$$\dot{\boldsymbol{x}}_i = \boldsymbol{A}_i \boldsymbol{x}_i + \boldsymbol{B}_i \boldsymbol{u}_i, \tag{4.57a}$$

$$\bar{\boldsymbol{m}}_i = \bar{\boldsymbol{C}}_i \boldsymbol{x}_i, \tag{4.57b}$$

where $\bar{\boldsymbol{m}}_i \in \mathbb{R}^{q_i - r_i}$ with $r_i = \operatorname{rank}(\boldsymbol{C}_{i,u} + \boldsymbol{C}_{i,y}(\boldsymbol{M}_i + \boldsymbol{S}_i))$.
Let (4.57) be strong detectable for each $i \in \mathbb{V}_{\mathrm{I}}$. Then, for each $i \in \mathbb{V}_{\mathrm{I}}$ we can*

find an asymptotic input-state-output observer of the form

$$\dot{z}_i = \boldsymbol{N}_i z_i + \boldsymbol{L}_i \bar{\boldsymbol{m}}_i, \quad z_i\big|_{t=0} = z_{0,i}, \tag{4.58a}$$

$$\hat{\boldsymbol{x}}_i = z_i - \boldsymbol{E}_i \bar{\boldsymbol{m}}_i, \tag{4.58b}$$

$$\hat{\boldsymbol{u}}_i = \left(\bar{\boldsymbol{C}}_i \boldsymbol{B}_i\right)^+ \left(\dot{\bar{\boldsymbol{m}}}_i - \bar{\boldsymbol{C}}_i \boldsymbol{A}_i \hat{\boldsymbol{x}}\right), \tag{4.58c}$$

$$\hat{\boldsymbol{y}}_i = \left(\boldsymbol{G}_i + \boldsymbol{P}_i\right)^\top \boldsymbol{Q}_i \hat{\boldsymbol{x}}_i + \left(\boldsymbol{M}_i + \boldsymbol{S}_i\right) \hat{\boldsymbol{u}}_i. \tag{4.58d}$$

Proof:
The proof is given for one subsystem $i \in \mathbb{V}_\mathrm{I}$ but translates to all other subsystems. Let Assumption 4.23 hold. Consider the PHS (4.55) with measurements (4.56). Assumption 4.23 implies that $\boldsymbol{u}_i$ is completely unknown. The rest follows from Theorem 4.8 for the special case without known inputs. □

The observers from Corollary 4.24 process only local measurement information to calculate reconstructions of the subsystems' inputs, states, and outputs. By this, the observers are fully distributed.

The design of the individual observers requires to determine the matrices $\boldsymbol{N}_i$, $\boldsymbol{L}_i$, and $\boldsymbol{E}_i$. These matrices can be automatically computed via a subsystem-wise application of Algorithm 4.10. For each subsystem, the respective execution of the algorithm requires only local model knowledge.

In the above, we considered a linear system where the exogenous variables of the subsystems are completely unknown. Instead, there have been measurements that are local with respect to the individual subsystems. In the following, we switch to a contrary case, viz. a class of nonlinear systems where the interactions between the subsystems are completely known without having additional measurement information from inside the system. Here, we apply the methods from Subsection 4.2.2 to propose an automated scheme for the design of distributed observers.

Design of a Distributed Observer with the Methods from Subsection 4.2.2

Suppose an interconnected system Σ consisting of N nonlinear subsystems. Let us consider the case in which the interactions amongst the subsystems and between the subsystems and the environment of Σ are fully available for measurement.

Assumption 4.25 (Known interaction)
For each $j \in \mathbb{B}$, the effort-flow pair $(\boldsymbol{e}_j, \boldsymbol{f}_j)$ is known or measured.

A key implication of Assumption 4.25 is that the input and the output of each subsystem model is known or measured. Thus, in this case all inputs are known and the measurement vector equals the output vector. Hence, no input and output reconstruction is required. As an example, we can think of a system consisting of subsystems with collocated sensors and actuators.

The following corollary particularizes the methods from Subsection 4.2.2 for an automated design of a distributed nonlinear observer:

Corollary 4.26 (Distributed observer under known interactions)
Given an interconnected system Σ described by a graph $\mathscr{G} = (\mathbb{V}, \mathbb{B})$. For each $i \in \mathbb{V}_{\mathrm{I}}$, let the subsystem Σ_i be described by an explicit PHS of the form

$$\dot{\boldsymbol{x}}_i = (\boldsymbol{J}_i(\boldsymbol{y}_i) - \boldsymbol{R}_i(\boldsymbol{y}_i)) \frac{\partial H_i}{\partial \boldsymbol{x}_i}(\boldsymbol{x}_i) + \boldsymbol{G}_i \boldsymbol{u}_i, \quad \boldsymbol{x}_i|_{t=0} = \boldsymbol{x}_{i,0}, \tag{4.59a}$$

$$\boldsymbol{y}_i = \boldsymbol{G}_i^\top \frac{\partial H_i}{\partial \boldsymbol{x}_i}(\boldsymbol{x}_i), \tag{4.59b}$$

with $\boldsymbol{u}_i \in \mathbb{U}_i \subset \mathbb{R}^{p_i}$, $\boldsymbol{x}_i \in \mathbb{X}_i \subset \mathbb{R}^{n_i}$, and $\boldsymbol{y}_i \in \mathbb{Y}_i \subset \mathbb{R}^{p_i}$ where $\mathbb{X}_i$ and $\mathbb{Y}_i$ are compact. The Hamiltonian of (4.59) is given by $H_i(\boldsymbol{x}_i) = \frac{1}{2}\boldsymbol{x}_i^\top \boldsymbol{Q}_i \boldsymbol{x}_i + \boldsymbol{N}_i(\boldsymbol{y}_i)$ where $\boldsymbol{Q}_i = \boldsymbol{Q}_i^\top \succ 0$ is a matrix and $\boldsymbol{N}_i : \mathbb{Y}_i \to \mathbb{R}$, $\boldsymbol{y}_i \mapsto \boldsymbol{N}_i(\boldsymbol{y}_i)$ is a positive semi-definite and twice differentiable function that may be nonlinear.
Let Assumption 4.25 hold. Moreover, let

$$\boldsymbol{R}_i(\boldsymbol{y}_i) + \boldsymbol{G}_i \boldsymbol{G}_i^\top \succ 0, \quad \forall \boldsymbol{y}_i \in \mathbb{Y}_i \tag{4.60}$$

be satisfied for all $i \in \mathbb{V}_\mathrm{I}$. Then, for each $i \in \mathbb{V}_\mathrm{I}$ an exponentially convergent state observer is given by

$$\dot{\hat{\boldsymbol{x}}}_i = (\boldsymbol{J}_i(\boldsymbol{y}_i) - \boldsymbol{R}_i(\boldsymbol{y}_i)) \frac{\partial H_i}{\partial \boldsymbol{x}_i}(\hat{\boldsymbol{x}}_i) + \boldsymbol{G}_i \left(\boldsymbol{u}_i + \boldsymbol{y}_i - \boldsymbol{G}_i^\top \frac{\partial H_i}{\partial \boldsymbol{x}_i}(\hat{\boldsymbol{x}}_i) \right), \tag{4.61}$$

with initial value $\hat{\boldsymbol{x}}_i|_{t=0} = \hat{\boldsymbol{x}}_{i,0}$.

Proof:
We particularize the results from Section 4.2.2 to the case where the measurement output equals the passive output.

Let (4.60) hold. The proof is given for the subsystem $i \in \mathbb{V}_\mathrm{I}$ but translates to all other subsystems. Consider the state reconstruction error $\boldsymbol{\varepsilon}_i = \boldsymbol{x}_i - \hat{\boldsymbol{x}}_i$. With (4.59a) and (4.61), the error dynamics read

$$\dot{\boldsymbol{\varepsilon}}_i = \left(\boldsymbol{J}_i(\boldsymbol{y}_i) - \boldsymbol{R}_i(\boldsymbol{y}_i) - \boldsymbol{G}_i \boldsymbol{G}_i^\top \right) \boldsymbol{Q}_i \boldsymbol{\varepsilon}_i. \tag{4.62}$$

Applying the Lyapunov function candidate $V_i(\boldsymbol{\varepsilon}_i) = \frac{1}{2} \boldsymbol{\varepsilon}_i^\top \boldsymbol{Q}_i \boldsymbol{\varepsilon}_i$ we obtain

$$\dot{V}(\boldsymbol{\varepsilon}_i) = -\boldsymbol{\varepsilon}_i^\top \underbrace{\boldsymbol{Q}_i \left(\boldsymbol{R}_i(\boldsymbol{y}_i) + \boldsymbol{G}_i \boldsymbol{G}_i^\top \right) \boldsymbol{Q}_i}_{\succ 0} \boldsymbol{\varepsilon}_i. \tag{4.63}$$

From this follows that $\boldsymbol{\varepsilon}_i \equiv \boldsymbol{0}$ is exponentially stable (cf. proof of Lemma 4.16). Thus, (4.61) is an exponentially convergent observer for (4.59). □

Remark 4.27 (PHS with feedthrough). Corollary 4.26 can be easily extended to a PHS with feedthrough (cf. Remark 4.14).

As can be seen, the observers from Corollary 4.26 require only knowledge of the interactions between the subsystems and are therewith distributed. Moreover, analogously to Subsection 4.2.2, the observers do not require a dedicated design. Instead, the observer parameters can be determined from the subsystems models.

This section showed that the centralized methods from subsections 4.2.1 and 4.2.2 can also be applied for an automated design of distributed observers based on local model information. It is important to note that the presented idea to distribute an observer design is not restricted to the two special cases from this subsection. The proposed principle may also be applied to several intermediate cases. For example, in Section 5.3 the methods from from Subsection 4.2.2 are applied to a nonlinear interconnected systems with partially known subsystem interaction. A theoretical examination of such intermediate cases, however, would not give new insights which is why we refrain from this step at this point.

This concludes the presentation of automated design methods for centralized and distributed observers for PHSs. In the next section, we discuss the presented methods with respect to the overall objectives of this thesis and previous results from the related literature.

4.3 Discussion

The first main results of this chapter is Theorem 4.8 in which an automatable design scheme for an input-state-output observer for linear PHSs is proposed. A sufficient existence condition for this observer requires (i) that there are more (independent) measurements than unknown inputs (i.e., $\tilde{q} > p_{\mathrm{u}}$) and (ii) that the state-space system (4.6) is strong* detectable. In comparison, the existence condition for the input-state-output estimator from the preliminary work authored by Pfeifer et al. [2019a] demands the number of measurements to be greater or equal to the number of unknown inputs plus the number of states, i.e., $\tilde{q} \geq p_{\mathrm{u}} + n$. Hence, subcondition (i) is less restrictive than the existence condition from Pfeifer et al. [2019a]. Hautus [1983, Theorem 1.12] showed that subcondition (ii) is necessary and sufficient for the existence of an asymptotic state observer for a system (4.6). Thus, for the presented approach, subcondition (ii) is inevitable.

An interesting special case of Theorem 4.8 appears for a PHS with known inputs. In this case, subcondition (i) is always fufilled. Moreover, subcondition (ii) reduces to the condition of the system being detectable. This indicates the existence conditions of Theorem 4.8 to be reasonable. For the special case of known inputs, the observer design from Theorem 4.8 yields a standard Luenberger observer extended by an output reconstruction.

The subconditions (i) and (ii) can be evaluated by a computer algebra system (cf. Remark 4.9). If both subconditions are fulfilled, we can automatically design an input-state-output observer by using Algorithm 4.10. Thereby, as for the modeling, the system size for a purely symbolic observer design is limited (cf. Section 3.3). The symbolic computation of the singular value decomposition, the generalized inverse, and the pole placement may lead to expressions of considerable size. Again, it is impossible to state an absolute limit for the system size as the feasibility of the symbolic calculations strongly depend on the specific structure of the matrices. For a numeric observer design with Algorithm 4.10 the computational complexity is determined by the complexity of the calculation of the inverse matrices required for the pole placement. Here, the upper bound is polynomial and given by $O(n^3)$ where n is the number of system states, see Lyche [2020, p. 65].[7]

The observer obtained from Algorithm 4.10 requires the first time-derivative of the measurement vector which, however, is unavoidable for an input reconstruction with a continuous observer [Hou and Patton, 1998]. To circumvent this problem, one can approach with a discontinuous sliding mode observer as from Edwards and Spurgeon [1994]. The design of sliding mode observers, however, involves more degrees of freedom which makes its automation delicate. The interested reader may refer to the thesis of Singer [2019].

Let us summarize that for linear PHSs with unknown inputs, states, and outputs we can design asymptotic observer in an automated manner. The existence conditions of the observer are reasonable.

The second main result of this chapter is Theorem 4.19. The theorem provides a sufficient condition for global exponential convergence of a state-output observer applicable to a class of nonlinear PHS. This class is quite general as it allows for state-dependent matrices and a possibly non-quadratic Hamiltonian. Venkatraman and van der Schaft [2010] consider an almost identical class of systems. A limitation of this class of PHSs is the assumption that those states which are responsible for the state-dependence of the PHSs matrices and which constitute the non-quadratic part of the Hamiltonian are measured. On the other hand, in practical systems this assumption may be satisfied by an appropriate sensor placement.

The observer from Theorem 4.19 obviates a dedicated "design" as it can be derived directly from the system model. This is in contrast to the observer design from Venkatraman and van der Schaft [2010] which requires the closed-form solution of a set of PDEs and algebraic equations. According to the underlying model, the observer resulting from Theorem 4.19 will be symbolic or numeric.

[7]The upper bound $O(n^3)$ is under the assumption that the number of states is greater than the number of measurements and greater than the number of unknown inputs.

The existence condition of the observer requires the error system to be sufficiently damped. Thereby, the damping consists of two parts, viz. the natural damping of the system and a virtual damping arising from the error injection. To ensure a fast convergence of all observer states, the error injection must access those states subject to no or weak natural dissipation. On the other hand, if the natural damping is sufficiently strong on all states (i.e., the system is strictly passive), one can completely omit the error injection in the observer (cf. Corollary 4.22).

The damping interpretation is closely related to well-known insights for the control of PHSs, see, e.g., Kugi [2001, Sec. 2.4] and van der Schaft [2017, Sec. 7.1]. By this relation, the results from Subsection 4.2.2 may also be applied for the design of controllers. As an example, in Appendix C.4 we use Lemma 4.17 for the automated design of an asymptotically stabilizing feedback controller for PHSs without feedthrough.

As an interim conclusion, for a class of nonlinear PHSs we can derive globally exponentially convergent observers directly from the system model.

A third considerable result from this chapter is given in corollaries 4.24 and 4.26. Corollary 4.24 shows that is is straightforward to reconsider the approach from Theorem 4.8 to design subsystem observers for linear interconnected systems on the basis of local models. The obtained observers only require measurement information which is local with respect to the subsystems. Even further, the observers from Corollary 4.24 obviate any information on the interactions between the subsystems. Hence, the observers (and their design) are fully distributed which is at the cost of a sufficiently large number of interior measurements in each subsystem.

Analogously to Corollary 4.24, Corollary 4.26 shows that for nonlinear interconnected systems with known interactions between subsystems, the design of the centralized observers from Theorem 4.19 translates into the design of a distributed observer. This claim is not restricted to the observer design from Theorem 4.19 but applies to many observer designs from the literature, see Section 4.1. Although this result is intuitive, it was not explicitly stated in the literature yet. The key is that known interaction variables imply the inputs and outputs of the local models to be known. Hence, based on the subsystem models, we can design distributed independently. To this end, the observer existence condition has to be fulfilled for each subsystem, e.g., by a suitable partitioning of the system into subsystems and/or an appropriate sensor placement.

It is noteworthy that, in general, the distributed observers resulting from corollaries 4.24 and 4.26 cannot be extracted from a centralized observer of the system. This can be explained by the fact that a local model of a subsystem is in general not equal to the corresponding submodel of a centralized model. For example, in Section 5.3, we consider an interconnected system in which the local models have feedthrough while the global model has not.

A natural limitation of the deterministic observer methods from this chapter is that they do not explicitly consider uncertainties. An analysis of the influence of measurement noise to the observers from this chapter will be provided in the numerical simulations that follow in Chapter 5.

Let us conclude the discussion of this section. For linear PHSs with unknown inputs, states, and outputs, we can design asymptotic observers in an automated manner. Moreover, for an important class of nonlinear PHSs we developed an approach to derive globally exponentially convergent observers directly from the system model. The observer methods can be also applied for an automated design of distributed observers based on local model knowledge.

4.4 Summary and Contributions

The extent of interconnected systems hampers a manual or partially computer-aided observer design. Existing automatable observer design techniques are purely numeric and disregard the physical background of a system. This chapter addressed this research gap by presenting methods and algorithms for an automated design of observers for PHSs. The original contributions of this research are:

(C2.1) an automatable design scheme for an asymptotic observer applicable to linear PHSs with unknown inputs, states, and outputs (Theorem 4.8 and Algorithm 4.10);

(C2.2) a globally exponentially convergent state-output observer for an important class of nonlinear PHSs; the observer exploits the natural damping of the system and obviates a dedicated "design" as it is directly obtained from the system model (Theorem 4.19);

(C2.3) two automated designs of distributed observers (corollaries 4.24 and 4.26) based on local model knowledge which are obtained from a reconsideration of the methods from C2.1 and C2.2.

The presented techniques are the first to enable an automated design of observers for a considerable class of PHSs and therewith reach research objective O2 from Section 1.3. The automated observer design methods were demonstrated for a linear and a nonlinear example system. Numerical simulations verified the convergence of the obtained observers.

Chapter 5

Proof of Principle

The aim of this thesis is to show that for a large class of interconnected systems the process of deriving explicit state-space models and the process of designing observers can be automated. This chapter provides a proof of principle towards this endeavor. To this end, in Section 5.1 the methods from the previous two chapters are integrated in a software prototype. The prototype enables an automated generation of models and observers for interconnected systems. In Section 5.2 and Section 5.3, the prototype is applied to two example systems, viz. an unbalanced power distribution system and a large-scale nonlinear system, respectively. The validity of the obtained models and observers is analyzed through numerical simulations. The proof of principle ends with a discussion and a summary of the contributions in Section 5.4 and Section 5.5, respectively.

5.1 Software Prototype

The prototype is named AMOTO which is an acronym for automatic model generation and observer design tool. AMOTO implements the methods from Chapter 3 (viz. algorithms 3.36 and 3.48) and Chapter 4 (viz. Algorithm 4.10 and Theorem 4.19). The development of AMOTO started in 2018. The first version of the program has been presented in Pfeiffer et al. [2019]. The current version is 1.1.4. The core of the tool is implemented in the Wolfram language. Hence, the main calculations are performed by the Mathematica kernel. A graphical user interface (GUI) for AMOTO was developed in Java. The Wolfram J/Link interface enables the communication between the Java GUI and the Mathematica core. AMOTO requires Wolfram Mathematica version 11 and Java runtime version 8 or respective later versions.

The workflow in AMOTO is as follows. First, the user enters a bond graph representation of the interconnected system under consideration. The constitutive relations of the bond graph elements are entered symbolically or numerically. Moreover, we specify in the bond graph model which variables are available for measurement. To this end, bond graph variables can be set to "measured". This step is optional and only required if we aim at an automated observer design.

Based on the bond graph model, the user can start the model generation in which AMOTO derives an explicit port-Hamiltonian model. The calculation runs fully automatic. After the calculation, the resulting PHS is displayed to the user. Afterwards, we may run the automated observer design. The obtained observer equations are also displayed to the user.

AMOTO features an export function in which the obtained models and observers can be written to external files in the formats of Mathematica, MATLAB, and LaTeX. This facilitates subsequent steps in the development of models and observers as, e.g., their simulation, analysis, documentation and, finally, their implementation in the system under consideration.

Figure 5.1 summarizes the workflow for the development of models and observers with AMOTO.

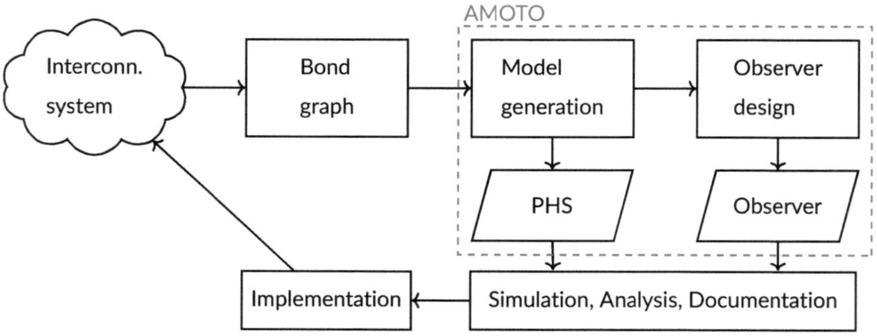

Figure 5.1: Illustration of the workflow with the software prototpye AMOTO

Figure 5.2 depicts the AMOTO GUI. The GUI consists of a menu bar, a graph panel, a type list, an element list, an element panel, a tool bar, and a log panel. In the sequel, the different parts of the GUI are briefly described.

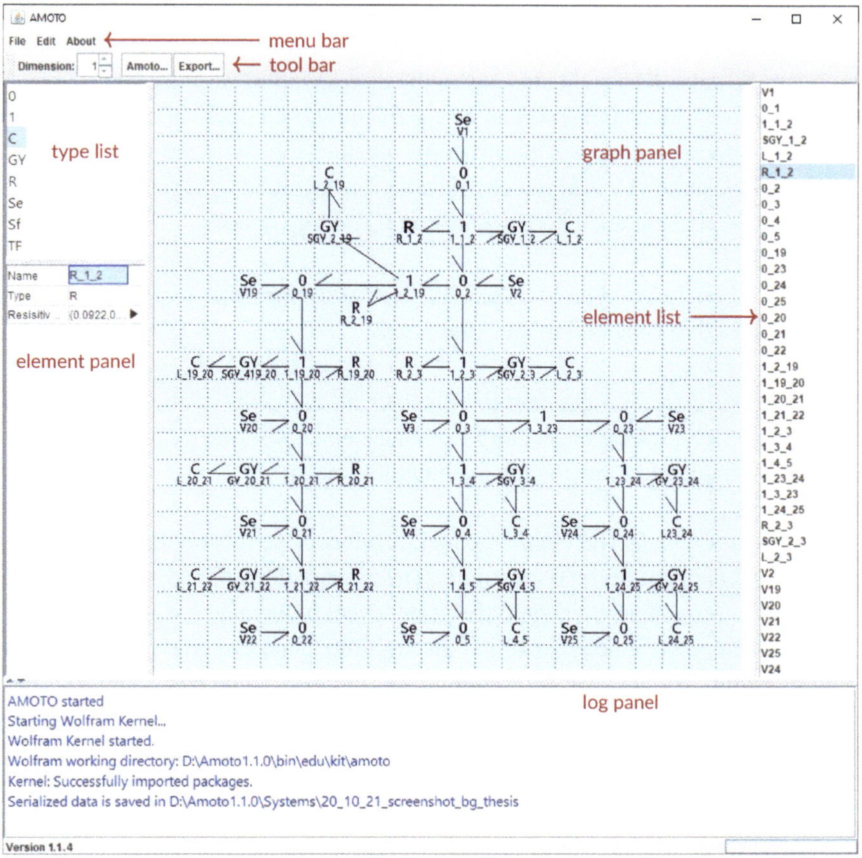

Figure 5.2: Screenshot of the AMOTO GUI with labeling of the panels, bars, and lists in red

- *Menu bar:* The menu bar consists of the three menu points “File”, “Edit”, and “About”. By clicking on “File” we can open a new blank model, save the current bond graph, or load a bond graph from an existing file. Under “Edit” the user may change program options as, e.g., the directory of the Mathematica kernel or the grid size in the graph panel. License information and the current software version are provided under “About”.

- *Graph panel:* The graph panel is the central playground to create bond graphs. With a right click on the graph panel, bond graph elements can be added. Moreover, by right clicking on an existing element, one can connect this element to another element by inserting a bond. Within the graph panel, elements and bonds can be moved with drag and drop. Elements and bonds can be deleted by right clicking on the corresponding instance and choosing "remove". Measurements can be specified by right clicking on a bond. For the selected bond, we then specify whether the flow, the effort or both are measured. Measurements of states are specified by right clicking on a C-type element and choosing the corresponding option.

- *Type list:* In the type list, the user can choose the type of element that is supposed to be added to the bond graph.

- *Element list:* The element list displays all elements of the current bond graph. By clicking on an entry from the element list, the respective element is highlighted in the graph panel in bold font.

- *Element panel:* The element panel displays basic information about a selected element of the bond graph. The information appears when clicking on the respective element in the graph panel or element list. For a selected element, the element panel first displays the name and type of the selected element. By clicking on the name, the element name can be edited. For elements of type TF and GY, the element panel provides an additional input field for the transformer and gyrator ratio, respectively. For elements of type R, an input field for the constitutive relation of the considered element is displayed. Likewise, for C-type elements, the energy state and the storage function can be specified. The ratios of TF- and GY-type elements as well as the constitutive relations of R- and C-type elements can be specified in the Mathematica language either symbolically or numerically.

- *Tool bar:* The tool bar provides access to the main functional features of AMOTO. It consists of three parts, viz. "Dimension", "AMOTO...", and "Export...". In the field "Dimension" we can specify the dimension of the bond graph. For the automated generation of the model and observer equations, we click on the button "AMOTO...". In the following user dialog, we select if the model equations, the observer equations or both are to be calculated. In the dialog, by clicking on "Run" we start the calculations, which are performed by the Mathematica kernel. The resulting equations are displayed in a new pop-up window. By clicking on "Export...", the obtained model and/or observer equations may be written to an external file.

For the export, one can choose between the file extensions *.wl, *.m, and *.tex, i.e., file formats for Mathematica, MATLAB, and LaTeX, respectively.

- *Log panel:* The log panel is used to document the session. The most important feature is that the log panel informs the user if any existence conditions are violated. Moreover, the log panel reports if there are problems with the program configuration.

This concludes the presentation of AMOTO. In the subsequent sections, AMOTO will be applied for the model generation and observer design in two interconnected systems.

5.2 Case Study 1: Unbalanced Distribution System

5.2.1 Study Objectives

Distribution systems are the final stage in the delivery of electric power. The distribution level plays a crucial role for the decarbonization of electric energy. In the control-oriented power systems research, the distribution system is one of the hot topics often addressed under the names "Smart grid" or "Microgrid", see, e.g., Schiffer et al. [2016]; Simpson-Porco et al. [2017]; Strehle et al. [2019].

A distribution system is a multi-phase power system that operates on a medium to low voltage level. Compared to transmission systems, the treatment of distribution systems is more intricate as their voltages and currents are unbalanced [Kersting, 2017]. In many cases, a distribution system comprises a high number of buses and lines and therewith involves high dimensional variable spaces.

State estimation techniques are of key importance for the monitoring and control of a distribution system. The state estimators currently used in distribution system control centers rely on the assumption that the network is in quasi-steady state [Zhao et al., 2019]. The quasi-steady state assumption simplifies the design of a state estimator for the network to the derivation of a static weighted least squares (WLS) estimator. However, as the IEEE Task Force on Power System Dynamic State and Parameter Estimation recently pointed out, in reality, power systems never operate in quasi-steady state as there are continuous variations of generation and demand [Zhao et al., 2019]. The situation is aggravated by the recent changes in the distribution system, viz. the extensive integration of distributed energy resources, complex loads, and demand-response technologies. In the distribution system, these changes induce faster transients and stronger harmonic distortions [IEEE Standards Association, 2014]. A WLS-based state estimator is unable to capture these transients and harmonics which motivates the development of new monitoring techniques.

Power system dynamic state estimation (PSDSE) is a promising approach to account for the recent changes in the distribution system. The survey papers from Primadianto and Lu [2017] and Zhao et al. [2019] outline the state of the art in PSDSE. In contrast to a WLS-based state estimation, PSDSE methods are based on a state-space dynamic model. Based on the dynamic model, an estimator is derived by designing a Luenberger observer, a Kalman filter, or extensions thereof. In existing PSDSE approaches, the system dynamics originate from a dynamic modeling of components that are connected to the network as for example synchronous generators or storage systems, see, e.g., Zhao et al. [2017]; Singh and Pal [2019]; Alhelou et al. [2019] and references therein. To our knowledge, in all existing approaches, the network is still described in quasi-steady state. In the literature, there are no previous reports on PSDSE approaches that reject the quasi-steady state assumption for the network.[1] In other words, the benefit of considering the network dynamics for the state estimation has been largely unexplored.

AMOTO is used to bridge this research gap. First, we derive a distribution system model which explicitly considers the line dynamics. Based on the model, we apply the automated observer design to generate a dynamic state estimator. The performance of the obtained model and estimator is analyzed through numerical simulations.

This study will demonstrate the capabilities of AMOTO. In particular, it will be shown that

- AMOTO is able to automatically generate a model and an observer for an unbalanced distribution system;
- the obtained model and observer produce plausible results;
- in distribution systems with fast transients, harmonic distortion, and measurement noise the observer extends the functionalities of a WLS-based state estimation.

[1] An exception is given by the class of transient state estimation methods, see, e.g., Watson and Yu [2008]. This class of methods, however, focuses on fault detection and isolation and cannot be used for a continuous system monitoring.

5.2.2 System Description

Consider a three-phase unbalanced distribution system with N buses. The three phases are denoted by A, B, and C. P of the N buses are connected to higher-level systems. The remaining $N-P$ buses are connected to $N-P$ uncontrollable loads. Thereby, each load is connected to exactly one bus. The N buses are connected by M three-phase lines. As can be seen, such a distribution system is clearly an interconnected system in the sense of Term 2.3: the set of inner vertices V_{I} contains the N buses; the set of boundary vertices $\mathbb{V}_{\mathrm{B}}$ contains the P higher-level systems and $N-P$ loads; the set of inner edges $\mathbb{B}_{\mathrm{I}}$ contains the M lines; and the set of boundary vertices $\mathbb{B}_{\mathrm{B}}$ contains the N (lossless) connections to the higher-level systems and loads. Let us assume that the open graph $\mathscr{G} = (\mathbb{V}, \mathbb{B})$ with $\mathbb{V} = \mathbb{V}_{\mathrm{I}} \cup \mathbb{V}_{\mathrm{B}}$ and $\mathbb{B} = \mathbb{B}_{\mathrm{I}} \cup \mathbb{B}_{\mathrm{B}}$ is connected.

Figure 5.3a shows an exemplary distribution system with $N = 33$, $M = 32$, and $P = 1$. The depicted system is the well-known IEEE 33 Bus System from Baran and Wu [1989]. The nominal voltage of this system is 12.66 kV; the system frequency is 50 Hz. The open graph representation of the IEEE 33 Bus System is illustrated in Figure 5.3b.

5.2.3 Bond Graph Model

This subsection presents a methodology for deriving a bond graph model of an unbalanced distribution system. To this end, we first describe the components of a distribution system separately. The overall distribution system model can then be obtained by assembling the component models. The component models are described in the following

A bus $i \in \mathbb{V}_{\mathrm{I}}$ is modeled as ideal Kirchhoff node, i.e., a 0-junction. An element $i \in \mathbb{V}_{\mathrm{B}}$ is modeled as three-phase voltage source, i.e., an Se-type element.[2] We describe an element $j \in \mathbb{B}_{\mathrm{B}}$ as an ideal (lossless) connection, i.e., an ordinary power bond. Finally, a line $j \in \mathbb{B}_{\mathrm{I}}$ is described by the π-section model in Figure 5.4. The π-section model considers phase resistances $R_{k,j}$, phase self inductances $L_{k,k,j}$ and line-line mutual inductances $L_{k,l,j}$ where $k,l \in \{\mathrm{A},\mathrm{B},\mathrm{C}\}$, $k \neq l$. The latter are necessary as distribution lines are in general untransposed [Kersting, 2017, p. 79]. The resistances and inductances are collected in the following resistance

[2]The parameters of loads are constantly changing during system operation and are unknown for the state estimation [Kersting, 2017, p. 28]. Hence, we describe the loads by the voltages across them.

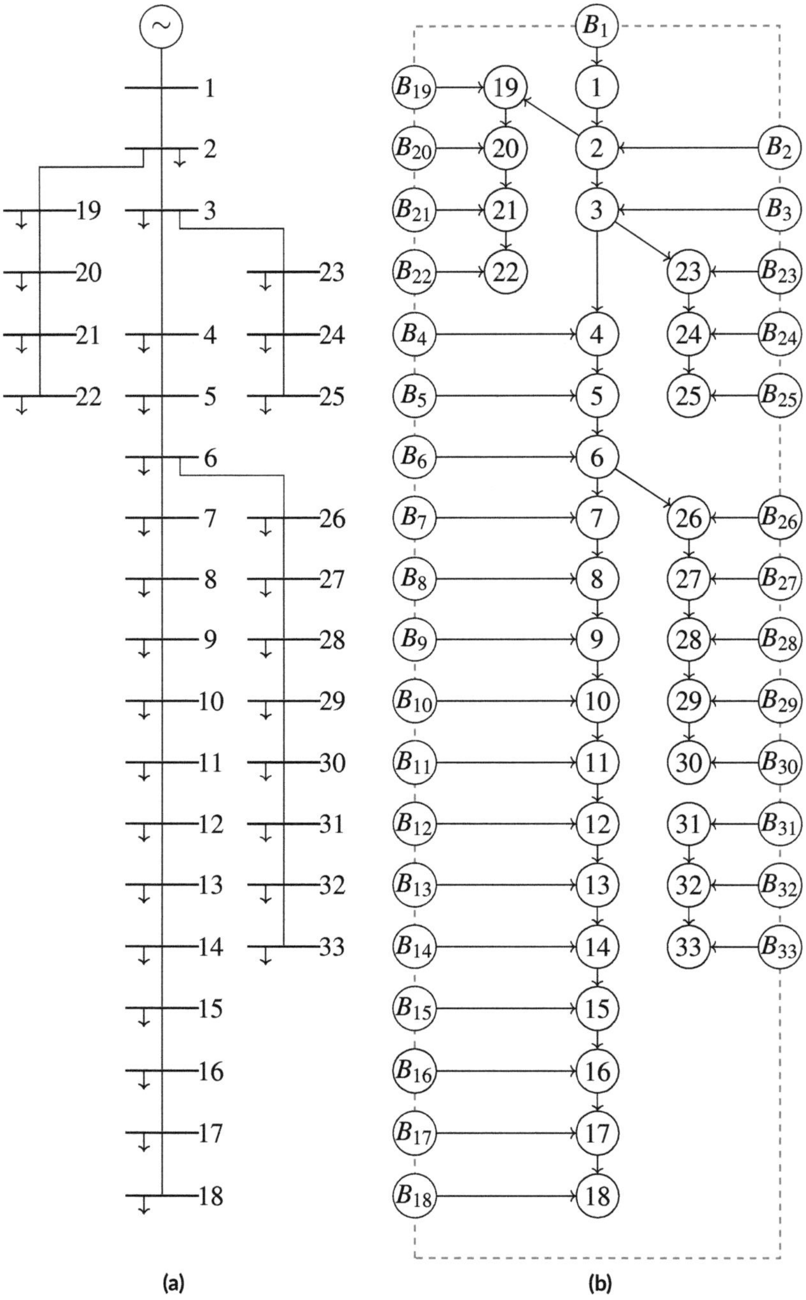

Figure 5.3: IEEE 33 Bus System: schematic diagram (a) and open graph representation (b)

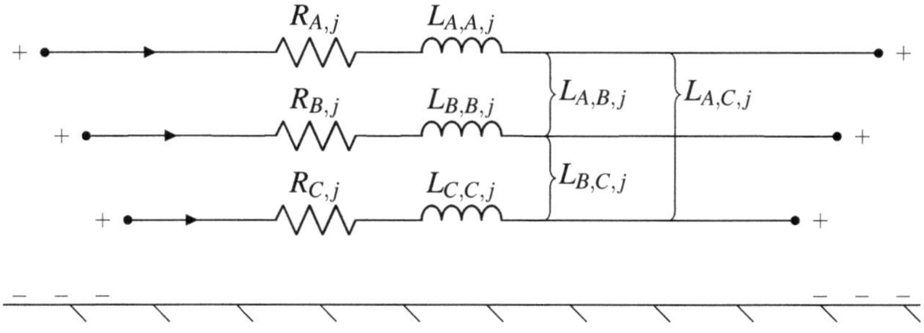

Figure 5.4: π-section equivalent circuit of a three-phase distribution line

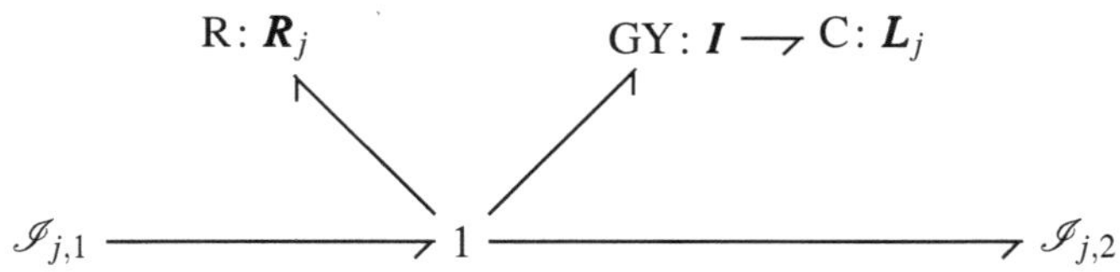

Figure 5.5: Bond graph representation of the π-section model from Figure 5.4

and inductance matrices:

$$\boldsymbol{R}_j = \begin{pmatrix} R_{A,j} & 0 & 0 \\ 0 & R_{B,j} & 0 \\ 0 & 0 & R_{C,j} \end{pmatrix}, \quad \text{and} \quad \boldsymbol{L}_j = \begin{pmatrix} L_{A,A,j} & L_{A,B,j} & L_{A,C,j} \\ L_{A,B,j} & L_{B,B,j} & L_{B,C,j} \\ L_{A,C,j} & L_{B,C,j} & L_{C,C,j} \end{pmatrix}, \tag{5.1}$$

respectively, with $j \in \mathbb{B}_\text{I}$. Figure 5.5 depicts the bond graph representation of the π-section model of line $j \in \mathbb{B}_\text{I}$ from Figure 5.4. The ports $\mathscr{I}_{1,j}$ and $\mathscr{I}_{2,j}$ are open ports for the interconnection to the incident buses.

By using the above component models, we may construct a bond graph model of an unbalanced distribution system. Figure 5.6 depicts the obtained bond graph model for the IEEE 33 Bus System from Figure 5.3. Note that each line model L_j consists of a bond graph as depicted in Figure 5.5 ($j \in \mathbb{B}_\text{I}$).

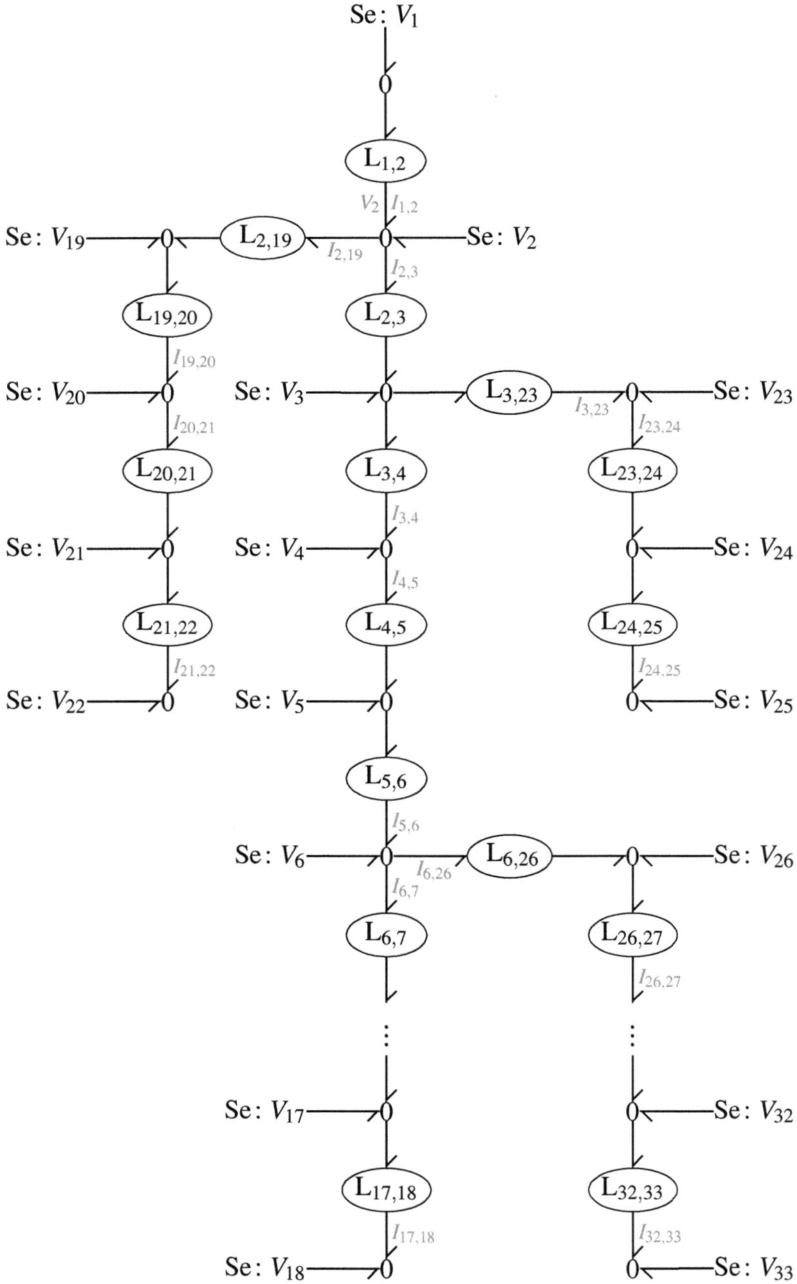

Figure 5.6: IEEE 33 Bus System: bond graph model with measured variables highlighted in green

5.2.4 Automated Model Generation

To demonstrate the automated model generation for the IEEE 33 Bus System, the bond graph from Figure 5.6 was entered into AMOTO. The system parameters were taken from Baran and Wu [1989].[3] Based on the entered bond graph, we then run the automated model generation of AMOTO. After approximately 19 seconds[4], AMOTO outputs a linear PHS of the form

$$\dot{\boldsymbol{x}} = -\boldsymbol{R}\boldsymbol{Q}\boldsymbol{x} + \boldsymbol{G}\boldsymbol{u}, \tag{5.2a}$$

$$\boldsymbol{y} = \boldsymbol{G}^\top \boldsymbol{Q}\boldsymbol{x}, \tag{5.2b}$$

with $\boldsymbol{x} \in \mathbb{R}^{96}$ and $\boldsymbol{u}, \boldsymbol{y} \in \mathbb{R}^{99}$.

The complete model is far too large to be displayed in the format of this thesis. Hence, we focus on structural properties of the model.

The state vector consists of the magnetic flux linkages ψ of the $M = 32$ lines. Each line is described by three states, viz. the magnetic flux linkages of the phases A, B, and C. Hence, the overall number of states is $3 \cdot 32 = 96$ and the state vector is given by $\boldsymbol{x} = ((\psi)_k)_j$ for all $k \in \{\mathrm{A,B,C}\}$ and $j \in \mathbb{B}_\mathrm{I}$.

The input vector of (5.2) consists of the $3 \cdot 33 = 99$ bus voltages, i.e., $\boldsymbol{u} = ((V)_k)_i$ for all $k \in \{\mathrm{A,B,C}\}$ and $i \in \mathbb{V}_\mathrm{I}$. The output vector contains the corresponding 99 bus currents, i.e., $\boldsymbol{y} = ((I)_k)_i$ for all $k \in \{\mathrm{A,B,C}\}$ and $i \in \mathbb{V}_\mathrm{I}$. These are the currents that flow between the buses on the one hand and the loads and higher level system on the other hand. As the bus voltages and currents are uncontrollable, the input vector is a pure disturbance vector; the output vector is the corresponding disturbance output vector.

The matrix $\boldsymbol{R}$ in (5.2) is given by $\boldsymbol{R} = \mathrm{blkdiag}\,(\boldsymbol{R}_j)$ with $j \in \mathbb{B}_\mathrm{I}$ where $\boldsymbol{R}_j$ is from (5.1). Likewise, the matrix $\boldsymbol{Q}$ is a block diagonal matrix that contains the inverse matrices of the inductance matrices from (5.1), i.e., $\boldsymbol{Q} = \mathrm{blkdiag}(\boldsymbol{L}_j^{-1})$. The matrix $\boldsymbol{G}$ is a sparse matrix. The non-zero entries in $\boldsymbol{G}$ appear in 3×3 diagonal blocks which are either $\mathrm{diag}(1,1,1)$ or $-\mathrm{diag}(1,1,1)$. The arrangement of the non-zero blocks reflects the incidence matrix of the connected subgraph $\mathcal{G}_\mathrm{I} = (\mathbb{V}_\mathrm{I}, \mathbb{B}_I)$. The remaining PHS matrices $\boldsymbol{J}$, $\boldsymbol{P}$, $\boldsymbol{M}$, and $\boldsymbol{S}$ are calculated to zero.

[3] Baran and Wu [1989] provide values only for the self-inductances of the lines. The ratio between the line self-inductances and the line-line mutual inductances is assumed to be 0.8.

[4] Calculated on a computer with Intel(R) Core(TM) i7-6600U CPU @ 2.60 GHz and 12 GB RAM.

5.2.5 Automated Observer Design

To demonstrate the automated observer design of AMOTO for the IEEE 33 Bus System, we first define a set of measurement buses $\mathbb{V}_M \subseteq \mathbb{V}_I$. We assume the following buses to be equipped with measurement units:

$$\mathbb{V}_M = \{2,4,6,8,10,12,14,16,18,20,22,23,25,27,29,31,33\}. \tag{5.3}$$

In each measurement bus $i \in \mathbb{V}_M$, sensors provide synchronized measurements of the three-phase network currents, i.e., the currents flowing through the incident lines. The load currents are assumed to be non-measured. For the entire system, we assume only one three-phase voltage measurement which is located at bus 2. This voltage measurement acts as reference for the voltage reconstructions. In Figure 5.6, the buses and quantities that are available for measurement are highlighted in green.

In the graph panel of AMOTO we enter the measurements to the bond graph by setting the appropriate efforts and flows to "measured". From the extended bond graph, AMOTO computes a measurement equation of the form $\bar{\boldsymbol{m}} = \bar{\boldsymbol{C}}\boldsymbol{y}$ with $\bar{\boldsymbol{m}} \in \mathbb{R}^{96}$ and $\bar{\boldsymbol{C}} \in \{0,-1\}^{96\times 99}$. Based on the measurement equation and the model (5.2), AMOTO calculates an input-state-output observer of the form

$$\dot{\boldsymbol{z}} = \boldsymbol{N}\boldsymbol{z}, \quad \boldsymbol{z}|_{t=0} = \boldsymbol{z}_0, \tag{5.4a}$$
$$\hat{\boldsymbol{x}} = \boldsymbol{z} - \boldsymbol{E}\bar{\boldsymbol{m}}, \tag{5.4b}$$
$$\hat{\boldsymbol{u}}_2 = \left(\bar{\boldsymbol{C}}\boldsymbol{G}_2\right)^+ \left(\dot{\bar{\boldsymbol{m}}} + \bar{\boldsymbol{C}}\boldsymbol{R}\hat{\boldsymbol{x}} - \bar{\boldsymbol{C}}\boldsymbol{G}_1\boldsymbol{u}_1\right), \tag{5.4c}$$
$$\hat{\boldsymbol{y}} = \boldsymbol{G}^\top \boldsymbol{Q}\hat{\boldsymbol{x}}, \tag{5.4d}$$

with $\boldsymbol{z} \in \mathbb{R}^{96}$. The vectors $\hat{\boldsymbol{x}} \in \mathbb{R}^{96}$ and $\hat{\boldsymbol{y}} \in \mathbb{R}^{99}$ are reconstructions of the magnetic flux linkages of the lines and the bus currents, respectively. The vector $\boldsymbol{u}_1 \in \mathbb{R}^3$ contains the known inputs, i.e., the three-phase bus voltage at bus 2. The vector $\hat{\boldsymbol{u}}_2 \in \mathbb{R}^{96}$ is a reconstruction of the unknown inputs, i.e., the three-phase bus voltages at the remaining buses. The matrices $\boldsymbol{G}_1$ and $\boldsymbol{G}_2$ can be obtained from splitting the matrix $\boldsymbol{G}$ from (5.2) according to the splitting of $\boldsymbol{u}$.

The matrix $\boldsymbol{N}$ is calculated as a diagonal matrix. The diagonal values of $\boldsymbol{N}$ are negative and lie in the complex plane left to the eigenvalues of the matrix $-\boldsymbol{R}\boldsymbol{Q}$ from (5.2). This is meaningful, as it ensures the error dynamics to be faster than the system dynamics. The matrix $\boldsymbol{E}$ is a sparse matrix. The arrangement of the non-zero matrix blocks again reflects the incidence matrix of the connected subgraph $\mathscr{G}_I = (\mathbb{V}_I, \mathbb{B}_I)$.

This concludes the presentation of the results obtained from the automated model generation and observer design of AMOTO for the IEEE 33 Bus System. Next, the obtained model and observer are validated through numerical simulations.

5.2.6 Simulation Setup

The aim of the simulation study is (i) to evaluate the validity of the model (5.2) and the observer (5.4) and (ii) to compare the performance of the observer with the performance of a WLS-based estimation. It is interesting to note that we can use AMOTO to calculate a WLS estimator for (ii). The detailed derivation of a WLS estimator for the IEEE 33 Bus System with AMOTO is illustrated in Appendix D.1.

As a ground truth we use the verified MATLAB/Simulink time-domain simulation model of the IEEE 33 Bus System from Wong [2020]. By default, this simulation model describes a scenario in which the system is balanced and in quasi-steady state. Hence, to account for the recent changes in distribution systems, the model from Wong [2020] is modified in three points:

- On the buses $\{4,21,27,30,31\}$, $\{7,17,22,25,26\}$, and $\{3,12,24,28,32\}$, load imbalances for the phases A, B, and C are introduced, respectively. For these buses, the load at the concerning phase is 30 % higher than the load at the other phases.
- At $t_1 = 1\,\mathrm{s}$, $t_2 = 2\,\mathrm{s}$, and $t_3 = 3\,\mathrm{s}$ we consider load transients in which the active and reactive power of the three-phase loads at the buses 14, 17, and 26, respectively, increase by the factor 2.
- The voltage at bus 1 is subject to harmonic distortion. We consider the 3rd, 5th, 7th and 11th harmonics of the fundamental frequency. The amplitudes of the harmonics are set to 2.5 % of the amplitude of the fundamental frequency.[5]

[5] This yields a total harmonic distortion of 5 % which is within the range of an allowed total harmonic distortion [IEEE Standards Association, 2014].

From now on, we denote to the modified model from Wong [2020] as benchmark model. For the evaluation of the model and observer, first the benchmark model has been simulated which gives us time series for all bus voltages, bus currents, line currents, and line flux linkages. The bus voltages are the inputs of the model (5.2). Hence, based on the time-series of the bus voltages one may simulate (5.2). Likewise, the time-series of the variables from the vectors $\bar{\boldsymbol{m}}$ and $\boldsymbol{u}_1$ obtained from the simulation of the benchmark model are collected. With these data one may simulate the observer (5.4).

All simulations were conducted in MATLAB/Simulink R2019a with a fixed-step solver at a simulation step size of 0.01 ms. The model, the observer, and the WLS estimator were written to MATLAB code by using the export function from AMOTO. The resulting files comprise more than 18000 lines of MATLAB code. The initial values of the models and the observer are chosen such that the simulation starts in quasi-steady state. The simulation time was set to $T = 4\,\mathrm{s}$.

5.2.7 Simulation Results

For the evaluation of the model (5.2) and observer (5.4) we use the relative error signal power (RESP).[6] For each signal obtained from the model and observer, an error signal based on the corresponding signal from the benchmark model is computed. The RESP is then the quotient of the signal power of the error signal and the signal power of the benchmark signal. Therewith, the RESP is a relative measure for the similarity of a signal to the corresponding benchmark signal. The formal definition of the RESP can be found in Appendix D.2.

First, we analyze the RESP of the bus currents that have been computed from the model (5.2) and the observer (5.4). Figure 5.7 shows the three-phase average RESP over the bus number for the model and the observer in blue and red, respectively. The underlying numerical data can be found in Table D.1 and Table D.2 in Appendix D.3.

As can be seen, for each of the buses the RESP of the model takes values equal or less than 0.25 %. Hence, we can say that the model (5.2) accurately reproduces the behavior of the benchmark model.[7] For the observer, we obtain

[6] Note that the deterministic, periodic nature of the signals obtained from the model (5.2) and observer (5.4) make the well-known similarity measures from statistics inappropriate for this study.

[7] The remaining difference between the models can be further decreased by choosing a smaller simulation step size.

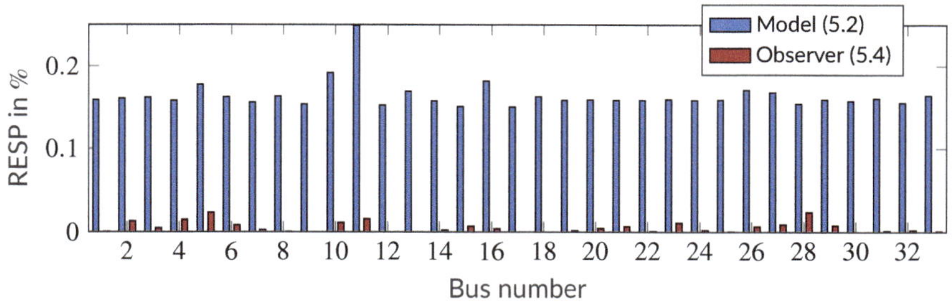

Figure 5.7: Three-phase average RESP of the bus currents for the model (5.2) (blue) and the observer (5.4) (red)

even smaller RESP values of less than 0.025 %. This can be explained by the observer error injection, which induces a robustness against model uncertainties.

Figure 5.8 depicts the time courses of the three-phase bus current at bus 17 for the benchmark model (solid, green), the model (5.2) (dashed, blue) and the observer (5.4) (dotdashed, red) for the time between 1.95 s and 2.10 s. We can clearly identify the fundamental frequency of 50 Hz. Due to the load imbalance at phase B of bus 17, the amplitude of the current at phase B is slightly higher than the amplitudes for the currents at the phases A and C. At $t = 2\,\mathrm{s}$ we can see the load jump in which the amplitude of the bus current increases approximately by the factor of 2. As can be seen, the model (5.2) and the observer (5.4) correctly reproduce the behavior of the benchmark model at this crucial point. Visually, we cannot distinguish between the obtained currents.

As an interim result, let us summarize that the model (5.2) accurately reflects the behavior of the benchmark model. Moreover, the observer (5.4) produces estimates that are very close to the values from the benchmark model. Hence, we can say that the observer is asymptotic.

Next, we compare the simulation results of the observer (5.4) with the results obtained for the WLS estimator (D.8). The RESPs for all bus voltage signals are given in Table D.3 and Table D.4 in Appendix D.3. For the observer, the mean RESP over all reconstructed bus voltages is 1.5×10^{-3} % (standard deviation: 9.2×10^{-4} %). The respective value for the WLS estimator is 5.1 % (standard deviation: 6.3×10^{-2} %). Hence, the observer significantly outperforms the WLS estimator. This can be explained by two reasons.

First, the WLS estimator cannot capture the harmonic distortion. This is illustrated in Figure 5.9 which shows one oscillation period of the voltage at phase A of bus 17 starting from 2 s. In contrast to the WLS estimator (dotdashed,

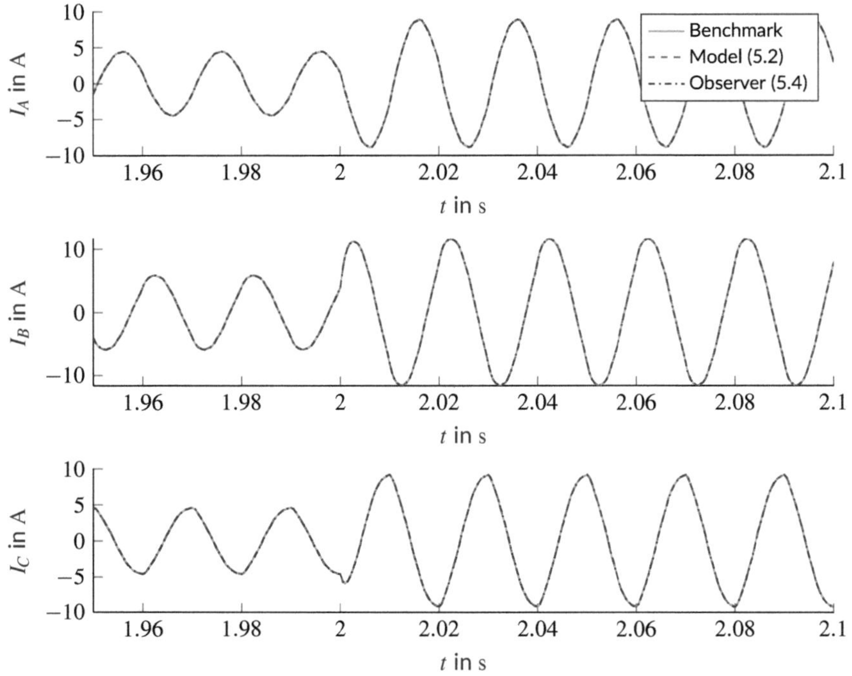

Figure 5.8: Bus current at bus 17 for the benchmark model (solid, green), the model (5.2) (dashed, blue) and the observer (5.4) (dotdashed, red) for the time between 1.95 s and 2.10 s

brown) the observer (dashed, red) is able to reconstruct the harmonic distortion from the benchmark model (solid, green).

Second, the WLS estimator cannot deal with abrupt changes in the load. This is shown in Figure 5.10 which depicts the estimated bus current for phase A of bus 17 between 1.95 s and 2.1 s. As can be seen, the WLS estimator (dotdashed, brown) requires about four periods to reach the amplitude of the current computed by the observer (dashed, red) and benchmark model (solid, green). Naturally, such errors in the bus current lead to an increase of the voltage RESP.

In the last part of this case study, we now analyze the validity of the reconstructions from the observer and the WLS estimator under measurement noise. To this end, the measurement equation is extended by noise, i.e., $\bar{\boldsymbol{m}} = \bar{\boldsymbol{C}}\boldsymbol{y} + \boldsymbol{\varepsilon}$ where $\boldsymbol{\varepsilon}$ is vector-valued Gaussian random process with zero mean and covariance matrix $\sigma^2 \boldsymbol{I}_{\bar{q}}$, $\sigma \in \mathbb{R}_{\geq 0}$. The above simulations of the observer and the WLS estimator are then repeated on the basis of the noisy measurements.

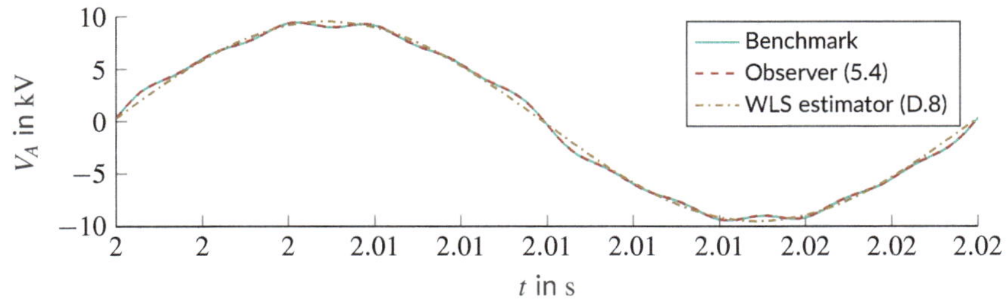

Figure 5.9: Bus voltage at phase A of bus 17 for the benchmark model (solid, green), the observer (5.4) (dashed, red), and the WLS estimator (D.8) (dotdashed, brown) for the time between 2 s and 2.02 s

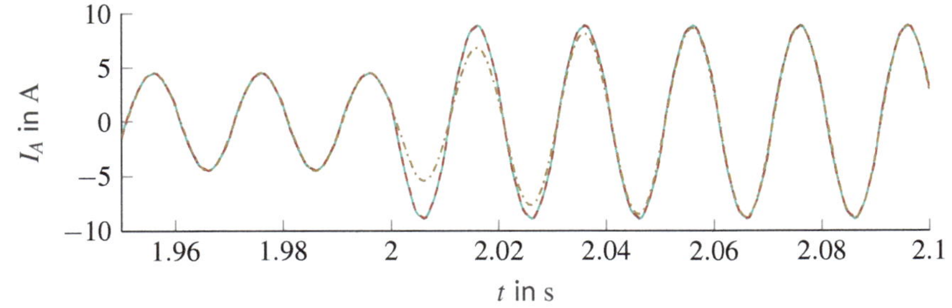

Figure 5.10: Bus current at phase A of bus 17 for the benchmark model (solid, green), the observer (5.4) (dashed, red), and the WLS estimator (D.8) (dotdashed, brown) for the time between 1.95 s and 2.1 s

The results are depicted in Figure 5.11. The figure shows the mean RESPs of the reconstructed bus voltages for the observer (5.4) (red) and the WLS estimator (D.8) (brown) for different noise variances σ^2. For the six considered noise variances 10^{-5}, 10^{-4}, 10^{-3}, 10^{-2}, 10^{-1}, and 10^{0}, the mean signal-to-noise ratios (SNRs) over all measurement signals is given by 98.0 dB, 86.5 dB, 75.0 dB, 63.4 dB, 51.9 dB, and 40.4 dB, respectively. As seen, starting on a low level, the mean RESP of the reconstructions from the observer increases with increasing noise variance. In contrast, the RESPs of the WLS estimator are almost constant over the different noise variances.

Figure 5.11 shows that for a variance smaller or equal to 10^{-2} (i.e., an SNR$\geq$ 63.4 dB), the observer gives adequate reconstructions which have a significantly lower mean RESP than the reconstructions obtained from the WLS estimator. However, an increasing noise level leads to an increasing deteriora-

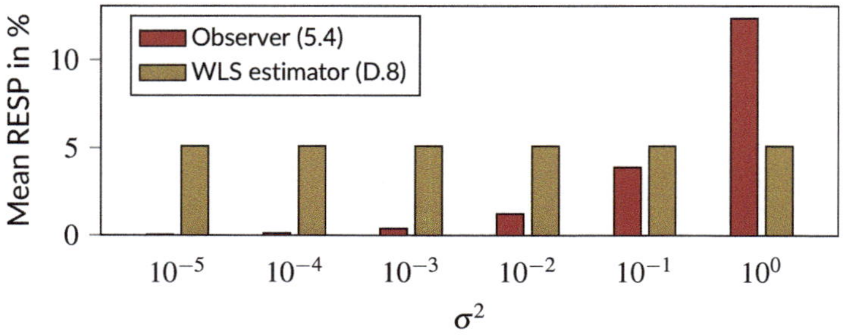

Figure 5.11: Mean RESPs over all voltages from the reconstructions of the observer (5.4) (red) and the WLS estimator (D.8) (brown) for different noise variances σ^2

tion of the reconstructions from the observer. In contrast, the WLS estimator remains almost unaffected by the measurement noise. This can be explained by the smoothing property of the phase-locked loops for the phasor computation in the WLS estimator.

In conclusion, the observer (5.4) captures harmonic distortion and thus allows for a bus-specific power quality assessment, e.g., to identify critical loads. Moreover, the algorithm immediately detects fast transients which is an important prerequisite for the application of new primary control schemes in low-inertia power systems (cf. Milano et al. [2018] and Strehle et al. [2019]). Therewith, the proposed observer extends the functionalities of a classical WLS-based power system state estimation. The basis for these functionalities are measurements with a low to moderate SNR. Provided such measurements are available, the proposed observer is a promising approach to deal with the challenges in the supervision of future power systems.

5.3 Case Study 2: Large-Scale Nonlinear System

5.3.1 Study Objectives

Compared to the previous subsection, we now turn our attention to a nonlinear system. We consider an interconnected system of academic nature which features nonlinearities in both, the interconnection structure and storages. Based on the system, it will be shown that

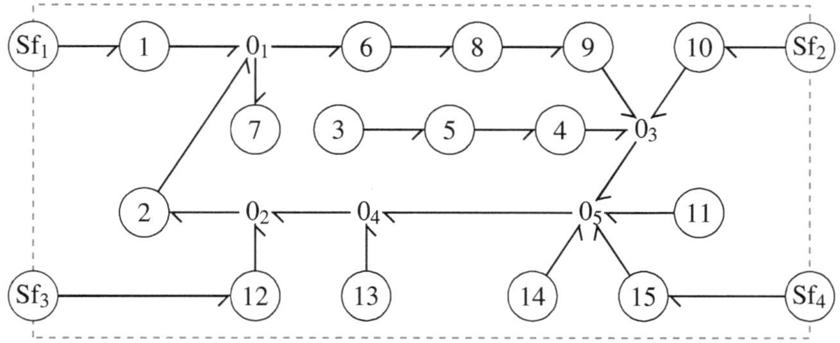

Figure 5.12: Large-scale nonlinear interconnected system

- AMOTO is able to automatically generate global and local models of a nonlinear interconnected system;
- AMOTO can be used to derive a centralized and a distributed observer based on local and global model knowledge, respectively;
- in numerical simulations, the centralized and the distributed observer obtained from AMOTO yield asymptotic reconstructions of the unknown system variables.

Moreover, we will discuss the convergence of the observers in the presence of measurement noise.

5.3.2 System Description

Let us consider the interconnected system from Figure 5.12. The system consists of 15 subsystems $\mathbb{V}_s = \{1,\dots,15\}$ and five zero junctions $\mathbb{V}_0 = \{0_1,\dots,0_5\}$. Following Definition 2.1, the set of inner vertices is given by $\mathbb{V}_\mathrm{I} = \mathbb{V}_s \cup \mathbb{V}_0$. The 20 inner vertices are connected by 20 inner edges $\mathbb{B}_\mathrm{I} = \{I_1,\dots,I_{20}\}$. The system contains four boundary vertices $\mathbb{V}_\mathrm{B} = \{\mathrm{Sf}_1,\dots,\mathrm{Sf}_4\}$ which determine the flow variables at the system boundary. The boundary vertices are connected to the subsystems $j = 1, 10, 12, 15$ by four boundary edges $\mathbb{B}_\mathrm{B} = \{B_1, B_2, B_3, B_4\}$.

Each subsystem $j \in \mathbb{V}_s$ consists of a four-dimensional bond graph as depicted in Figure 5.13. The bond graph contains two energy-storing elements with the

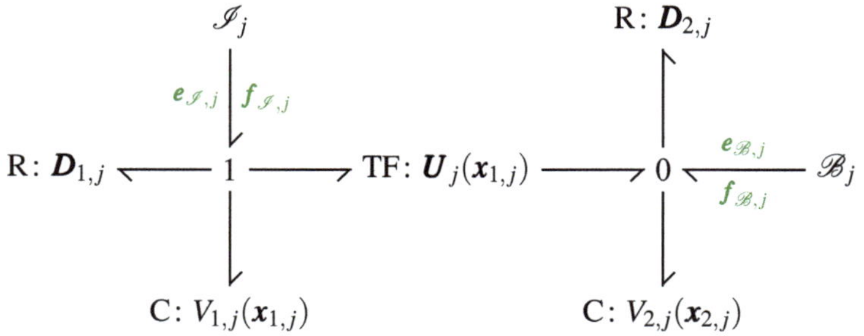

Figure 5.13: Interior structure of the subsystems $j \in \mathbb{V}_s$ from Figure 5.12

following storage functions

$$V_{1,j}(\boldsymbol{x}_{1,j}) = \frac{1}{2}\boldsymbol{x}_{1,j}^\top \boldsymbol{Q}_{1,j}\boldsymbol{x}_{1,j} + \frac{||\boldsymbol{x}_{1,j}||^4}{||\boldsymbol{x}_{1,j}||^2 + \cos\left(||\boldsymbol{x}_{1,j}||^2 + 1\right)}, \tag{5.5a}$$

$$V_{2,j}(\boldsymbol{x}_{2,j}) = \frac{1}{2}\boldsymbol{x}_{2,j}^\top \boldsymbol{Q}_{2,j}\boldsymbol{x}_{2,j}, \tag{5.5b}$$

where $\boldsymbol{x}_{1,j}, \boldsymbol{x}_{2,j} \in \mathbb{R}^4$ and $\boldsymbol{Q}_{1,j}, \boldsymbol{Q}_{2,j} \in \mathbb{R}^{4\times 4}$ symmetric, positive-definite matrices ($j \in \mathbb{V}_s$). The two energy-dissipating elements are specified by the symmetric, positive-definite matrices $\boldsymbol{D}_{1,j}, \boldsymbol{D}_{2,j} \in \mathbb{R}^{4\times 4}$ ($j \in \mathbb{V}_s$). The junction structure consists of a 1-junction, a 0-junction, and a state-modulated transformer with transformation ratio

$$\boldsymbol{U}_j(\boldsymbol{x}_{1,j}) = \exp\left(\operatorname{diag}\left(\boldsymbol{x}_{1,j}\right)\right) \in \mathbb{R}^{4\times 4}, \tag{5.6}$$

where $\exp(\cdot)$ is the matrix exponential ($j \in \mathbb{V}_s$). The port $\mathscr{I}$ is an open port for the connection to the inner edges. Likewise, the port $\mathscr{B}$ is an open port for the connection to boundary edges and hence only relevant for the subsystems $j = 1, 10, 12, 15$.

5.3.3 Automated Model Generation

This subsection initially presents a global model obtained from AMOTO for the system depicted in Figure 5.12. Afterwards, a local model for the subsystems $j \in \mathbb{V}_s$ is described.

Global Model

For the global model, we first set up a bond graph model of the overall system. To this end, we substitute the subsystems $j \in \mathbb{V}_s$ in Figure 5.12 by their bond graph representation from Figure 5.13. The resulting overall bond graph is entered into AMOTO. The dissipation matrices $\boldsymbol{D}_{1,j}, \boldsymbol{D}_{2,j}$ are specified with the numerical values as given in Appendix D.4 ($j \in \mathbb{V}_s$).

AMOTO outputs a nonlinear PHS of the form

$$\dot{\boldsymbol{x}} = -\boldsymbol{R}(\boldsymbol{x}')\frac{\partial H}{\partial \boldsymbol{x}} + \boldsymbol{G}\boldsymbol{u}, \tag{5.7}$$

$$\boldsymbol{y} = \boldsymbol{G}^{\top}\frac{\partial H}{\partial \boldsymbol{x}}, \tag{5.8}$$

with $\boldsymbol{x} \in \mathbb{R}^{120}$ and $\boldsymbol{u}, \boldsymbol{y} \in \mathbb{R}^{16}$.

The vector $\boldsymbol{x}' \in \mathbb{R}^{60}$ is the subvector of $\boldsymbol{x}$ which contains the states of the storage elements with the storage functions $V_{1,j}$, i.e., $\boldsymbol{x}' = (\boldsymbol{x}_{1,j})$ for all $j \in \mathbb{V}_s$. The input vector consists of the flows that are determined by the boundary vertices, i.e., $\boldsymbol{u} = (\boldsymbol{f}_{\mathscr{B},i})$ for $i \in \mathbb{B}_\mathrm{B}$. The output vector contains the conjugated variables, i.e., the efforts, and is thus given by $\boldsymbol{y} = (\boldsymbol{e}_{\mathscr{B},i})$ for $i \in \mathbb{B}_\mathrm{B}$.

The Hamiltonian in (5.7) is the sum of the individual storage functions from (5.5):

$$H(\boldsymbol{x}) = \sum_{j=1}^{15} \left(V_{1,j}(\boldsymbol{x}_{1,j}) + V_{2,j}(\boldsymbol{x}_{2,j})\right). \tag{5.9}$$

The matrix $\boldsymbol{R}(\boldsymbol{x}')$ consists of 30×30 non-zero blocks of the size 4×4. This reveals the damping of the energy-dissipating elements to act across subsystem boundaries. On the other hand, $\boldsymbol{R}(\boldsymbol{x}')$ neither reflects the incidence nor the adjacency relations of the subsystems or alike. In contrast, the matrix $\boldsymbol{G}$ is a sparse matrix in which the only non-zero blocks are four identity matrices $\boldsymbol{I}_4$. These identity matrices are located in the lines corresponding to $\dot{\boldsymbol{x}}_{2,j}$ for

$j = 1, 10, 12, 15$. This makes the matrix $\boldsymbol{G}$ transparent with respect to the system structure. The remaining matrices of the PHS are calculated to zero.

Local Model

Next we use AMOTO to derive a local model of the subsystems $j \in \mathbb{V}_s$. To this end, the bond graph from Figure 5.13 is entered into AMOTO and the automated model generation is executed.[8] From this, we obtain the following PHS for subsystem $j \in \mathbb{V}_s$:

$$\underbrace{\frac{\mathrm{d}}{\mathrm{d}t}\begin{pmatrix} \boldsymbol{x}_{1,j} \\ \boldsymbol{x}_{2,j} \end{pmatrix}}_{=\boldsymbol{x}_j} = -\underbrace{\begin{pmatrix} \boldsymbol{0} & \boldsymbol{0} \\ \boldsymbol{0} & \boldsymbol{D}_{2,j} \end{pmatrix}}_{=\boldsymbol{R}_j} \frac{\partial H_j}{\partial \boldsymbol{x}_j}(\boldsymbol{x}_j) + \underbrace{\begin{pmatrix} \boldsymbol{0} & \boldsymbol{I} \\ \boldsymbol{I} & \boldsymbol{U}(\boldsymbol{x}_{1,j}) \end{pmatrix}}_{=\boldsymbol{G}_j(\boldsymbol{x}_{1,j})} \underbrace{\begin{pmatrix} \boldsymbol{f}_{\mathscr{B},j} \\ \boldsymbol{f}_{\mathscr{I},j} \end{pmatrix}}_{=\boldsymbol{u}_j}, \tag{5.10a}$$

$$\underbrace{\begin{pmatrix} \boldsymbol{e}_{\mathscr{B},j} \\ \boldsymbol{e}_{\mathscr{I},j} \end{pmatrix}}_{=\boldsymbol{y}_j} = \underbrace{\begin{pmatrix} \boldsymbol{0} & \boldsymbol{I} \\ \boldsymbol{I} & \boldsymbol{U}(\boldsymbol{x}_{1,j}) \end{pmatrix}}_{=\boldsymbol{G}_j^\top(\boldsymbol{x}_{1,j})} \frac{\partial H_j}{\partial \boldsymbol{x}_j}(\boldsymbol{x}_j) + \underbrace{\begin{pmatrix} \boldsymbol{0} & \boldsymbol{0} \\ \boldsymbol{0} & \boldsymbol{D}_{1,j}^{-1} \end{pmatrix}}_{=\boldsymbol{S}_j} \underbrace{\begin{pmatrix} \boldsymbol{f}_{\mathscr{B},j} \\ \boldsymbol{f}_{\mathscr{I},j} \end{pmatrix}}_{=\boldsymbol{u}_j}, \tag{5.10b}$$

where $\boldsymbol{u}, \boldsymbol{x}, \boldsymbol{y} \in \mathbb{R}^8$ and $H_j(\boldsymbol{x}_j) = V_{1,j}(\boldsymbol{x}_{1,j}) + V_{2,j}(\boldsymbol{x}_{2,j})$ with the storage functions from (5.5). As can be seen, the PHS is passive but not strictly passive (cf. Proposition 2.26). Moreover, in contrast to the global model, the local model has feedthrough. Keep in mind that the input $\boldsymbol{f}_{\mathscr{B},j}$ is only relevant for the subsystems $j = 1, 10, 12, 15$. For the remaining subsystems, the input $\boldsymbol{f}_{\mathscr{B},j}$ can be neglected.

5.3.4 Automated Observer Design

First, we apply AMOTO to design a centralized observer based on the global model. Subsequently, we use the local model to design a distributed observer.

Centralized Observer Based on the Global Model

Consider the global model (5.7). For the centralized observer design, suppose that $\boldsymbol{u}$ and $\boldsymbol{x}'$ are known. Hence, the measurement vector reads $\boldsymbol{m} = \boldsymbol{x}' = \left(\boldsymbol{x}_{1,j}\right)$ for

[8] The open ports $\mathscr{B}$ and $\mathscr{I}$ of the bond graph in Figure 5.13 are terminated with two Sf-type elements as AMOTO does not allow for open ports.

$j \in \mathbb{V}_s$. In AMOTO, we set the variables from $\boldsymbol{u}$ and $\boldsymbol{x}'$ to "measured" and execute the automated observer design. The program first calculates a measurement equation of the form (4.31):

$$\boldsymbol{m} = \boldsymbol{x}' = \underbrace{\begin{pmatrix} \boldsymbol{Q}_{1,1}^{-1} & \boldsymbol{0} & \boldsymbol{0} & \boldsymbol{0} & \boldsymbol{0} & \boldsymbol{0} & \cdots \\ \boldsymbol{0} & \boldsymbol{0} & \boldsymbol{Q}_{1,2}^{-1} & \boldsymbol{0} & \boldsymbol{0} & \boldsymbol{0} & \cdots \\ \boldsymbol{0} & \boldsymbol{0} & \boldsymbol{0} & \boldsymbol{0} & \boldsymbol{Q}_{1,3}^{-1} & \boldsymbol{0} & \cdots \\ \vdots & \vdots & \vdots & \vdots & \vdots & \ddots \end{pmatrix}}_{=\boldsymbol{C} \in \mathbb{R}^{60 \times 120}} \boldsymbol{Q}\boldsymbol{x}, \tag{5.11}$$

where $\boldsymbol{Q} = \operatorname{blkdiag}\left(\boldsymbol{Q}_{1,j}, \boldsymbol{Q}_{2,j}\right)$ for all $j \in \mathbb{V}_s$. Moreover, AMOTO reports that the observer existence condition $\boldsymbol{R}(\boldsymbol{x}') + \boldsymbol{C}^\top \boldsymbol{C}$ is met (cf. Theorem 4.19). The obtained centralized nonlinear state-output observer reads

$$\dot{\hat{\boldsymbol{x}}} = -\boldsymbol{R}(\boldsymbol{x}') \frac{\partial H}{\partial \boldsymbol{x}}(\hat{\boldsymbol{x}}) + \boldsymbol{G}\boldsymbol{u} + \alpha \boldsymbol{C}^\top \left(\boldsymbol{m} - \boldsymbol{C}\boldsymbol{Q}\hat{\boldsymbol{x}}\right), \tag{5.12a}$$

$$\hat{\boldsymbol{y}} = \boldsymbol{G}^\top \frac{\partial H}{\partial \boldsymbol{x}}(\hat{\boldsymbol{x}}), \tag{5.12b}$$

where $\hat{\boldsymbol{x}} \in \mathbb{R}^{120}$, $\hat{\boldsymbol{y}} \in \mathbb{R}^{16}$. The parameter $\alpha \in \mathbb{R}_{>0}$ is a convergence parameter (see Remark 4.18).

Distributed Observer Based on the Local Model

Now consider the local model (5.10). For the design of a distributed observer, assume $\boldsymbol{u}_j$ and $\boldsymbol{x}_{1,j}$ to be known ($j \in \mathbb{V}_s$). Note that from a global point of view we have now more known variables than for the centralized observer. In the local bond graph model in AMOTO, the variables $\boldsymbol{u}_j$ and $\boldsymbol{x}_{1,j}$ are set to "measured". Afterwards, we run the automated observer design. AMOTO outputs the local measurement equation

$$\boldsymbol{m}_j = \boldsymbol{x}_{1,j} = \underbrace{\begin{pmatrix} \boldsymbol{Q}_{1,j}^{-1} & \boldsymbol{0} \end{pmatrix}}_{=\boldsymbol{C}_j \in \mathbb{R}^{4 \times 8}} \boldsymbol{Q}_j \boldsymbol{x}_j, \tag{5.13}$$

where $\boldsymbol{Q}_j = \operatorname{blkdiag}\left(\boldsymbol{Q}_{1,j}, \boldsymbol{Q}_{2,j}\right)$ $(j \in \mathbb{V}_s)$. The observer existence condition $\boldsymbol{R}_j + \boldsymbol{C}_j^\top \boldsymbol{C}_j \succ 0$ is fulfilled for all $j \in \mathbb{V}_s$. The obtained state-output observer for the subsystems $j \in \mathbb{V}_s$ read:

$$\dot{\hat{\boldsymbol{x}}}_j = -\boldsymbol{R}_j \frac{\partial H_j}{\partial \boldsymbol{x}_j}(\hat{\boldsymbol{x}}_j) + \boldsymbol{G}_j(\boldsymbol{x}_{1,j})\boldsymbol{u}_j + \alpha \boldsymbol{C}_j^\top \left(\boldsymbol{m}_j - \boldsymbol{C}_j \boldsymbol{Q}_j \hat{\boldsymbol{x}}_j\right), \tag{5.14a}$$

$$\hat{\boldsymbol{y}}_j = \boldsymbol{G}_j^\top(\boldsymbol{x}_{1,j}) \frac{\partial H_j}{\partial \boldsymbol{x}}(\hat{\boldsymbol{x}}_j) + \boldsymbol{S}_j \boldsymbol{u}_j, \tag{5.14b}$$

where $\hat{\boldsymbol{x}}_j, \hat{\boldsymbol{y}}_j \in \mathbb{R}^8$ and $\alpha \in \mathbb{R}_{>0}$. All subsystem observers (5.14) together form the distributed observer.

5.3.5 Simulation Setup

The objective of the simulation study is to analyze the convergence of the reconstructions produced by the centralized and distributed observer. To this end, the global model (5.7), the centralized observer (5.12), and the distributed observer (5.14) are exported to MATLAB/Simulink. It is noteworthy that the MATLAB code only for the nonlinear expression $\boldsymbol{R}(\boldsymbol{x}')$ from (5.7) comprises about 5000 lines of MATLAB code.

For the simulation, the matrices $\boldsymbol{D}_{1,j}$, $\boldsymbol{D}_{2,j}$, $\boldsymbol{Q}_{1,j}$, and $\boldsymbol{Q}_{2,j}$ are chosen as diagonal matrices. The diagonal entries were specified by random numbers between 0.1 and 10. The obtained matrices can be found in Appendix D.4. For the observers (5.12) and (5.14), the convergence parameter is set to $\alpha = 10$.

For the specification of the input of the model (5.7), we split $\boldsymbol{u} \in \mathbb{R}^{16}$ into $\boldsymbol{u} = \left(\boldsymbol{u}_1^\top \; \boldsymbol{u}_2^\top \; \boldsymbol{u}_3^\top \; \boldsymbol{u}_4^\top\right)^\top$. The subvectors $\boldsymbol{u}_k \in \mathbb{R}^4$ are chosen as

$$\boldsymbol{u}_k = 100 \cdot \begin{pmatrix} 1 & 1 & 1 & 1 \end{pmatrix}^\top \operatorname{pulse}_{0.2\,\mathrm{s}}^{5\,\%}(t - t_{0,k}), \tag{5.15}$$

for $k = 1, \ldots, 4$. The function $\operatorname{pulse}_T^w(t)$ is a periodic pulse excitation of period T and pulse width w:

$$\operatorname{pulse}_T^w(t) = \begin{cases} 1, & (i-1)T \leq t < (i-1+w)T, \quad i \in \mathbb{N}_{\geq 1} \\ 0, & \text{otherwise.} \end{cases} \tag{5.16}$$

The initial state of the system (5.7) is set to $\boldsymbol{x}_0 = (2\dots2)^\top \in \mathbb{R}^{120}$. The centralized observer (5.12) is initialized with $\hat{\boldsymbol{x}}_0 = (1\dots1)^\top \in \mathbb{R}^{120}$. Accordingly, the initial values of the distributed observer (5.14) are set to $\hat{\boldsymbol{x}}_{0,j} = (1\dots1)^\top \in \mathbb{R}^8$ for $j \in \mathbb{V}_s$.

The simulation was carried out in MATLAB / Simulink R2019a by using the automatic variable-step solver selection from Simulink. The simulation time was set to 20 seconds.

5.3.6 Simulation Results

In order to analyze the convergence of the centralized and the distributed observer let us introduce a variable $t_{\delta,i}$. Consider a state x_i and a corresponding reconstruction $\hat{x}_i$. For a constant $\delta \in \mathbb{R}_{>0}$, the time $t_{\delta,i}$ is defined as the smallest time $t \geq 0$ such that $|x_i - \hat{x}_i| < \delta$ for all $t > t_{\delta,i}$. The time $t_{\delta,i}$ is hence a measure for the advance of the convergence of a reconstruction towards the true state. In the sequel, we set $\delta = 0.01$.

Figure 5.14 shows the values of $t_{0.01,i}$ for the $i = 1,\dots,120$ states as obtained from the centralized observer (5.12) (red) and the distributed observer (5.14) (brown). The corresponding numerical data can be found in Appendix D.5. For both observers, in the majority of states, the time $t_{0.01,i}$ is smaller than one second (see lowest help line in Figure 5.14). For some states, $t_{0.01,i}$ is higher than 1 s but still less than 5 s, e.g., for $i = 32, 64, 78$. An outlier is given for the 87th state from the distributed observer by $t_{0.01,87} \approx 12.8$ s. This outlier can be explained by two reasons: (i) the state x_{87} is subject to a weak natural damping (see the third element in the matrix $\boldsymbol{D}_{2,11}$ in Appendix D.4); (ii) the error injection in the observer has no access to this observer state. The centralized observer is robust against (i) as it makes use of the damping of all subsystems and not only of one single subsystem (cf. matrix $\boldsymbol{R}(\boldsymbol{x}')$ in (5.7)). From Figure 5.14 we can also see that none of the observers outperforms the other over all states. This might be unintuitive at a first glance as a centralized observer is usually expected to produce better results as a distributed observer. However, this does not apply in our case as the distributed observer has more known variables available than the centralized observer.

Figure 5.15 depicts the time courses of the first, fifth, and 32nd state of the model (5.7), the centralized observer (5.12), and the distributed observer (5.14), respectively. The reconstructions of the first and fifth state quickly converge to the true values. The convergence of the 32nd state is slower which confirms the corresponding value from Figure 5.14.

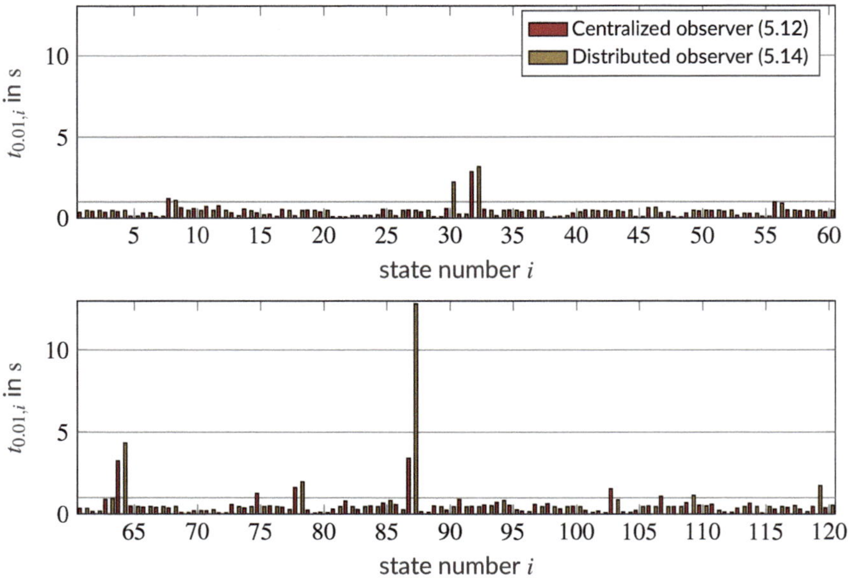

Figure 5.14: Time $t_{\delta,i}$ for $\delta = 0.01$ over the state number i for the centralized and distributed observer

Finally, let us discuss the convergence of the observers under the influence of measurement noise. To this end, a zero-mean Gaussian white noise is added to the measurements. Afterwards, the simulations of the centralized observer (5.12) and the distributed observer (5.14) are rerun on the basis of the noisy measurements.

The simulation setup and the results are described in Appendix D.6. It turns out that those observer states whose convergence is ascribed only to the natural damping of the system remain unaffected by the measurement noise. In contrast, observer states which are influenced by the measurement error injection lose the property of asymptotic convergence.What can be stated positively, however, is that these reconstructions reach with increasing time a tolerance band around the value of the variable to be reconstructed. The width of this tolerance band depends on the noise level. In practice, it has to be considered whether such a tolerance band is acceptable or if additional measures for noise suppression are to be implemented.

Let us conclude the insights from this subsection. AMOTO was shown to be able to generate a model, a centralized observer, and a distributed observer for a large-scale nonlinear system. The simulation results verify the theoretical result

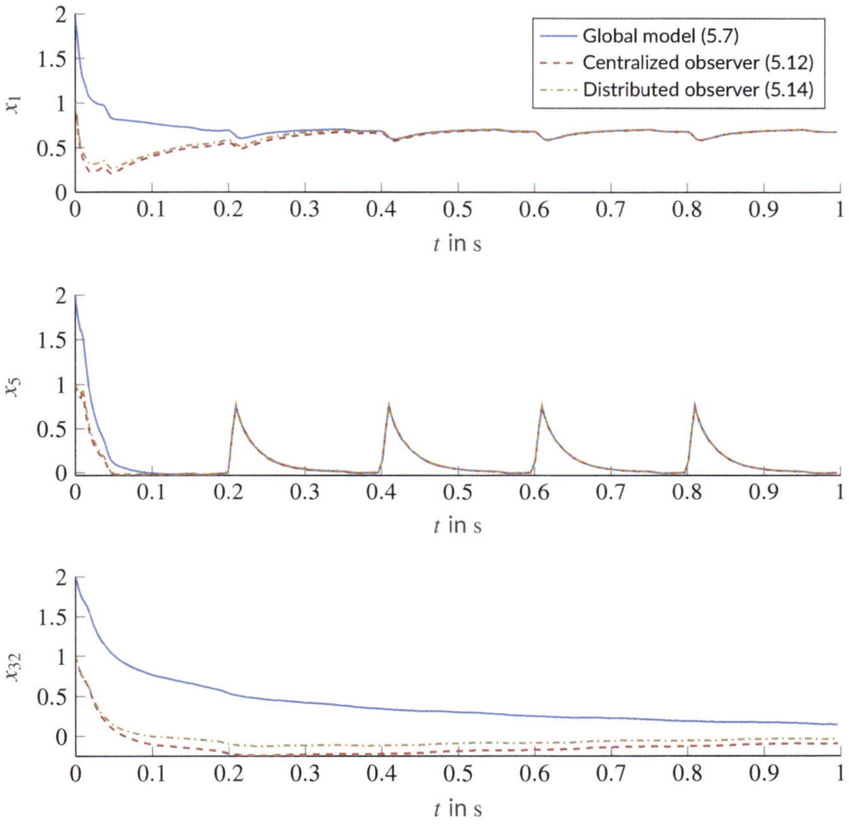

Figure 5.15: System state x_i from the model (5.7) and reconstructions $\hat{x}_i$ from the centralized observer (5.12) and the distributed observer (5.14) for $i = 1, 5, 32$ and the time between $0\,\mathrm{s}$ and $1\,\mathrm{s}$

that the obtained observers are asymptotic. In the case of noisy measurements, the property of asymptotic convergence reduces to a convergence into a tolerance band around the value to be reconstructed.

5.4 Discussion

The previous two sections demonstrated the capabilities of the software prototype AMOTO. It was shown that the program can be used for the automated model generation and observer design in linear and nonlinear interconnected systems with 100 states and more. The results from this chapter have been confirmed in many other studies as, e.g., by Muller [2018] in which AMOTO is applied to the modeling of a wind turbine system with doubly fed induction generator.

In numerical simulations, the obtained models produced plausible results. Section 5.2 showed that the model obtained from AMOTO yields simulation results that are almost identical to the results produced by a verified benchmark model from the literature. This is remarkable as the AMOTO model is a causal ODE model while the benchmark model is an acausal, component-based Simscape model. Numerical simulations verified the observers obtained from AMOTO to be asymptotic. In the presence of measurement noise, the property of asymptotic convergence reduces to a convergence into a tolerance band whose width depends on the noise level. In this context, a pre-processing of the measurements may be good choice to reduce the influence of the noise to the observer.

In AMOTO, the user exclusively operates on an intuitive graphical level. As was shown, this prevents the engineer from dealing with hundreds of equations. Therewith, AMOTO enables a transparent and comfortable modeling and observer design for interconnected systems. The program fits seamlessly in the workflow for developing control systems. The bond graph for AMOTO can be generated with well-known computer tools that feature extensive component libraries, e.g., Modelica or 20-sim. Moreover, AMOTO's export function allows to write the obtained models and observers to the file formats of MATLAB, Mathematica, and LaTeXfor subsequent steps in the development of control systems as, e.g., their simulation, analysis, and documentation.

In both case studies, the time required to calculate the models and observers is well under a minute. It is noteworthy, however, that the models and observers obtained from AMOTO may exceed the simulation capabilities of MATLAB/Simulink. The bottleneck is not the simulation itself but the compilation of a model or observer. Large expressions that involve several thousand lines of MATLAB code may lead to the situation where the compilation cannot be finished in a reasonable time.

In conclusion, this proof of principle demonstrated that AMOTO allows for an intuitive, time-efficient, and error-resistant model generation and observer design for large-scale interconnected systems.

5.5 Summary and Contributions

This chapter presented a proof of principle for the automated model generation and observer design. To this end, the software prototype AMOTO integrating the methods and algorithms from Chapter 3 and Chapter 4 was introduced. In a first case study, AMOTO was applied to design a time-domain dynamic state estimator for an unbalanced power distribution system. In a second case study, we applied AMOTO to a large-scale nonlinear interconnected system. In both case studies, numerical simulations illustrated the validity of the models and observers that were generated by AMOTO. The main achievements of this chapter, including contributions are:

(C3.1) a software prototype for an automated model generation and observer design which fits seamlessly in the workflow for the development of control systems;

(C3.2) a time-domain dynamic state estimator which extends the functionalities of a static WLS-based state estimation and which is, to the best of our knowledge, the first power system state estimator that explicitly considers the line dynamics;

(C3.3) a demonstration that we can automatically generate a model, a centralized observer, and a distributed observer for a nonlinear interconnected system.

These contributions verify that the process of deriving explicit state-space models and the process of designing model-based observers for interconnected systems can be indeed automated. Moreover, they emphasize the practical usefulness of such an approach.

Chapter 6

Conclusion

The model derivation in interconnected systems is time- and cost-intensive and may require experts from different engineering fields. Moreover, for such systems, the curse of dimensionality leads to extensive equations which makes a manual or partially computer-aided model development prone to error. Besides the modeling, the curse of dimensionality hampers many subsequent steps in a model-based control system development as, e.g., the design of observers. Hence, to reduce complexity, engineers may be forced to apply unjustified simplifications in the description of the systems' physics.

This thesis outlined a new approach to handle the complexity of interconnected systems. The idea is to consistently *automate* the model derivation and observer design in the physically unifying framework of port-Hamiltonian systems (PHSs). Following this idea, this work has made innovative methodological contributions in two areas: first, methods and algorithms for an automated generation of explicit port-Hamiltonian models from bond graphs have been presented; second, methods and algorithms for an automated PHS-based design of observers have been developed. By these contributions, we reach the research objectives stated in Chapter 1. The approach is intuitive, efficient, and error-resistant and allows to treat a wide class of interconnected systems covering the electric, mechanic, hydraulic, thermal, and chemical domain.

The methods and algorithms from this work are implemented in a software prototype named AMOTO. AMOTO was successfully demonstrated for two interconnected systems, viz. an unbalanced power distribution network and a large-scale nonlinear system. For the former, one obtains a dynamic state estimator which significantly extends the functionalities of the static WLS estimator that is used in nowadays power system control centers. Various numerical simulations verify the validity of the models and observers generated by AMOTO.

A natural direction for future research is to consider the automated design of controllers. Appendix C.4 provides a first result towards this direction. Moreover, the thesis at hand follows an approach based on explicit models and observers. Two well-known limitations of an explicit approach are (i) that it disallows to consider systems with irreversible thermodynamic processes, and (ii) that it imposes an input-state-output causality on the system which may reduce the physical transparency in the obtained models and observers. This motivates to reconsider the research questions from this thesis in the framework of implicit PHSs—a class of systems that has been recently proposed, e.g., by van der Schaft and Maschke [2018] and Beattie et al. [2018].

As a final conclusion, this thesis is the first to consistently automate the processes of a physics-based state-space model derivation and a model-based observer design for a wide class of interconnected systems.

Appendix A

Appendices for Chapter 2

A.1 Proof of Proposition 2.26

Proposition 2.26 claims that explicit PHSs are passive or even strictly passive as a consequence of their system formulation. The proof follows here.

Proof:
The proof is oriented on van der Schaft [2017, p. 114]. Consider the explicit PHS from Definition 2.23. As the Hamiltonian H is bounded from below, we always find a constant $c \in \mathbb{R}_{\geq 0}$ such that $V(\boldsymbol{x}) = H(\boldsymbol{x}) + c$ is a non-negative function. The time derivative of V is calculated as

$$\begin{aligned}
\dot{V}(\boldsymbol{x}) &= \left(\frac{\partial V}{\partial \boldsymbol{x}}(\boldsymbol{x})\right)^{\top} \dot{\boldsymbol{x}} = \left(\frac{\partial H}{\partial \boldsymbol{x}}(\boldsymbol{x})\right)^{\top} \dot{\boldsymbol{x}} \\
&\overset{(2.12a)}{=} \left(\frac{\partial H}{\partial \boldsymbol{x}}(\boldsymbol{x})\right)^{\top} \left(\left(\boldsymbol{J}(\boldsymbol{x}) - \boldsymbol{R}(\boldsymbol{x})\right) \frac{\partial H}{\partial \boldsymbol{x}}(\boldsymbol{x}) + \left(\boldsymbol{G}(\boldsymbol{x}) - \boldsymbol{P}(\boldsymbol{x})\right) \boldsymbol{u} \right) \\
&= -\left(\frac{\partial H}{\partial \boldsymbol{x}}(\boldsymbol{x})\right)^{\top} \boldsymbol{R}(\boldsymbol{x}) \frac{\partial H}{\partial \boldsymbol{x}}(\boldsymbol{x}) + \left(\frac{\partial H}{\partial \boldsymbol{x}}(\boldsymbol{x})\right)^{\top} \left(\boldsymbol{G}(\boldsymbol{x}) - \boldsymbol{P}(\boldsymbol{x})\right) \boldsymbol{u}. \qquad \text{(A.1)}
\end{aligned}$$

With (2.12b), the last term in (A.1) can be written as

$$\begin{aligned}
&\left(\frac{\partial H}{\partial \boldsymbol{x}}(\boldsymbol{x})\right)^{\top} \left(\boldsymbol{G}(\boldsymbol{x}) - \boldsymbol{P}(\boldsymbol{x})\right) \boldsymbol{u} \\
&\qquad = \boldsymbol{y}^{\top} \boldsymbol{u} - 2 \left(\frac{\partial H}{\partial \boldsymbol{x}}(\boldsymbol{x})\right)^{\top} \boldsymbol{P}(\boldsymbol{x}) \boldsymbol{u} - \boldsymbol{u}^{\top} \left(\boldsymbol{M}(\boldsymbol{x}) + \boldsymbol{S}(\boldsymbol{x})\right)^{\top} \boldsymbol{u}. \qquad \text{(A.2)}
\end{aligned}$$

Inserting (A.2) into (A.1) and recalling that $\boldsymbol{M}(\boldsymbol{x})$ is skew-symmetric yields

$$\dot{V}(\boldsymbol{x}) = -\left(\frac{\partial H}{\partial \boldsymbol{x}}(\boldsymbol{x})\right)^{\top} \boldsymbol{R}(\boldsymbol{x})\frac{\partial H}{\partial \boldsymbol{x}}(\boldsymbol{x}) + \boldsymbol{y}^{\top}\boldsymbol{u} - 2\left(\frac{\partial H}{\partial \boldsymbol{x}}(\boldsymbol{x})\right)^{\top} \boldsymbol{P}(\boldsymbol{x})\boldsymbol{u} - \boldsymbol{u}^{\top}\boldsymbol{S}^{\top}(\boldsymbol{x})\boldsymbol{u}. \tag{A.3}$$

Equation (A.3) can be rearranged as

$$\dot{V}(\boldsymbol{x}) = \boldsymbol{u}^{\top}\boldsymbol{y} - \left(\left(\tfrac{\partial H}{\partial \boldsymbol{x}}(x)\right)^{\top} \quad \boldsymbol{u}^{\top}\right) \underbrace{\begin{pmatrix} \boldsymbol{R}(\boldsymbol{x}) & \boldsymbol{P}(\boldsymbol{x}) \\ \boldsymbol{P}^{\top}(\boldsymbol{x}) & \boldsymbol{S}(\boldsymbol{x}) \end{pmatrix}}_{=\boldsymbol{\Theta}(x)} \begin{pmatrix} \frac{\partial H}{\partial \boldsymbol{x}}(x) \\ \boldsymbol{u} \end{pmatrix} \overset{(2.13)}{\leq} \boldsymbol{u}^{\top}\boldsymbol{y}, \quad \forall t \geq 0. \tag{A.4}$$

Hence, the PHS is passive. If $\boldsymbol{\Theta}(\boldsymbol{x})$ is positive-definite for all $\boldsymbol{x} \in \mathbb{X}$, we have $\dot{V} < \boldsymbol{u}^{\top}\boldsymbol{y}$ for all non-equilibrium points and the system is strictly passive. □

Appendix **B**

Appendices for Chapter 3

B.1 Discussion on Assumption 3.22

In Subsection 3.2.5, it was argued that the source elements of a bond graph are independent if and only if Assumption 3.22 is met. In the following, we prove this statement formally. To this end, let us first introduce another matrix representation of a Dirac structure, viz. the *image representation*.

Definition B.1 (Image representation)
An image representation of a Dirac structure $\mathbb{D}(\boldsymbol{x}) \subset \mathbb{R}^n \times \mathbb{R}^n$ *with* $\boldsymbol{x} \in \mathbb{X}$ *is*

$$\mathbb{D}(\boldsymbol{x}) = \{(\boldsymbol{f}, \boldsymbol{e}) \in \mathbb{R}^n \times \mathbb{R}^n \mid \exists \lambda \in \mathbb{R}^n \text{ s.t. } \boldsymbol{f} = \boldsymbol{E}^\top(\boldsymbol{x})\lambda, \boldsymbol{e} = \boldsymbol{F}^\top(\boldsymbol{x})\lambda\}. \tag{B.1}$$

As with the kernel representation from Definition 2.18, the matrices $\boldsymbol{F}(\boldsymbol{x})$ *and* $\boldsymbol{E}(\boldsymbol{x})$ *in* (B.1) *satisfy* (2.5)*; the power balance is given by* (2.6).

Now for the dependent sources.

Proposition B.2 (Dependent sources)
Given a bond graph as in Definition 3.5 whose junction structure is described by a Dirac structure of the form (3.27)*. The source elements of the bond graph are independent for all* $\boldsymbol{x} \in \mathbb{X}$ *if and only if Assumption 3.22 is satisfied.*

Proof:
We first show that a violation of Assumption 3.22 implies that the bond graph has dependent source elements. To this end, we use the idea from Golo et al. [2003, Remark 1].

Let Assumption 3.22 be violated, i.e.,

$$\exists \boldsymbol{x} \in \mathbb{X}: \ \mathrm{rank}\left(\boldsymbol{E}_{\mathrm{Sf}}(\boldsymbol{x}) \quad \boldsymbol{F}_{\mathrm{Se}}(\boldsymbol{x})\right) < K\left(N_{\mathrm{Sf}} + N_{\mathrm{Se}}\right). \tag{B.2}$$

For such an $\boldsymbol{x}$, the matrix $\left(\boldsymbol{E}_{\mathrm{Sf}}(\boldsymbol{x}) \ \boldsymbol{F}_{\mathrm{Se}}(\boldsymbol{x})\right)$ has not full rank and there exists a non-zero matrix $\left(\boldsymbol{U}(\boldsymbol{x}) \ \boldsymbol{V}(\boldsymbol{x})\right)$ such that

$$\left(\boldsymbol{U}(\boldsymbol{x}) \quad \boldsymbol{V}(\boldsymbol{x})\right)\left(\boldsymbol{E}_{\mathrm{Sf}}(\boldsymbol{x}) \quad \boldsymbol{F}_{\mathrm{Se}}(\boldsymbol{x})\right)^{\top} = \boldsymbol{0}. \tag{B.3}$$

By post-multiplying (B.3) with a non-zero $\boldsymbol{\lambda} \in \mathbb{R}^{K(N_{\mathrm{Sf}}+N_{\mathrm{Se}})}$ we obtain

$$\boldsymbol{U}(\boldsymbol{x})\underbrace{\boldsymbol{E}_{\mathrm{Sf}}^{\top}\boldsymbol{\lambda}}_{\overset{(\mathrm{B.1})}{=}\boldsymbol{f}_{\mathrm{Sf}}} + \boldsymbol{V}(\boldsymbol{x})\underbrace{\boldsymbol{F}_{\mathrm{Se}}^{\top}\boldsymbol{\lambda}}_{\overset{(\mathrm{B.1})}{=}\boldsymbol{e}_{\mathrm{Se}}} = \boldsymbol{0}. \tag{B.4}$$

Equation (B.4) reveals the junction structure to imply an algebraic dependency between $\boldsymbol{f}_{\mathrm{Sf}}$ and $\boldsymbol{e}_{\mathrm{Se}}$. This proves Assumption 3.22 to be a necessary condition for the independence of sources. The proof for sufficiency is obtained from the backwards calculation of the above formulas from (B.4) over (B.3) to (B.2). □

B.2 Analysis of the Input-Output Matrix

At the end of the proof of Lemma 3.23, we apply the insights from the proof of Theorem 4 from Bloch and Crouch [1999] for the construction of the skew-symmetric matrix $\boldsymbol{Z}(\boldsymbol{x})$. In their proof, Bloch and Crouch [1999] consider the Dirac structure on abstract (finite-dimensional) vector spaces. This is in contrast to most practical cases in which a Dirac structure is considered on Euclidean vector spaces. In the following proposition, we elaborate the calculation law of the matrix $\boldsymbol{Z}(\boldsymbol{x})$ from Lemma 3.23 for Dirac structures on Euclidean vector spaces:

Proposition B.3 (Caculation law for $Z(x)$)

The calculation law (3.38) *ensures the equation systems in* (3.36) *and* (3.27) *to describe the same relations. Moreover,* (3.38) *yields a matrix* $\boldsymbol{Z}(\boldsymbol{x})$ *that is skew-symmetric.*

Proof:

We neglect the argument $\boldsymbol{x}$ and the supplement "for all $\boldsymbol{x} \in \mathbb{X}$" in this proof.

Using the splitting of $\boldsymbol{F}_\mathrm{C}$, $\boldsymbol{F}_\mathrm{R}$, $\boldsymbol{E}_\mathrm{C}$, $\boldsymbol{E}_\mathrm{R}$ and likewise the splitting of $\boldsymbol{f}_\mathrm{C}$, $\boldsymbol{f}_\mathrm{R}$, $\boldsymbol{e}_\mathrm{C}$, $\boldsymbol{e}_\mathrm{R}$, the condition in (3.27) reads

$$\begin{aligned}
&-\boldsymbol{F}_{\mathrm{C},1}\boldsymbol{f}_{\mathrm{C},1}-\boldsymbol{F}_{\mathrm{C},2}\boldsymbol{f}_{\mathrm{C},2}-\boldsymbol{F}_{\mathrm{R},1}\boldsymbol{f}_{\mathrm{R},1}-\boldsymbol{F}_{\mathrm{R},2}\boldsymbol{f}_{\mathrm{R},2}+\boldsymbol{F}_{\mathrm{Sf}}\boldsymbol{f}_{\mathrm{Sf}}+\boldsymbol{F}_{\mathrm{Se}}\boldsymbol{f}_{\mathrm{Se}}\\
&\quad+\boldsymbol{E}_{\mathrm{C},1}\boldsymbol{e}_{\mathrm{C},1}+\boldsymbol{E}_{\mathrm{C},2}\boldsymbol{e}_{\mathrm{C},2}+\boldsymbol{E}_{\mathrm{R},1}\boldsymbol{e}_{\mathrm{R},1}+\boldsymbol{E}_{\mathrm{R},2}\boldsymbol{e}_{\mathrm{R},2}+\boldsymbol{E}_{\mathrm{Sf}}\boldsymbol{e}_{\mathrm{Sf}}+\boldsymbol{E}_{\mathrm{Se}}\boldsymbol{e}_{\mathrm{Se}}=\boldsymbol{0}. \quad \text{(B.5)}
\end{aligned}$$

Combining and gathering the different matrices we can write this as

$$\begin{aligned}
&\begin{pmatrix}\boldsymbol{F}_{\mathrm{C},1} & \boldsymbol{E}_{\mathrm{C},2} & \boldsymbol{F}_{\mathrm{R},1} & \boldsymbol{E}_{\mathrm{R},2} & \boldsymbol{E}_{\mathrm{Sf}} & \boldsymbol{F}_{\mathrm{Se}}\end{pmatrix}\begin{pmatrix}-\boldsymbol{f}_{\mathrm{C},1}\\ \boldsymbol{e}_{\mathrm{C},2}\\ -\boldsymbol{f}_{\mathrm{R},1}\\ \boldsymbol{e}_{\mathrm{R},2}\\ \boldsymbol{e}_{\mathrm{Sf}}\\ \boldsymbol{f}_{\mathrm{Se}}\end{pmatrix}+\\
&\qquad\begin{pmatrix}\boldsymbol{E}_{\mathrm{C},1} & \boldsymbol{F}_{\mathrm{C},2} & \boldsymbol{E}_{\mathrm{R},1} & \boldsymbol{F}_{\mathrm{R},2} & \boldsymbol{F}_{\mathrm{Sf}} & \boldsymbol{E}_{\mathrm{Se}}\end{pmatrix}\begin{pmatrix}\boldsymbol{e}_{\mathrm{C},1}\\ -\boldsymbol{f}_{\mathrm{C},2}\\ \boldsymbol{e}_{\mathrm{R},1}\\ -\boldsymbol{f}_{\mathrm{R},2}\\ \boldsymbol{f}_{\mathrm{Sf}}\\ \boldsymbol{e}_{\mathrm{Se}}\end{pmatrix}=\boldsymbol{0}, \quad \text{(B.6)}
\end{aligned}$$

which is equivalent to

$$\begin{pmatrix} -\boldsymbol{f}_{\mathrm{C},1} \\ \boldsymbol{e}_{\mathrm{C},2} \\ -\boldsymbol{f}_{\mathrm{R},1} \\ \boldsymbol{e}_{\mathrm{R},2} \\ \boldsymbol{e}_{\mathrm{Sf}} \\ \boldsymbol{f}_{\mathrm{Se}} \end{pmatrix} = -\begin{pmatrix} \boldsymbol{F}_{\mathrm{C},1} & \boldsymbol{E}_{\mathrm{C},2} & \boldsymbol{F}_{\mathrm{R},1} & \boldsymbol{E}_{\mathrm{R},2} & \boldsymbol{E}_{\mathrm{Sf}} & \boldsymbol{F}_{\mathrm{Se}} \end{pmatrix}^{-1} \cdot \begin{pmatrix} \boldsymbol{E}_{\mathrm{C},1} & \boldsymbol{F}_{\mathrm{C},2} & \boldsymbol{E}_{\mathrm{R},1} & \boldsymbol{F}_{\mathrm{R},2} & \boldsymbol{F}_{\mathrm{Sf}} & \boldsymbol{E}_{\mathrm{Se}} \end{pmatrix} \begin{pmatrix} \boldsymbol{e}_{\mathrm{C},1} \\ -\boldsymbol{f}_{\mathrm{C},2} \\ \boldsymbol{e}_{\mathrm{R},1} \\ -\boldsymbol{f}_{\mathrm{R},2} \\ \boldsymbol{f}_{\mathrm{Sf}} \\ \boldsymbol{e}_{\mathrm{Se}} \end{pmatrix}, \tag{B.7}$$

i.e., equivalent to the condition in (3.36). To prove the skew-symmetry of $\boldsymbol{Z}$ we again follow the idea of Bloch and Crouch [1999, Theorem 4]. Equation (2.5a) can be written as

$$\begin{aligned} \mathbf{0} &= \boldsymbol{E}\boldsymbol{F}^\top + \boldsymbol{F}\boldsymbol{E}^\top \\ &= \begin{matrix} \boldsymbol{E}_{\mathrm{C},1}\boldsymbol{F}_{\mathrm{C},1}^\top + \boldsymbol{E}_{\mathrm{C},2}\boldsymbol{F}_{\mathrm{C},2}^\top + \boldsymbol{E}_{\mathrm{R},1}\boldsymbol{F}_{\mathrm{R},1}^\top + \boldsymbol{E}_{\mathrm{R},2}\boldsymbol{F}_{\mathrm{R},2}^\top + \boldsymbol{E}_{\mathrm{Sf}}\boldsymbol{F}_{\mathrm{Sf}}^\top + \boldsymbol{E}_{\mathrm{Se}}\boldsymbol{F}_{\mathrm{Se}}^\top + \\ \boldsymbol{F}_{\mathrm{C},1}\boldsymbol{E}_{\mathrm{C},1}^\top + \boldsymbol{F}_{\mathrm{C},2}\boldsymbol{E}_{\mathrm{C},2}^\top + \boldsymbol{F}_{\mathrm{R},1}\boldsymbol{E}_{\mathrm{R},1}^\top + \boldsymbol{F}_{\mathrm{R},2}\boldsymbol{E}_{\mathrm{R},2}^\top + \boldsymbol{F}_{\mathrm{Sf}}\boldsymbol{E}_{\mathrm{Sf}}^\top + \boldsymbol{F}_{\mathrm{Se}}\boldsymbol{E}_{\mathrm{Se}}^\top. \end{matrix} \end{aligned} \tag{B.8}$$

Resorting and combining the matrices we can write this as

$$\begin{aligned} \mathbf{0} = &\begin{pmatrix} \boldsymbol{F}_{\mathrm{C},1} & \boldsymbol{E}_{\mathrm{C},2} & \boldsymbol{F}_{\mathrm{R},1} & \boldsymbol{E}_{\mathrm{R},2} & \boldsymbol{E}_{\mathrm{Sf}} & \boldsymbol{F}_{\mathrm{Se}} \end{pmatrix} \\ &\quad \cdot \begin{pmatrix} \boldsymbol{E}_{\mathrm{C},1} & \boldsymbol{F}_{\mathrm{C},2} & \boldsymbol{E}_{\mathrm{R},1} & \boldsymbol{F}_{\mathrm{R},2} & \boldsymbol{F}_{\mathrm{Sf}} & \boldsymbol{E}_{\mathrm{Se}} \end{pmatrix}^\top \\ &\quad + \begin{pmatrix} \boldsymbol{E}_{\mathrm{C},1} & \boldsymbol{F}_{\mathrm{C},2} & \boldsymbol{E}_{\mathrm{R},1} & \boldsymbol{F}_{\mathrm{R},2} & \boldsymbol{F}_{\mathrm{Sf}} & \boldsymbol{E}_{\mathrm{Se}} \end{pmatrix} \\ &\quad \cdot \begin{pmatrix} \boldsymbol{F}_{\mathrm{C},1} & \boldsymbol{E}_{\mathrm{C},2} & \boldsymbol{F}_{\mathrm{R},1} & \boldsymbol{E}_{\mathrm{R},2} & \boldsymbol{E}_{\mathrm{Sf}} & \boldsymbol{F}_{\mathrm{Se}} \end{pmatrix}^\top. \end{aligned} \tag{B.9}$$

Inverting the first and last matrix we get the equation of the skew-symmetry of $\boldsymbol{Z}$:

$$\begin{aligned}
-\boldsymbol{Z}^\top &= \begin{pmatrix}\boldsymbol{E}_{\mathrm{C},1} & \boldsymbol{F}_{\mathrm{C},2} & \boldsymbol{E}_{\mathrm{R},1} & \boldsymbol{F}_{\mathrm{R},2} & \boldsymbol{F}_{\mathrm{Sf}} & \boldsymbol{E}_{\mathrm{Se}}\end{pmatrix}^\top \\
&\quad \cdot \left(\begin{pmatrix}\boldsymbol{F}_{\mathrm{C},1} & \boldsymbol{E}_{\mathrm{C},2} & \boldsymbol{F}_{\mathrm{R},1} & \boldsymbol{E}_{\mathrm{R},2} & \boldsymbol{E}_{\mathrm{Sf}} & \boldsymbol{F}_{\mathrm{Se}}\end{pmatrix}^{-1}\right)^\top \\
&= -\begin{pmatrix}\boldsymbol{F}_{\mathrm{C},1} & \boldsymbol{E}_{\mathrm{C},2} & \boldsymbol{F}_{\mathrm{R},1} & \boldsymbol{E}_{\mathrm{R},2} & \boldsymbol{E}_{\mathrm{Sf}} & \boldsymbol{F}_{\mathrm{Se}}\end{pmatrix}^{-1} \\
&\quad \cdot \begin{pmatrix}\boldsymbol{E}_{\mathrm{C},1} & \boldsymbol{F}_{\mathrm{C},2} & \boldsymbol{E}_{\mathrm{R},1} & \boldsymbol{F}_{\mathrm{R},2} & \boldsymbol{F}_{\mathrm{Sf}} & \boldsymbol{E}_{\mathrm{Se}}\end{pmatrix} \\
&= \boldsymbol{Z}.
\end{aligned} \tag{B.10}$$

□

B.3 Discussion on Assumption 3.30

In Subsection 3.3, it was argued that Assumption 3.30 is always fulfilled if all resistive elements from the bond graph are truly dissipative. In the following proposition, we elaborate this statement more in detail:

Proposition B.4 (Truly dissipative resistive elements)
Given a bond graph according to Definition 3.5. Let all resistive elements $i \in \mathbb{V}_{\mathrm{R}}$ be truly dissipative, i.e., they obey the constitutive relation

$$\boldsymbol{f}_j = \boldsymbol{D}_i(\boldsymbol{x})\boldsymbol{e}_j, \tag{B.11}$$

with $j \in \mathbb{B}(i)$, $\boldsymbol{D}_i(\boldsymbol{x}) \in \mathbb{R}^{K \times K}$, and $\boldsymbol{D}_i(\boldsymbol{x}) = \boldsymbol{D}_i(\boldsymbol{x})^\top \succ 0$, for all $\boldsymbol{x} \in \mathbb{X}$. Note that in contrast to the general case from Subsection 3.2.2, we now assume $\boldsymbol{D}_i(\boldsymbol{x})$ to be positive-definite. Let the junction structure of the bond graph be described by a Dirac structure in kernel form (3.27) which satisfies Assumption 3.22. Then, we always find an input-output representation (3.36) of (3.27) with a splitting of resistive variables that fulfills Assumption 3.30.

Proof:
We neglect the argument $\boldsymbol{x}$ and the supplement "for all $\boldsymbol{x} \in \mathbb{X}$" in this proof.

Consider a Dirac structure in kernel form (3.27) which satisfies Assumption 3.22. From Lemma 3.23 and the properties of a causal bond graph we know that we always find an input-output representation of (3.27) in the form (3.36) where for each $j \in \mathbb{B}(i)$, $i \in \mathbb{V}_{\mathrm{R}}$ the pair $(\boldsymbol{f}_j, \boldsymbol{e}_j)$ lies either completely in $(\boldsymbol{f}_{\mathrm{R},1}, \boldsymbol{e}_{\mathrm{R},1})$ or completely in $(\boldsymbol{f}_{\mathrm{R},2}, \boldsymbol{e}_{\mathrm{R},2})$. Possibly after permutations, we may thus write the resistive relations (B.11) as

$$\begin{pmatrix} \boldsymbol{f}_{\mathrm{R},1} \\ \boldsymbol{f}_{\mathrm{R},2} \end{pmatrix} = \underbrace{\begin{pmatrix} \boldsymbol{D}_{11} & \boldsymbol{0} \\ \boldsymbol{0} & \boldsymbol{D}_{22} \end{pmatrix}}_{=\boldsymbol{D}} \begin{pmatrix} \boldsymbol{e}_{\mathrm{R},1} \\ \boldsymbol{e}_{\mathrm{R},2} \end{pmatrix}, \tag{B.12}$$

where $\boldsymbol{D}_{11} = \boldsymbol{D}_{11}^{\top} \succ 0$ and $\boldsymbol{D}_{22} = \boldsymbol{D}_{22}^{\top} \succ 0$ are block diagonal matrices. Equation (B.12) may be rewritten as

$$\underbrace{\begin{pmatrix} \boldsymbol{e}_{\mathrm{R},1} \\ -\boldsymbol{f}_{\mathrm{R},2} \end{pmatrix}}_{=\boldsymbol{u}_{\mathrm{R}}} = -\underbrace{\begin{pmatrix} \boldsymbol{D}_{11}^{-1} & \boldsymbol{0} \\ \boldsymbol{0} & \boldsymbol{D}_{22} \end{pmatrix}}_{=\tilde{\boldsymbol{R}}} \underbrace{\begin{pmatrix} -\boldsymbol{f}_{\mathrm{R},1} \\ \boldsymbol{e}_{\mathrm{R},2} \end{pmatrix}}_{=\boldsymbol{y}_{\mathrm{R}}}, \tag{B.13}$$

where $\tilde{\boldsymbol{R}} = \tilde{\boldsymbol{R}}^{\top} \succ 0$. Equation (B.13) is of the form (3.66) which is why Assumption 3.30 is fulfilled. □

B.4 Port-Hamiltonian Systems with Nonlinear Dissipation

In this section, it is shown that the modeling approach from Chapter 3 can be generalized to systems with resistive relations which are nonlinear in the power variables.

Consider the Dirac structure obtained from Corollary 3.28:

$$\mathbb{D}(\boldsymbol{x}) = \{(\begin{pmatrix} \boldsymbol{f}_{\mathrm{C}} \\ \boldsymbol{f}_{\mathrm{R}} \\ \boldsymbol{f}_{\mathrm{Sf}} \\ \boldsymbol{f}_{\mathrm{Se}} \end{pmatrix}, \begin{pmatrix} \boldsymbol{e}_{\mathrm{C}} \\ \boldsymbol{e}_{\mathrm{R}} \\ \boldsymbol{e}_{\mathrm{Sf}} \\ \boldsymbol{e}_{\mathrm{Se}} \end{pmatrix}) \in \mathbb{R}^{KN_{\mathrm{E}}} \times \mathbb{R}^{KN_{\mathrm{E}}} \mid \begin{pmatrix} \boldsymbol{y}_{\mathrm{C}} \\ \boldsymbol{y}_{\mathrm{R}} \\ \boldsymbol{y}_{\mathrm{P}} \end{pmatrix} = \underbrace{\begin{pmatrix} \boldsymbol{Z}_{\mathrm{CC}}(\boldsymbol{x}) & -\boldsymbol{Z}_{\mathrm{CR}}(\boldsymbol{x}) & -\boldsymbol{Z}_{\mathrm{CP}}(\boldsymbol{x}) \\ \boldsymbol{Z}_{\mathrm{CR}}^{\top}(\boldsymbol{x}) & \boldsymbol{Z}_{\mathrm{RR}}(\boldsymbol{x}) & -\boldsymbol{Z}_{\mathrm{RP}}(\boldsymbol{x}) \\ \boldsymbol{Z}_{\mathrm{CP}}^{\top}(\boldsymbol{x}) & \boldsymbol{Z}_{\mathrm{RP}}^{\top}(\boldsymbol{x}) & \boldsymbol{Z}_{\mathrm{PP}}(\boldsymbol{x}) \end{pmatrix}}_{\boldsymbol{Z}(\boldsymbol{x})} \begin{pmatrix} \boldsymbol{u}_{\mathrm{C}} \\ \boldsymbol{u}_{\mathrm{R}} \\ \boldsymbol{u}_{\mathrm{P}} \end{pmatrix} \}, \tag{B.14}$$

where $\boldsymbol{Z}(\boldsymbol{x}) = -\boldsymbol{Z}^{\top}(\boldsymbol{x}) \in \mathbb{R}^{KN_{\mathrm{E}} \times KN_{\mathrm{E}}}$ for all $\boldsymbol{x} \in \mathbb{X}$. The inputs and outputs in (B.14) are given by $\boldsymbol{u}_{\mathrm{C}} = \boldsymbol{e}_{\mathrm{C}}$, $\boldsymbol{u}_{\mathrm{R}} = (\boldsymbol{e}_{\mathrm{R},1}^{\top}\ \boldsymbol{f}_{\mathrm{R},2}^{\top})^{\top}$, $\boldsymbol{u}_{\mathrm{P}} = (\boldsymbol{f}_{\mathrm{Sf}}^{\top}\ \boldsymbol{e}_{\mathrm{Se}}^{\top})^{\top}$, $\boldsymbol{y}_{\mathrm{C}} = -\boldsymbol{f}_{\mathrm{C}}$, $\boldsymbol{y}_{\mathrm{R}} = (-\boldsymbol{f}_{\mathrm{R},1}^{\top}\ \boldsymbol{e}_{\mathrm{R},2}^{\top})^{\top}$, and $\boldsymbol{y}_{\mathrm{P}} = (\boldsymbol{e}_{\mathrm{Sf}}^{\top}\ \boldsymbol{f}_{\mathrm{Se}}^{\top})^{\top}$. With the following assumption, we exclude the presence of causal paths between resistive elements:

Assumption B.5 (Matrix block in (B.14))
In (B.14), we have $\boldsymbol{Z}_{\mathrm{RR}}(\boldsymbol{x}) = \boldsymbol{0}$ for all $\boldsymbol{x} \in \mathbb{X}$.

Recall the constitutive relations of storages from (3.65):

$$\boldsymbol{y}_{\mathrm{C}} = -\boldsymbol{f}_{\mathrm{C}} = -\dot{\boldsymbol{x}}, \qquad \boldsymbol{u}_{\mathrm{C}} = \boldsymbol{e}_{\mathrm{C}} = \frac{\partial V}{\partial \boldsymbol{x}}(\boldsymbol{x}), \tag{B.15}$$

and define $\boldsymbol{z} := \boldsymbol{u}_{\mathrm{C}}$. For resistive elements, suppose nonlinear constitutive relations which are expressed as the graph of an input-output map (cf. van der Schaft and Jeltsema [2014, p. 24]):

$$\boldsymbol{u}_{\mathrm{R}} = \boldsymbol{\Phi}(\boldsymbol{y}_{\mathrm{R}}, \boldsymbol{x}, \boldsymbol{z}, \boldsymbol{u}), \tag{B.16}$$

where $\boldsymbol{y}_{\mathrm{R}}^{\top}\boldsymbol{u}_{\mathrm{R}} \leq 0$. Now we have everything prepared for a port-Hamiltonian formulation of a system with nonlinear dissipation.

Proposition B.6 (PHS of a system with nonlinear dissipation)
Given an explicit Dirac structure (B.14) which satisfies Assumption B.5. Let the constitutive relations of storage elements and resistive elements be given in the forms (B.15) and (B.16), respectively. Equations (B.14), (B.15),

and (B.16) can be written as explicit input-state-output PHSs of the form

$$\dot{\boldsymbol{x}} = \boldsymbol{J}(\boldsymbol{x})\,\boldsymbol{z} - \mathscr{R}(\boldsymbol{x},\boldsymbol{z},\boldsymbol{u}) + \boldsymbol{G}(\boldsymbol{x})\,\boldsymbol{u}, \tag{B.17a}$$

$$\boldsymbol{y} = \boldsymbol{G}^\top(\boldsymbol{x})\,\boldsymbol{z} + \mathscr{P}(\boldsymbol{x},\boldsymbol{z},\boldsymbol{u}) + \boldsymbol{M}(\boldsymbol{x})\,\boldsymbol{u}, \tag{B.17b}$$

with mappings $\mathscr{R}(\boldsymbol{x},\cdot,\cdot) : \mathbb{R}^n \to \mathbb{R}^n$, $\mathscr{P}(\boldsymbol{x},\cdot,\cdot) : \mathbb{R}^n \to \mathbb{R}^p$ *and* $\boldsymbol{u} = \boldsymbol{u}_{\mathrm{P}}$, $\boldsymbol{y} = \boldsymbol{y}_{\mathrm{P}}$. *In (B.17), the matrices and mappings satisfy* $\boldsymbol{J}(\boldsymbol{x}) = -\boldsymbol{J}^\top(\boldsymbol{x})$, $\boldsymbol{M}(\boldsymbol{x}) = -\boldsymbol{M}^\top(\boldsymbol{x})$, *and*

$$\begin{pmatrix} \boldsymbol{z} \\ \boldsymbol{u} \end{pmatrix}^\top \begin{pmatrix} \mathscr{R}(\boldsymbol{x},\boldsymbol{z},\boldsymbol{u}) & \boldsymbol{0} \\ 0 & \mathscr{P}(\boldsymbol{x},\boldsymbol{z},\boldsymbol{u}) \end{pmatrix} \geq 0, \quad \forall \boldsymbol{x} \in \mathbb{X}, \boldsymbol{z} \in \mathbb{R}^n, \boldsymbol{u} \in \mathbb{U}. \tag{B.18}$$

The matrices can be obtained from $\boldsymbol{J}(\boldsymbol{x}) = -\boldsymbol{Z}_{\mathrm{CC}}(\boldsymbol{x})$, $\boldsymbol{G}(\boldsymbol{x}) = \boldsymbol{Z}_{\mathrm{CP}}(\boldsymbol{x})$, *and* $\boldsymbol{M}(\boldsymbol{x}) = \boldsymbol{Z}_{\mathrm{PP}}(\boldsymbol{x})$; *the mappings are calculated as*

$$\mathscr{R}(\boldsymbol{x},\boldsymbol{z},\boldsymbol{u}) = -\boldsymbol{Z}_{\mathrm{CR}}(\boldsymbol{x})\,\boldsymbol{\Phi}\left(\boldsymbol{Z}_{\mathrm{CR}}^\top(\boldsymbol{x})\,\boldsymbol{z} - \boldsymbol{Z}_{\mathrm{CP}}(\boldsymbol{x})\,\boldsymbol{u}, \boldsymbol{x}, \boldsymbol{z}, \boldsymbol{u}\right), \tag{B.19a}$$

$$\mathscr{P}(\boldsymbol{x},\boldsymbol{z},\boldsymbol{u}) = \boldsymbol{Z}_{\mathrm{RP}}^\top(\boldsymbol{x})\,\boldsymbol{\Phi}\left(\boldsymbol{Z}_{\mathrm{CR}}^\top(\boldsymbol{x})\,\boldsymbol{z} - \boldsymbol{Z}_{\mathrm{CP}}(\boldsymbol{x})\,\boldsymbol{u}, \boldsymbol{x}, \boldsymbol{z}, \boldsymbol{u}\right). \tag{B.19b}$$

Proof:

Inserting $\boldsymbol{u}_{\mathrm{C}} = \boldsymbol{z}$ and (B.16) into the first line of the equation system from (B.14) yields

$$\boldsymbol{y}_{\mathrm{C}} = \boldsymbol{Z}_{\mathrm{CC}}(\boldsymbol{x})\,\boldsymbol{u}_{\mathrm{C}} - \boldsymbol{Z}_{\mathrm{CR}}(\boldsymbol{x})\,\boldsymbol{\Phi}(\boldsymbol{y}_{\mathrm{R}},\boldsymbol{x},\boldsymbol{z},\boldsymbol{u}) - \boldsymbol{Z}_{\mathrm{CP}}(\boldsymbol{x})\,\boldsymbol{u}_{\mathrm{P}}. \tag{B.20}$$

For the second term from the right side we write

$$-\boldsymbol{Z}_{\mathrm{CR}}(\boldsymbol{x})\,\boldsymbol{\Phi}(\boldsymbol{y}_{\mathrm{R}},\boldsymbol{x},\boldsymbol{z},\boldsymbol{u}) \overset{\text{(B.16)}}{=} \underbrace{-\boldsymbol{Z}_{\mathrm{CR}}(\boldsymbol{x})\,\boldsymbol{\Phi}\left(\boldsymbol{Z}_{\mathrm{CR}}^\top(\boldsymbol{x})\,\boldsymbol{z} - \boldsymbol{Z}_{\mathrm{CP}}(\boldsymbol{x})\,\boldsymbol{u}, \boldsymbol{x}, \boldsymbol{z}, \boldsymbol{u}\right)}_{=\mathscr{R}(\boldsymbol{x},\boldsymbol{z},\boldsymbol{u})}. \tag{B.21}$$

By inserting (B.21) into (B.20) we obtain (B.17a):

$$\begin{aligned} & \boldsymbol{y}_{\mathrm{C}} = \boldsymbol{Z}_{\mathrm{CC}}(\boldsymbol{x})\,\boldsymbol{z} + \mathscr{R}(\boldsymbol{x},\boldsymbol{z},\boldsymbol{u}) - \boldsymbol{Z}_{\mathrm{CP}}(\boldsymbol{x})\,\boldsymbol{u}_{\mathrm{P}} \\ \overset{(B.15)}{\Leftrightarrow}\quad & -\dot{\boldsymbol{x}} = \boldsymbol{Z}_{\mathrm{CC}}(\boldsymbol{x})\,\boldsymbol{z} + \mathscr{R}(\boldsymbol{x},\boldsymbol{z},\boldsymbol{u}) - \boldsymbol{Z}_{\mathrm{CP}}(\boldsymbol{x})\,\boldsymbol{u}_{\mathrm{P}} \\ \Leftrightarrow\quad & \dot{\boldsymbol{x}} = \underbrace{-\boldsymbol{Z}_{\mathrm{CC}}(\boldsymbol{x})}_{=\boldsymbol{J}(\boldsymbol{x})}\boldsymbol{z} - \mathscr{R}(\boldsymbol{x},\boldsymbol{z},\boldsymbol{u}) + \underbrace{\boldsymbol{Z}_{\mathrm{CP}}(\boldsymbol{x})}_{=\boldsymbol{G}(\boldsymbol{x})}\underbrace{\boldsymbol{u}_{\mathrm{P}}}_{=\boldsymbol{u}}. \end{aligned} \tag{B.22}$$

Now for the output equation. From the third line of the equation system in (B.14) and with $\boldsymbol{u} = \boldsymbol{u}_{\mathrm{P}}$, $\boldsymbol{y} = \boldsymbol{y}_{\mathrm{P}}$ we obtain

$$\boldsymbol{y} = \underbrace{\boldsymbol{Z}_{\mathrm{CP}}^{\top}(\boldsymbol{x})}_{=\boldsymbol{G}^{\top}(\boldsymbol{x})}\boldsymbol{z} + \boldsymbol{Z}_{\mathrm{RP}}^{\top}(\boldsymbol{x})\,\boldsymbol{u}_{\mathrm{R}} + \underbrace{\boldsymbol{Z}_{\mathrm{PP}}^{\top}(\boldsymbol{x})}_{=\boldsymbol{M}(\boldsymbol{x})}\boldsymbol{u}. \tag{B.23}$$

For the second term from the right side we write

$$\begin{aligned} \boldsymbol{Z}_{\mathrm{RP}}^{\top}(\boldsymbol{x})\,\boldsymbol{u}_{\mathrm{R}} & \overset{(B.16)}{=} \boldsymbol{Z}_{\mathrm{RP}}^{\top}(\boldsymbol{x})\,\boldsymbol{\Phi}(\boldsymbol{y}_{\mathrm{R}},\boldsymbol{x},\boldsymbol{z},\boldsymbol{u}) \\ & \overset{(B.14)}{=} \underbrace{\boldsymbol{Z}_{\mathrm{RP}}^{\top}(\boldsymbol{x})\,\boldsymbol{\Phi}\left(\boldsymbol{Z}_{\mathrm{CR}}^{\top}(\boldsymbol{x})\,\boldsymbol{z} - \boldsymbol{Z}_{\mathrm{CP}}(\boldsymbol{x})\,\boldsymbol{u},\boldsymbol{x},\boldsymbol{z},\boldsymbol{u}\right)}_{=\mathscr{P}(\boldsymbol{x},\boldsymbol{z},\boldsymbol{u})}. \end{aligned} \tag{B.24}$$

Inserting (B.24) into (B.23) yields (B.17b).

Next, we show that (B.18) holds. By multiplying (B.17a) from the right side with $\boldsymbol{z}^{\top}$ we obtain

$$\boldsymbol{z}^{\top}\dot{\boldsymbol{x}} = -\boldsymbol{z}^{\top}\mathscr{R}(\boldsymbol{x},\boldsymbol{z},\boldsymbol{u}) + \boldsymbol{z}^{\top}\boldsymbol{G}(\boldsymbol{x})\,\boldsymbol{u}. \tag{B.25}$$

On the other hand, by the power balance of the Dirac structure (B.14) we have

$$\boldsymbol{z}^{\top}\dot{\boldsymbol{x}} \overset{(B.15)}{=} -\boldsymbol{u}_{\mathrm{C}}^{\top}\boldsymbol{y}_{\mathrm{C}} \overset{(2.9)}{=} \boldsymbol{u}_{\mathrm{R}}^{\top}\boldsymbol{y}_{\mathrm{R}} + \boldsymbol{u}_{\mathrm{P}}^{\top}\boldsymbol{y}_{\mathrm{P}}. \tag{B.26}$$

Equating (B.25) and (B.26) yields

$$-z^\top \mathscr{R}(x,z,u) + z^\top G(x)\,u = u_R^\top y_R + u_P^\top y_P. \tag{B.27}$$

The last term reads

$$\begin{aligned}
u_P^\top y_P &\overset{\text{(B.14)}}{=} u_P^\top \left(Z_{CP}^\top(x)\,u_C + Z_{RP}^\top(x)\,u_R + Z_{PP}(x)\,u_P\right) \\
&= u_P^\top Z_{CP}^\top(x)\,u_C + u_P^\top Z_{RP}^\top(x)\,u_R \\
&\overset{\text{(B.16)}}{=} u_P^\top Z_{CP}^\top(x)\,u_C + u_P^\top Z_{RP}^\top(x)\,\Phi(y_R, x, z, u) \\
&\overset{\text{(B.14)}}{=} u_P^\top Z_{CP}^\top(x)\,u_C + u_P^\top Z_{RP}^\top(x)\,\Phi\left(Z_{CR}^\top(x)\,z - Z_{CP}(x)\,u, x, z, u\right).
\end{aligned} \tag{B.28}$$

With (B.19a) and $u = u_P$, $y = y_P$ we write (B.28) as

$$u_P^\top y_P = u^\top \underbrace{Z_{CP}^\top(x)}_{=G^\top(x)} z + u^\top \mathscr{P}(x,z,u). \tag{B.29}$$

By inserting (B.29) into (B.27) we prove (B.18):

$$\begin{aligned}
& -z^\top \mathscr{R}(x,z,u) + \cancel{z^\top G(x)\,u} = u_R^\top y_R + \cancel{u^\top G^\top(x)\,z} + u^\top \mathscr{P}(x,z,u) \\
\Leftrightarrow \quad & -z^\top \mathscr{R}(x,z,u) - u^\top \mathscr{P}(x,z,u) = u_R^\top y_R \overset{\text{(B.16)}}{\leq} 0.
\end{aligned} \tag{B.30}$$

□

To the best of our knowledge, the explicit PHS (B.17) has not been presented in the literature so far. The next proposition shows that this PHS is passive.

Proposition B.7 (Passivity of a PHS with nonlinear dissipation)
The explicit port-Hamiltonian model from (B.17) *is passive.*

Proof:
Recall that $z = \frac{\partial V}{\partial x}(x)$ where $V(x)$ is a storage function that is bounded from below. We always find a constant $c \in \mathbb{R}_{\geq 0}$ such that $\tilde{V}(x) = V(x) + c$ is a non-negative function. The derivative of $\tilde{V}(x)$ with respect to time reads

$$\begin{aligned}\dot{\tilde{V}}(x) = \left(\frac{\partial \tilde{V}}{\partial x}(x)\right)^{\top} \dot{x} = \left(\frac{\partial V}{\partial x}(x)\right)^{\top} \dot{x} \\ = z^{\top}\left(J(x)z - \mathscr{R}(x,z,u) + G(x)u\right) \\ = -z^{\top}\mathscr{R}(x,z,u) + z^{\top}G(x)u. \qquad \text{(B.31)}\end{aligned}$$

Transposing (B.17b) and multiplying with u from the right gives

$$\begin{aligned} y^{\top}u = z^{\top}G(x)u + \mathscr{P}^{\top}(x,z,u)u \\ \Leftrightarrow \quad z^{\top}G(x)u = y^{\top}u - u^{\top}\mathscr{P}(x,z,u). \qquad \text{(B.32)}\end{aligned}$$

Inserting (B.32) into (B.31) then yields

$$\begin{aligned}\dot{\tilde{V}}(x) &= -z^{\top}\mathscr{R}(x,z,u) + y^{\top}u - u^{\top}\mathscr{P}(x,z,u) \\ &= y^{\top}u - \underbrace{\begin{pmatrix} z \\ u \end{pmatrix}^{\top}\begin{pmatrix}\mathscr{R}(x,z,u) & 0 \\ 0 & \mathscr{P}(x,z,u)\end{pmatrix}}_{\geq 0} \leq y^{\top}u. \qquad \text{(B.33)}\end{aligned}$$

□

For the case of no feedthrough we obtain from Proposition B.6 the "input-state-output PHS with nonlinear resistive structure" introduced by van der Schaft [2017, Def. 6.1.4]. This special case is outlined in the subsequent corollary.

Corollary B.8 (PHS with nonlinear dissipation from van der Schaft [2017])

Given an explicit Dirac structure (B.14) *which satisfies Assumption B.5. Let* $Z_{\mathrm{RP}}(x) = 0$ *and* $Z_{\mathrm{PP}}(x) = 0$ *for all* $x \in \mathbb{X}$. *Equations* (B.14), (B.15), *and* (B.16) *can be written as explicit input-state-output PHS of the form*

$$\dot{\boldsymbol{x}} = \boldsymbol{J}(\boldsymbol{x})\boldsymbol{z} - \mathscr{R}(\boldsymbol{x},\boldsymbol{z},\boldsymbol{u}) + \boldsymbol{G}(\boldsymbol{x})\boldsymbol{u}, \tag{B.34a}$$

$$\boldsymbol{y} = \boldsymbol{G}^{\top}(\boldsymbol{x})\boldsymbol{z}, \tag{B.34b}$$

where $\boldsymbol{J}(\boldsymbol{x}) = -\boldsymbol{J}^{\top}(\boldsymbol{x})$ *and* $\boldsymbol{z}^{\top}\mathscr{R}(\boldsymbol{x},\boldsymbol{z},\boldsymbol{u}) \geq 0$ *for all* $\boldsymbol{x} \in \mathbb{X}$, $\boldsymbol{z} \in \mathbb{R}^n$, $\boldsymbol{u} \in \mathbb{U}$.

Proof:
The proof follows from Proposition B.6 under $\boldsymbol{Z}_{\mathrm{RP}}(\boldsymbol{x}) = 0$ and $\boldsymbol{Z}_{\mathrm{PP}}(\boldsymbol{x}) = 0$. □

B.5 Notes to Section 3.3

The following notes provide some additional information to the discussion from Section 3.3, especially with regard to the related literature:

Theorems 3.35 and 3.47 are the first to give rigorous existence conditions for the complete transfer from a bond graph to an explicit PHS. In the literature, there exist some existence conditions for intermediate steps of this transfer. Golo et al. [2000] and Golo et al. [2003] show that the junction structure of a bond graph can always be related to a Dirac structure in implicit form. This is in line with our findings from Theorem 3.35.Donaire and Junco [2009] provide sufficient conditions for the transfer from a Dirac structure in explicit form to a non-feedthrough PHS. Theorems 3.35 and 3.47 put the conditions of Golo et al. [2000], Golo et al. [2003], and Donaire and Junco [2009] into the perspective of an explicit port-Hamiltonian formulation of bond graphs. It turns out, that the most crucial step for such a formulation is the transfer of the Dirac structure from an implicit to an explicit representation. Concerning this transfer, Theorem 3.35 is the first to provide necessary and sufficient conditions.

The sufficient existence condition from Theorem 3.35 makes the well-known conditions for the derivation of an explicit PHS mathematically traceable. In the literature, there exist three well-known prerequisites under which a system can be formulated as an explicit PHS, see, e.g., van der Schaft [2009, p. 70] or van der Schaft and Jeltsema [2014, p. 53]. The class of explicit input-state-output

PHS occurs if "(1) the external port variables can be split into input and output variables, (2) there are no algebraic constraints between the state variables, and (3) the resistive structure is linear and of input-output form"[1] [van der Schaft and Jeltsema, 2014, p. 53]. These conditions are formulated only verbally which makes them mathematically untraceable and hampers their practical evaluation. For the particular case of bond graph systems, Theorem 3.35 provides remedy as it gives mathematically traceable representations of the three conditions (1), (2), and (3), viz. assumptions 3.22, 3.27, and 3.30, respectively. Based on these assumptions, computer algebra systems can evaluate if an explicit PHS exists or not.

The existence conditions from Theorem 3.35 are related to the existence condition for an implicit port-Hamiltonian formulation of a bond graph. Golo et al. [2003] shows that each well-posed bond graph[2] permits an implicit port-Hamiltonian formulation. Assumption 3.22 (i.e., the independence of sources) is a necessary condition for a bond graph to be well-posed. Moreover, Assumption 3.27 (i.e., the independence of sources and the independence of storages) is sufficient for the well-posedness. In consequence, each explicit PHS obtained from Theorem 3.35 permits an implicit representation. This is intuitive as the class of implicit PHSs is more general than the class of explicit PHSs. Lopes [2016] gives conditions under which an implicit PHS can be transferred into an explicit PHS [Lopes, 2016, Propriété 6]. However, the results from Theorem 3.35 show that the conditions from Lopes [2016] are unnecessarily strict.

Assumptions 3.22 and 3.27 can be interpreted by means of bond graph causality. The transfer of the Dirac structure from an implicit to an explicit form can be interpreted as the causalization of the bond graph.[3] A first indication of this interpretation has already been given by Lopes [2016]. A causality-based interpretation of assumptions 3.22 and 3.27 reads "in the causal bond graph, all sources have proper causality" and "in the causal bond graph, all sources have proper causality and all storages are in preferred (integral) causality", respectively.

Theorem 3.47 generalizes the results from previous literature on state-space models of systems with dependent storages. In bond graph theory, dependent storages are well-known to lead in general to models in the form of DAEs (cf. Borutzky [2010, pp. 142–143]). Nevertheless, Theorem 3.47 states that

[1] Compared to the original texts, the points (1) and (2) have been switched in the order.
[2] For the well-posedness of a bond graph the reader is asked to refer to Golo et al. [2003, Def. 2].
[3] For the concept of bond graph causality refer to Borutzky [2010, pp. 92ff.].

(under the given conditions) we can obtain a PHS in the form of an ODE. This insight is in line with some previous results from the literature. For linear systems, Wellstead [1979] has shown that if Assumption 3.37 is satisfied we can always obtain an explicit state-space model in spite of the presence of dependent storages. Theorem 3.47 generalizes the results from Wellstead [1979] to nonlinearities in the junction structure. Moreover, the insights from Theorem 3.47 are in line with the findings of Donaire and Junco [2009]. Compared to Donaire and Junco [2009], however, Theorem 3.47 is more general as it allows for systems that possibly contain feedthrough. Najnudel et al. [2018] stated conditions under which the dependent storages can be expressed by independent storages. However, the work of Najnudel et al. [2018] is restricted to dependent storages that occur from the parallel or serial interconnection of storage elements. Theorem 3.47 generalizes these results to more subtle structures which lead to the dependence between storage elements.

Algorithms 3.36 and 3.48 are the first to enable a fully automated generation of explicit port-Hamiltonian models. In the literature, there exists various algorithms that are related to algorithms 3.36 and 3.48. Golo et al. [2000] present an algorithm which determines a Dirac structure in explicit form based on a bond graph [Golo et al., 2000, Algorithm 3]. The algorithm is based on graphical manipulations. Dai [2016] provides an algorithm for the generation of a Dirac structure in implicit form [Dai, 2016, Algorithm 1]. Moreover, the author provides an algorithm for transferring a Dirac structure from an implicit to an explicit form [Dai, 2016, Algorithm 4].Donaire and Junco [2009] derive calculation rules for the matrices of a PHS based on a Dirac structure in explicit form [Donaire and Junco, 2009, Theorem 1]. An algorithm for the determination of an implicit PHS from an analog circuit was proposed by Falaize and Hélie [2016]. The algorithm is implemented in a Python program package named PyPHS, see Falaize and Hélie [2019]. As can be seen, different aspects concerning an automated generation of explicit port-Hamiltonian models have been addressed in the literature. However, algorithms 3.36 and 3.48 are the first to fully automate the derivation of an explicit PHS.

Appendix C

Appendices for Chapter 4

C.1 Detectability, Strong Detectability, and Strong* Detectability

This section gives a brief introduction into the concepts of detectability, strong detectability, and strong* detectability. The latter is of particular importance for the observer design in Subsection 4.2.1.

Consider a linear state-space system in standard form:

$$\dot{\boldsymbol{x}} = \boldsymbol{A}\boldsymbol{x} + \boldsymbol{B}\boldsymbol{u}, \quad \boldsymbol{x}|_{t=0} = \boldsymbol{x}_0, \tag{C.1a}$$

$$\boldsymbol{y} = \boldsymbol{C}\boldsymbol{x} + \boldsymbol{D}\boldsymbol{u}. \tag{C.1b}$$

The following definition is inspired by Sontag [1998, p. 329]:

Definition C.1 (Detectability)
The system (C.1) is detectable if in the autonomous system $\boldsymbol{y} \equiv \boldsymbol{0}$ *implies* $\lim_{t\to\infty} \boldsymbol{x} \to \boldsymbol{0}$*, for all* $\boldsymbol{x}_0$.

In Definition C.1, we assume the input $\boldsymbol{u}$ of the system to be known. Hence, the restriction to the autonomous system (i.e., the case where $\boldsymbol{u} \equiv \boldsymbol{0}$) is without loss of generality (cf. Ludyk [1995, p. 7]). In the following, the concept of detectability is extended to systems with unknown inputs. This leads us to the definitions of *strong detectability*, and *strong* detectability* as introduced by Hautus [1983].

Definition C.2 (Strong detectability)
The system (C.1) *is strong detectable if* $\boldsymbol{y} \equiv \mathbf{0}$ *implies* $\lim_{t\to\infty} \boldsymbol{x} \to \mathbf{0}$, *for all* $\boldsymbol{x}_0$ *and* $\boldsymbol{u}$.

Definition C.3 (Strong* detectability)
The system (C.1) *is strong* detectable if* $\lim_{t\to\infty} \boldsymbol{y} \to \mathbf{0}$ *implies* $\lim_{t\to\infty} \boldsymbol{x} \to \mathbf{0}$.

Note that strong* detectability implies strong detectability.

C.2 Specification of the Eigenvalues for the Error Dynamics Matrix

The state reconstruction error dynamics of the observer proposed in Theorem 4.8 obey $\dot{\boldsymbol{\varepsilon}} = \boldsymbol{N}\boldsymbol{\varepsilon}$ (i.e., (4.13) under (4.14)). The design of the observer is conducted with Algorithm 4.10. In code line 21 of Algorithm 4.10, we have to specify the eigenvalues of the matrix $\boldsymbol{N}$. Here, an automatable procedure for this is proposed.

Recall (4.18), i.e,

$$\boldsymbol{N} = \boldsymbol{K}\boldsymbol{A} - \boldsymbol{Z}\bar{\boldsymbol{C}} \tag{C.2}$$

and define $\bar{\boldsymbol{A}} := \boldsymbol{K}\boldsymbol{A}$. Let Assumption 4.7 hold. Then the pair $(\bar{\boldsymbol{A}}, \bar{\boldsymbol{C}})$ is detectable (cf. Darouach et al. [1994, Theorem 2] and Hautus [1983, Theorem 1.5]).

First, we apply a Kalman decomposition with respect to observability to $(\bar{\boldsymbol{A}}, \bar{\boldsymbol{C}})$. Let $\boldsymbol{O} \in \mathbb{R}^{\bar{q}n\times n}$ denote the observability matrix of $(\bar{\boldsymbol{A}}, \bar{\boldsymbol{C}})$ with $r = \operatorname{rank}(\boldsymbol{O}) (\leq n)$. Collect r linear independent rows of $\boldsymbol{O}$ in a matrix $\boldsymbol{T}_1 \in \mathbb{R}^{r\times n}$. The rows of $\boldsymbol{T}_1$ span the observable subspace. Collect the vectors of a basis of $\ker(\boldsymbol{O})$ in a matrix $\boldsymbol{T}_2 \in \mathbb{R}^{(n-r)\times n}$. The rows of $\boldsymbol{T}_2$ span the orthogonal complement of the observable subspace.

Conjugating (C.2) with $\boldsymbol{T} := \left(\boldsymbol{T}_1^\top \ \boldsymbol{T}_2^\top\right)^\top$ gives

$$\underbrace{\boldsymbol{T}\boldsymbol{N}\boldsymbol{T}^{-1}}_{=:\boldsymbol{N}^\dagger} = \underbrace{\boldsymbol{T}\bar{\boldsymbol{A}}\boldsymbol{T}^{-1}}_{=:\bar{\boldsymbol{A}}^\dagger} - \underbrace{\boldsymbol{T}\boldsymbol{Z}}_{=:\boldsymbol{Z}^\dagger} \underbrace{\bar{\boldsymbol{C}}\boldsymbol{T}^{-1}}_{=:\bar{\boldsymbol{C}}^\dagger}, \tag{C.3}$$

where

$$\bar{\boldsymbol{A}}^\dagger = \begin{pmatrix} \bar{\boldsymbol{A}}_{11}^\dagger & \mathbf{0} \\ \bar{\boldsymbol{A}}_{21}^\dagger & \bar{\boldsymbol{A}}_{22}^\dagger \end{pmatrix}, \quad \bar{\boldsymbol{C}}^\dagger = \begin{pmatrix} \bar{\boldsymbol{C}}_1^\dagger & \mathbf{0} \end{pmatrix}, \tag{C.4}$$

with $\bar{\boldsymbol{A}}_{11}^{\dagger} \in \mathbb{R}^{r\times r}$, $\bar{\boldsymbol{A}}_{21}^{\dagger} \in \mathbb{R}^{(n-r)\times r}$, $\bar{\boldsymbol{A}}_{22}^{\dagger} \in \mathbb{R}^{(n-r)\times(n-r)}$, and $\bar{\boldsymbol{C}}_{1}^{\dagger} \in \mathbb{R}^{\bar{q}\times r}$. Note that the eigenvalues are invariant under conjugation, i.e., the eigenvalues of $\boldsymbol{N}^{\dagger}$ are equal to the eigenvalues of $\boldsymbol{N}$. Moreover, as $(\bar{\boldsymbol{A}}, \bar{\boldsymbol{C}})$ is detectable, $\bar{\boldsymbol{A}}_{22}^{\dagger}$ is a Hurwitz matrix.

Now we have everything prepared to specify the eigenvalues of $\boldsymbol{N}$. The first r eigenvalues $\lambda_i(\boldsymbol{N})$ are placed at

$$\lambda_i(\boldsymbol{N}) = \min\left\{\mathrm{Re}\left(\lambda_i(\bar{\boldsymbol{A}}_{11}^{\dagger})\right) - \beta, \gamma\right\}, \tag{C.5}$$

where $\beta \in \mathbb{R}_{\geq 0}$, $\gamma \in \mathbb{R}_{<0}$, and $\lambda_i(\bar{\boldsymbol{A}}_{11}^{\dagger})$ the i-th eigenvalue of the matrix $\bar{\boldsymbol{A}}_{11}^{\dagger}$ $(i = 1, \ldots, r)$. The remaining $n-r$ eigenvalues are set to

$$\lambda_{i+r}(\boldsymbol{N}) = \lambda_i(\bar{\boldsymbol{A}}_{22}^{\dagger}), \tag{C.6}$$

where $\lambda_i(\bar{\boldsymbol{A}}_{22}^{\dagger})$ is the i-th eigenvalue of the matrix $\bar{\boldsymbol{A}}_{22}^{\dagger}$ $(i = 1, \ldots, n-r)$. This choice is motivated by (C.3) and (C.4) which reveal the (stable) eigenvalues of $\bar{\boldsymbol{A}}_{22}^{\dagger}$ are also eigenvalues of $\boldsymbol{N}^{\dagger}$ and therewith of $\boldsymbol{N}$.

With the chosen eigenvalues for $\boldsymbol{N}^{\dagger}$ and (C.3) we can calculate the matrix $\boldsymbol{Z}^{\dagger}$ by using well-established pole placement techniques. Finally, by $\boldsymbol{Z} = \boldsymbol{T}^{-1}\boldsymbol{Z}^{\dagger}$ we obtain the matrix $\boldsymbol{Z}$ in the original coordinates.

C.3 Additional Statements for the Proof of Lemma 4.16

In the proof of Lemma 4.16, we applied Lyapunov's direct method to prove $\boldsymbol{0}$ to be a globally exponentially stable equilibrium point of an error system. In this proof, we made use of the following two propositions:

Proposition C.4 ($\dot{V}$ for an autonomous system)
Consider the autonomous system

$$\dot{\boldsymbol{x}} = \boldsymbol{A}(\boldsymbol{s})\boldsymbol{Q}\boldsymbol{x}, \tag{C.7}$$

where $\boldsymbol{x} \in \mathbb{R}^n$, $\boldsymbol{A}(\boldsymbol{s}) \in \mathbb{R}^{n\times n}$, and $\boldsymbol{Q} \in \mathbb{R}^{n\times n}$ with $\boldsymbol{Q} = \boldsymbol{Q}^\top \succ 0$ for some parameter $\boldsymbol{s} \in \mathbb{S}$. In order to analyze the stability of the equilibrium $\boldsymbol{x} \equiv \boldsymbol{0}$ suppose the Lyapunov function candidate

$$V(\boldsymbol{x}) = \frac{1}{2}\boldsymbol{x}^\top \boldsymbol{Q}\boldsymbol{x}. \tag{C.8}$$

The derivative of (C.8) with respect to time can be expressed as

$$\dot{V}(\boldsymbol{x}) = \boldsymbol{x}^\top \boldsymbol{Q}\left(\tfrac{1}{2}\left(\boldsymbol{A}(\boldsymbol{s}) + \boldsymbol{A}^\top(\boldsymbol{s})\right)\right)\boldsymbol{Q}\boldsymbol{x}. \tag{C.9}$$

Equation (C.9) depends only on the symmetric part of the matrix $\boldsymbol{A}(\boldsymbol{s})$, i.e., $\dot{V}(\boldsymbol{x})$ it is independent of the skew-symmetric part of $\boldsymbol{A}(\boldsymbol{s})$.

Proof:
The derivative of (C.8) reads

$$\begin{aligned}
\dot{V}(\boldsymbol{x}) &= \frac{1}{2}\dot{\boldsymbol{x}}^\top \boldsymbol{Q}\boldsymbol{x} + \frac{1}{2}\boldsymbol{x}^\top \boldsymbol{Q}\dot{\boldsymbol{x}} \\
&\overset{(C.7)}{=} \frac{1}{2}\left(\boldsymbol{A}(\boldsymbol{s})\boldsymbol{Q}\boldsymbol{x}\right)^\top \boldsymbol{Q}\boldsymbol{x} + \frac{1}{2}\boldsymbol{x}^\top \boldsymbol{Q}\boldsymbol{A}(\boldsymbol{s})\boldsymbol{Q}\boldsymbol{x} \\
&= \frac{1}{2}\boldsymbol{x}^\top \boldsymbol{Q}\boldsymbol{A}^\top(\boldsymbol{s})\boldsymbol{Q}\boldsymbol{x} + \frac{1}{2}\boldsymbol{x}^\top \boldsymbol{Q}\boldsymbol{A}(\boldsymbol{s})\boldsymbol{Q}\boldsymbol{x} \\
&= \boldsymbol{x}^\top \boldsymbol{Q}\left(\tfrac{1}{2}\left(\boldsymbol{A}(\boldsymbol{s}) + \boldsymbol{A}^\top(\boldsymbol{s})\right)\right)\boldsymbol{Q}\boldsymbol{x}.
\end{aligned} \tag{C.10}$$

□

Proposition C.5 (Lower and upper bound)

Given a vector $\boldsymbol{x} \in \mathbb{R}^n$ and a family of symmetric, positive-definite matricies $\boldsymbol{D}(\boldsymbol{s}) \in \mathbb{R}^{n \times n}$ depending continuously on some parameter $\boldsymbol{s} \in \mathbb{S}$ with $\mathbb{S}$ compact. Then, there exist positive constants $k_1, k_2 \in \mathbb{R}_{>0}$ such that

$$k_1 \|\boldsymbol{x}\|^2 \leq \boldsymbol{x}^\top \boldsymbol{D}(\boldsymbol{s})\boldsymbol{x} \leq k_2 \|\boldsymbol{x}\|^2, \quad \forall \boldsymbol{s} \in \mathbb{S}, \forall \boldsymbol{x} \in \mathbb{R}^n. \tag{C.11}$$

Proof:

We first show that, without loss of generality, $\boldsymbol{D}(\boldsymbol{s})$ can be assumed to be diagonal.

As $\boldsymbol{D}(\boldsymbol{s})$ is symmetric there exists a continuous family of orthogonal matrices $\boldsymbol{T}(\boldsymbol{s})$ such that

$$\boldsymbol{x}^\top \boldsymbol{D}(\boldsymbol{s})\boldsymbol{x} = \boldsymbol{x}^\top \boldsymbol{T}^\top(\boldsymbol{s}) \underbrace{\boldsymbol{T}(\boldsymbol{s})\boldsymbol{D}(\boldsymbol{s})\boldsymbol{T}^\top(\boldsymbol{s})}_{=:\tilde{\boldsymbol{D}}(\boldsymbol{s})} \boldsymbol{T}(\boldsymbol{s})\boldsymbol{x}, \tag{C.12}$$

for all $\boldsymbol{s} \in \mathbb{S}$ and for all $\boldsymbol{x} \in \mathbb{R}^n$ where $\tilde{\boldsymbol{D}}(\boldsymbol{s})$ is a diagonal matrix with the (positive) eigenvalues of $\boldsymbol{D}(\boldsymbol{s})$ on its diagonal. By defining $\boldsymbol{y} := \boldsymbol{T}(\boldsymbol{s})\boldsymbol{x}$ we may rewrite (C.11) as

$$k_1 \|\boldsymbol{y}\|^2 \leq \boldsymbol{y}^\top \tilde{\boldsymbol{D}}(\boldsymbol{s})\boldsymbol{y} \leq k_2 \|\boldsymbol{y}\|^2, \quad \forall \boldsymbol{s} \in \mathbb{S}, \forall \boldsymbol{y} \in \mathbb{R}^n. \tag{C.13}$$

In (C.13), we use that

$$\|\boldsymbol{y}\|^2 = \|\boldsymbol{T}(\boldsymbol{s})\boldsymbol{x}\|^2 = \langle \boldsymbol{T}(\boldsymbol{s})\boldsymbol{x}, \boldsymbol{T}(\boldsymbol{s})\boldsymbol{x} \rangle = \langle \boldsymbol{x}, \boldsymbol{T}^\top(\boldsymbol{s})\boldsymbol{T}(\boldsymbol{s})\boldsymbol{x} \rangle = \|\boldsymbol{x}\|^2, \tag{C.14}$$

for all $\boldsymbol{s} \in \mathbb{S}$, i.e., the invariance of the Euclidean norm under orthogonal transformations. Equation (C.13) shows that, without loss of generality, we may assume $\boldsymbol{D}(\boldsymbol{s})$ to be diagonal.

Now for the claim from the proposition. Let $\boldsymbol{D}(\boldsymbol{s})$ be a positive-definite and diagonal matrix for all $\boldsymbol{s} \in \mathbb{S}$. Recall that $\boldsymbol{D}(\boldsymbol{s})$ depends continuously on $\boldsymbol{s}$. Hence, the eigenvalues $\lambda_i(\boldsymbol{s})$ of $\boldsymbol{D}(\boldsymbol{s})$ are also continuous in $\boldsymbol{s}$ for $i = 1, \dots, n$. From the positive definiteness of $\boldsymbol{D}(\boldsymbol{s})$ and the compactness of $\mathbb{S}$ we conclude that all eigenvalues $\lambda_i(\boldsymbol{s})$ are contained in a compact subset of $\mathbb{R}_{>0}$. Thus, there exist positive constants $k_1, k_2 \in \mathbb{R}_{>0}$ with $k_1 \leq \lambda_i(\boldsymbol{s}) \leq k_2$ for all $\boldsymbol{s} \in \mathbb{S}$ and $i = 1, \dots, n$. Such constants then fulfill (C.11) as

$$k_1 \boldsymbol{x}^\top \boldsymbol{I} \boldsymbol{x} \leq \boldsymbol{x}^\top \boldsymbol{D}(\boldsymbol{s}) \boldsymbol{x} \leq k_2 \boldsymbol{x}^\top \boldsymbol{I} \boldsymbol{x}, \quad \forall \boldsymbol{s} \in \mathbb{S}, \forall \boldsymbol{x} \in \mathbb{R}^n. \tag{C.15}$$

□

C.4 Automated Design of a Feedback Controller

Subsection 4.2.2 aims at the design of observers. The resulting methods, however, can also be used for the design of controllers. In the following proposition, we use Lemma 4.17 for the automated design of an asymptotically stabilizing feedback controller for nonlinear PHSs without feedthrough:

Proposition C.6 (Automated design of a feedback controller)
Consider the nonlinear PHS without feedthrough from Definition 2.24:

$$\dot{\boldsymbol{x}} = (\boldsymbol{J}(\boldsymbol{x}) - \boldsymbol{R}(\boldsymbol{x})) \frac{\partial H}{\partial \boldsymbol{x}}(\boldsymbol{x}) + \boldsymbol{G}(\boldsymbol{x})\boldsymbol{u}, \tag{C.16a}$$

$$\boldsymbol{y} = \boldsymbol{G}^\top(\boldsymbol{x}) \frac{\partial H}{\partial \boldsymbol{x}}(\boldsymbol{x}). \tag{C.16b}$$

Suppose a feedback controller of the form

$$\boldsymbol{u} = -\boldsymbol{K}(\boldsymbol{x}) \frac{\partial H}{\partial \boldsymbol{x}}(\boldsymbol{x}). \tag{C.17}$$

The condition

$$\boldsymbol{R}(\boldsymbol{x}) + \boldsymbol{G}(\boldsymbol{x})\boldsymbol{K}(\boldsymbol{x}) \succ 0, \quad \forall \boldsymbol{x} \in \mathbb{X}, \tag{C.18}$$

is a sufficient condition for $\boldsymbol{x} \equiv \boldsymbol{0}$ being an asymptotically stable equilibrium of the closed-loop system. Moreover, (C.18) can be fulfilled if and only if it is *fulfilled for $\boldsymbol{K}(\boldsymbol{x}) = \boldsymbol{G}^\top(\boldsymbol{x})$, that is the output feedback:*

$$\boldsymbol{u} = -\boldsymbol{G}^\top(\boldsymbol{x}) \frac{\partial H}{\partial \boldsymbol{x}}(\boldsymbol{x}) = -\boldsymbol{y}. \tag{C.19}$$

Proof:
Inserting (C.17) into (C.16a) gives the closed-loop dynamics:

$$\dot{\boldsymbol{x}} = (\boldsymbol{J}(\boldsymbol{x}) - \boldsymbol{R}(\boldsymbol{x}) - \boldsymbol{G}(\boldsymbol{x})\boldsymbol{K}(\boldsymbol{x})) \frac{\partial H}{\partial \boldsymbol{x}}(\boldsymbol{x}). \tag{C.20}$$

As $H(\boldsymbol{x})$ is bounded from below, we always find a constant $c \in \mathbb{R}_{\geq 0}$ such that $\tilde{H}(\boldsymbol{x}) = H(\boldsymbol{x}) + c$ is a positive definite function. Let us apply $\tilde{H}(\boldsymbol{x})$ as Lyapunov function candidate to analyze the stability of the equilibrium $\boldsymbol{x} \equiv \boldsymbol{0}$. The derivative of $\tilde{H}(\boldsymbol{x})$ with respect to time reads:

$$\begin{aligned}\dot{\tilde{H}}(\boldsymbol{x}) &= \left(\frac{\partial \tilde{H}}{\partial \boldsymbol{x}}(\boldsymbol{x})\right)^\top \dot{\boldsymbol{x}} = \left(\frac{\partial H}{\partial \boldsymbol{x}}(\boldsymbol{x})\right)^\top \dot{\boldsymbol{x}} \\ &\overset{(C.20)}{=} \left(\frac{\partial H}{\partial \boldsymbol{x}}(\boldsymbol{x})\right)^\top (\boldsymbol{J}(\boldsymbol{x}) - \boldsymbol{R}(\boldsymbol{x}) - \boldsymbol{G}(\boldsymbol{x})\boldsymbol{K}(\boldsymbol{x})) \frac{\partial H}{\partial \boldsymbol{x}}(\boldsymbol{x}) \\ &= -\left(\frac{\partial H}{\partial \boldsymbol{x}}(\boldsymbol{x})\right)^\top (\boldsymbol{R}(\boldsymbol{x}) + \boldsymbol{G}(\boldsymbol{x})\boldsymbol{K}(\boldsymbol{x})) \frac{\partial H}{\partial \boldsymbol{x}}(\boldsymbol{x}).\end{aligned} \tag{C.21}$$

Let (C.18) hold. Then, we have $\dot{\tilde{H}} < 0$ for all $\boldsymbol{x} \neq \boldsymbol{0}$ and $\dot{\tilde{H}} = 0$ for $\boldsymbol{x} = \boldsymbol{0}$. Hence, (C.18) is a sufficient condition for $\boldsymbol{x} \equiv \boldsymbol{0}$ being asymptotically stable. The rest follows from Lemma 4.17. □

Appendix D

Appendices for Chapter 5

D.1 Design of a WLS Estimator for Section 5.2

Section 5.2 addresses the observation of the IEEE 33 Bus System. We compare the performance of an observer generated with AMOTO with the performance of a WLS estimator. Here, it is illustated that this WLS estimator can also be generated with AMOTO.

Initially, we revise the bond graph model of the IEEE 33 Bus System from Section 5.2 under the assumption that the system is in quasi-steady state. From the resulting static bond graph model a static PHS is generated by using AMOTO. Moreover, we apply AMOTO to calculate a measurement equation for the static model. Based on the static PHS and the measurement equation we then derive a WLS estimator.

D.1.1 Static Bond Graph Model

In the bond graph model from Section 5.2, the elements $i \in \mathbb{V}_\mathrm{I}$, $i \in \mathbb{V}_\mathrm{B}$, and $j \in \mathbb{B}_\mathrm{B}$ are already described with static submodels. Hence, it suffices to reconsider the submodel of the lines $j \in \mathbb{B}_\mathrm{I}$.

Let the system be in quasi-steady state. For the lines $j \in \mathbb{B}_\mathrm{I}$, the π-equivalent circuit from Figure 5.4 then translates into the circuit in Figure D.1. The impedances from Figure D.1 can be collected in an impedence matrix:

$$\boldsymbol{Z}_j = \begin{pmatrix} Z_{j,A,A} & Z_{j,A,B} & Z_{j,A,C} \\ Z_{j,A,B} & Z_{j,B,B} & Z_{j,B,C} \\ Z_{j,A,C} & Z_{j,B,C} & Z_{j,C,C} \end{pmatrix}, \quad j \in \mathbb{B}_\mathrm{I}. \tag{D.1}$$

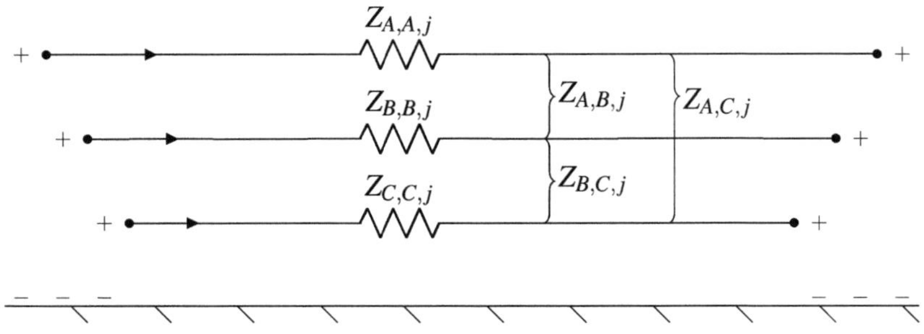

Figure D.1: Quasi-steady state π-section equivalent circuit of a three-phase distribution line

The impedance matrix for the line $j \in \mathbb{B}_\mathrm{I}$ is calculated from the resistance and inductance matrices from (5.1):

$$\boldsymbol{Z}_j = \begin{pmatrix} R_{j,A} & 0 & 0 \\ 0 & R_{j,B} & 0 \\ 0 & 0 & R_{j,C} \end{pmatrix} + \mathrm{i}2\pi f \begin{pmatrix} L_{j,A,A} & L_{j,A,B} & L_{j,A,C} \\ L_{j,A,B} & L_{j,B,B} & L_{j,B,C} \\ L_{j,A,C} & L_{j,B,C} & L_{j,C,C} \end{pmatrix}, \tag{D.2}$$

where i is the imaginary unit and $f = 50\,\mathrm{Hz}$ is the system frequency.

A bond graph representation of Figure D.1 is shown in Figure D.2. As can be seen, from a bond graph point of view the impedance $\boldsymbol{Z}_j$ is a resistive element. Hence, the line model from Figure D.2 is static. Therewith, all subsystem models are static. For the IEEE 33 Bus System, a static bond graph model is given by Figure 5.6 by replacing each line model L_j with the bond graph from Figure D.2 ($j \in \mathbb{B}_\mathrm{I}$).

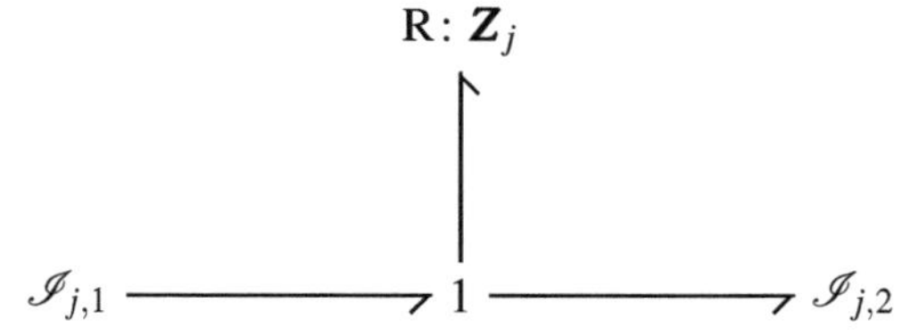

Figure D.2: Bond graph representation of the π-section model from Figure D.1

D.1.2 Automated Model Generation

The static bond graph model of the IEEE 33 Bus System is entered into AMOTO and the automated model generation is executed. AMOTO calculates a static port-Hamiltonian model consisting only of an output equation:

$$\boldsymbol{y} = \boldsymbol{S}\boldsymbol{u}, \tag{D.3}$$

where $\boldsymbol{u}, \boldsymbol{y} \in \mathbb{C}^{99}$ and $\boldsymbol{S} \in \mathbb{C}^{99 \times 99}$ with $\boldsymbol{S} = \boldsymbol{S}^\top$. As for the dynamic model (5.2), the input vector and the output vector from (D.3) contain the bus voltages and currents, respectively. However, in the static model these quantities are not in the time-domain but in the phasor domain. The matrix $\boldsymbol{S}$ reflects the system topology. Moreover, we can identify the inverses of the impedance matrices (D.1) in the non-zero matrix blocks of $\boldsymbol{S}$.

D.1.3 Estimator Design

For the estimator design, we assume the same measured variables as for the observer design in Section 5.2. The measured variables are collected in a vector $\boldsymbol{m} \in \mathbb{C}^{99}$. With AMOTO we calculate a measurement equation which relates the measurements to the inputs and outputs of (D.3):

$$\boldsymbol{m} = \boldsymbol{C}_u \boldsymbol{u} + \boldsymbol{C}_y \boldsymbol{y}, \tag{D.4}$$

where $\boldsymbol{C}_u, \boldsymbol{C}_y \in \{-1, 0, 1\}^{99 \times 99}$. Inserting the static model (D.3) into (D.4) yields

$$\boldsymbol{m} = \underbrace{(\boldsymbol{C}_u + \boldsymbol{C}_y \boldsymbol{S})}_{=\boldsymbol{C}} \boldsymbol{u}. \tag{D.5}$$

The matrix $\boldsymbol{C} \in \mathbb{C}^{99\times 99}$ is quadratic and of full rank. Hence, a static input-output estimator for the IEEE 33 Bus System can be easily calculated to

$$\hat{\boldsymbol{u}} = \boldsymbol{C}^{-1}\boldsymbol{m}, \tag{D.6}$$

$$\hat{\boldsymbol{y}} = \boldsymbol{S}\hat{\boldsymbol{u}}, \tag{D.7}$$

where $\hat{\boldsymbol{u}} \in \mathbb{C}^{99}$ and $\hat{\boldsymbol{y}} \in \mathbb{C}^{99}$ are reconstructions of $\boldsymbol{u}$ and $\boldsymbol{y}$, respectively.

Remark D.1 (WLS estimator). In general, the matrix $\boldsymbol{C}$ is a rectangular matrix with more rows than columns. Hence, from (D.5) we in general obtain

$$\hat{\boldsymbol{u}} = \left(\boldsymbol{C}^\top\boldsymbol{C}\right)^{-1}\boldsymbol{C}^\top\boldsymbol{m}. \tag{D.8}$$

This is the well known WLS estimator from the power systems literature under unit covariance.

D.2 Definition of the Relative Error Signal Power

In Subsection 5.2.7, we use the relative error signal power (RESP) as a measure for similarity of a signal to a benchmark signal. In this section, the RESP is formally introduced.

Consider a benchmark time-signal $s_{\mathrm{bm}}\colon [0,T] \to \mathbb{R}$, $t \mapsto s_{\mathrm{bm}}(t)$. The mean power of the signal $s_{\mathrm{bm}}(t)$ is

$$p_{\mathrm{bm}} = \frac{1}{T}\int_0^T s_{\mathrm{bm}}^2(t)\mathrm{d}t. \tag{D.9}$$

Furthermore, consider a second signal $s_{\mathrm{ap}}(t)\colon [0,T] \to \mathbb{R}$, $t \mapsto s_{\mathrm{ap}}(t)$ which represents an approximation of the signal $s_{\mathrm{bm}}(t)$. This approximation may stem, e.g., from a model or an observer. We define the *error signal* as $\varepsilon(t) := s_{\mathrm{bm}}(t) - s_{\mathrm{ap}}(t)$.

The (mean) *error signal power* is defined as

$$p_\varepsilon := \frac{1}{T}\int_0^T \varepsilon(t)^2 \mathrm{d}t = \frac{1}{T}\int_0^T \left(s_{\mathrm{bm}}(t) - s_{\mathrm{ap}}(t)\right)^2 \mathrm{d}t. \tag{D.10}$$

The RESP is defined as the quotient of (D.9) and (D.10):

$$r_\varepsilon := \frac{p_\varepsilon}{p_{\mathrm{bm}}}. \tag{D.11}$$

D.3 Numeric Data for the Simulation Results from Section 5.2

This section provides the numeric data from the simulation results in Subsection 5.2.7. Tables D.1 and D.2 display the numeric values for the RESP of the bus currents for the model (5.2) and the observer (5.4), respectively. For the model and the observer, the data underlying Figure 5.7 are given in the last column of the respective table. Likewise, Table D.3 and Table D.4 show the RESP values of the bus voltages for the observer (5.4) and the WLS estimator (D.8), respectively.

Table D.1: RESP of the bus currents computed from the model (5.2) in %

Bus no.	Phase A	Phase B	Phase C	Mean
1	0.1587	0.1587	0.1587	0.1587
2	0.1644	0.1587	0.1591	0.1607
3	0.1641	0.1632	0.1579	0.1617
4	0.1567	0.1564	0.1605	0.1579
5	0.1827	0.1779	0.1720	0.1776
6	0.1178	0.1739	0.1949	0.1622
7	0.1625	0.1588	0.1462	0.1558
8	0.1595	0.1640	0.1661	0.1632
9	0.1512	0.1489	0.1605	0.1535
10	0.2112	0.1867	0.1776	0.1918
11	0.3662	0.1441	0.2361	0.2488
12	0.1354	0.1763	0.1451	0.1523
13	0.1615	0.1643	0.1814	0.1690
14	0.1519	0.1615	0.1588	0.1574
15	0.1700	0.1640	0.1180	0.1506
16	0.1696	0.1443	0.2321	0.1820
17	0.1461	0.1597	0.1448	0.1502
18	0.1649	0.1616	0.1615	0.1627
19	0.1575	0.1587	0.1590	0.1584
20	0.1587	0.1590	0.1587	0.1588
21	0.1583	0.1583	0.1583	0.1583
22	0.1584	0.1583	0.1583	0.1583
23	0.1575	0.1591	0.1613	0.1593
24	0.1583	0.1583	0.1582	0.1583
25	0.1587	0.1586	0.1586	0.1586
26	0.1698	0.1975	0.1451	0.1708
27	0.1837	0.1428	0.1768	0.1678
28	0.1478	0.1544	0.1603	0.1542
29	0.1605	0.1617	0.1553	0.1592
30	0.1583	0.1561	0.1577	0.1574
31	0.1601	0.1621	0.1599	0.1607
32	0.1545	0.1537	0.1580	0.1554
33	0.1674	0.1681	0.1571	0.1642
Mean	0.1668	0.1615	0.1641	0.1641

Table D.2: RESP of the bus currents computed from the observer (5.4) in %

Bus no.	Phase A	Phase B	Phase C	Mean
1	0.0004	0.0004	0.0005	0.0004
2	0.0134	0.0107	0.0155	0.0132
3	0.0047	0.0058	0.0040	0.0048
4	0.0129	0.0162	0.0162	0.0151
5	0.0232	0.0229	0.0246	0.0236
6	0.0034	0.0101	0.0120	0.0085
7	0.0028	0.0025	0.0030	0.0028
8	0.0006	0.0007	0.0006	0.0006
9	0.0003	0.0003	0.0003	0.0003
10	0.0116	0.0114	0.0114	0.0114
11	0.0159	0.0161	0.0160	0.0160
12	0.0003	0.0004	0.0002	0.0003
13	0.0004	0.0001	0.0003	0.0003
14	0.0023	0.0023	0.0023	0.0023
15	0.0066	0.0076	0.0067	0.0070
16	0.0038	0.0047	0.0038	0.0041
17	0.0003	0.0004	0.0003	0.0004
18	0.0001	0.0001	0.0001	0.0001
19	0.0019	0.0019	0.0017	0.0018
20	0.0047	0.0047	0.0043	0.0046
21	0.0057	0.0077	0.0069	0.0068
22	0.0011	0.0010	0.0011	0.0010
23	0.0107	0.0115	0.0115	0.0112
24	0.0023	0.0025	0.0020	0.0023
25	0.0009	0.0006	0.0009	0.0008
26	0.0104	0.0043	0.0055	0.0067
27	0.0071	0.0101	0.0108	0.0093
28	0.0272	0.0243	0.0200	0.0239
29	0.0082	0.0079	0.0083	0.0081
30	0.0001	0.0003	0.0004	0.0003
31	0.0012	0.0013	0.0020	0.0015
32	0.0027	0.0027	0.0022	0.0025
33	0.0011	0.0011	0.0011	0.0011
Mean	0.0057	0.0059	0.0060	0.0059

Table D.3: RESP of the bus voltages computed from the observer (5.4) in %

Bus no.	Phase A	Phase B	Phase C	Mean
1	0.0004	0.0004	0.0004	0.0004
2	0.0000	0.0000	0.0000	0.0000
3	0.0004	0.0004	0.0004	0.0004
4	0.0006	0.0006	0.0005	0.0006
5	0.0007	0.0008	0.0006	0.0007
6	0.0014	0.0014	0.0011	0.0013
7	0.0016	0.0017	0.0014	0.0016
8	0.0017	0.0018	0.0015	0.0017
9	0.0019	0.0020	0.0017	0.0019
10	0.0021	0.0023	0.0019	0.0021
11	0.0021	0.0023	0.0019	0.0021
12	0.0021	0.0023	0.0019	0.0021
13	0.0024	0.0026	0.0022	0.0024
14	0.0025	0.0028	0.0023	0.0025
15	0.0025	0.0029	0.0024	0.0026
16	0.0026	0.0030	0.0024	0.0026
17	0.0027	0.0032	0.0025	0.0028
18	0.0027	0.0032	0.0025	0.0028
19	0.0000	0.0000	0.0000	0.0000
20	0.0001	0.0001	0.0001	0.0001
21	0.0002	0.0002	0.0001	0.0001
22	0.0002	0.0002	0.0001	0.0002
23	0.0004	0.0005	0.0005	0.0005
24	0.0005	0.0008	0.0008	0.0007
25	0.0005	0.0010	0.0009	0.0008
26	0.0015	0.0014	0.0012	0.0013
27	0.0015	0.0015	0.0012	0.0014
28	0.0020	0.0017	0.0015	0.0017
29	0.0024	0.0019	0.0017	0.0020
30	0.0026	0.0019	0.0017	0.0021
31	0.0027	0.0020	0.0020	0.0022
32	0.0027	0.0020	0.0020	0.0023
33	0.0027	0.0020	0.0020	0.0023
Mean	0.0015	0.0015	0.0013	0.0015

Table D.4: RESP of the bus voltages computed from the WLS estimator (D.8) in %

Bus no.	Phase A	Phase B	Phase C	Mean
1	4.9917	4.9873	4.9897	4.9896
2	4.9997	4.9956	4.9957	4.9970
3	5.0374	5.0351	5.0249	5.0325
4	5.0555	5.0554	5.0416	5.0508
5	5.0734	5.0755	5.0581	5.0690
6	5.1040	5.1130	5.0824	5.0998
7	5.1020	5.1090	5.0769	5.0960
8	5.1173	5.1256	5.0929	5.1119
9	5.1313	5.1407	5.1065	5.1262
10	5.1450	5.1553	5.1201	5.1401
11	5.1481	5.1586	5.1233	5.1433
12	5.1536	5.1644	5.1291	5.1490
13	5.1681	5.1786	5.1437	5.1635
14	5.1710	5.1806	5.1465	5.1660
15	5.1732	5.1819	5.1487	5.1679
16	5.1765	5.1843	5.1521	5.1710
17	5.1797	5.1834	5.1555	5.1729
18	5.1810	5.1848	5.1567	5.1742
19	5.0005	4.9963	4.9967	4.9978
20	5.0061	5.0019	5.0042	5.0041
21	5.0068	5.0025	5.0056	5.0050
22	5.0082	5.0021	5.0067	5.0057
23	5.0461	5.0433	5.0318	5.0404
24	5.0617	5.0570	5.0426	5.0538
25	5.0705	5.0616	5.0505	5.0609
26	5.1083	5.1178	5.0871	5.1044
27	5.1140	5.1241	5.0933	5.1105
28	5.1295	5.1448	5.1110	5.1284
29	5.1407	5.1595	5.1250	5.1417
30	5.1488	5.1680	5.1339	5.1502
31	5.1552	5.1783	5.1401	5.1579
32	5.1570	5.1804	5.1405	5.1593
33	5.1574	5.1808	5.1409	5.1597
Mean	5.1036	5.1099	5.0865	5.1000

D.4 Numeric System Parameters for the Simulation in Section 5.3

In Section 5.3, we consider a large-scale nonlinear interconnected system depending on some symmetric, positive-definite (4×4) matrices $\boldsymbol{D}_{1,j}$, $\boldsymbol{D}_{2,j}$, $\boldsymbol{Q}_{1,j}$, and $\boldsymbol{Q}_{1,j}$ ($j \in \mathbb{V}_s = \{1, \ldots, 15\}$). For the simulation of the system, we have to specify these matrices numerically. To this end, the matrices $\boldsymbol{D}_{1,j}$, $\boldsymbol{D}_{2,j}$, $\boldsymbol{Q}_{1,j}$, and $\boldsymbol{Q}_{1,j}$ were chosen as diagonal matrices. The diagonal entries were specified by random numbers between 0.1 and 10. The obtained matrices are as follows:

$$\boldsymbol{D}_{1,1} = \operatorname{diag}(7.58, 6.22, 2.18, 6.67) \qquad \boldsymbol{D}_{2,1} = \operatorname{diag}(6.84, 2.99, 5.66, 9.08) \tag{D.12}$$

$$\boldsymbol{D}_{1,2} = \operatorname{diag}(6.10, 7.05, 5.50, 4.40) \qquad \boldsymbol{D}_{2,2} = \operatorname{diag}(4.53, 5.16, 8.50, 6.57) \tag{D.13}$$

$$\boldsymbol{D}_{1,3} = \operatorname{diag}(6.23, 3.57, 5.72, 3.58) \qquad \boldsymbol{D}_{2,3} = \operatorname{diag}(9.75, 5.64, 5.86, 8.54) \tag{D.14}$$

$$\boldsymbol{D}_{1,4} = \operatorname{diag}(2.41, 3.34, 7.82, 1.85) \qquad \boldsymbol{D}_{2,4} = \operatorname{diag}(8.52, 0.27, 2.65, 8.52) \tag{D.15}$$

$$\boldsymbol{D}_{1,5} = \operatorname{diag}(8.02, 7.25, 0.71, 5.62) \qquad \boldsymbol{D}_{2,5} = \operatorname{diag}(5.72, 7.34, 3.65, 4.90) \tag{D.16}$$

$$\boldsymbol{D}_{1,6} = \operatorname{diag}(8.26, 6.99, 5.52, 8.56) \qquad \boldsymbol{D}_{2,6} = \operatorname{diag}(9.09, 3.63, 2.50, 8.78) \tag{D.17}$$

$$\boldsymbol{D}_{1,7} = \operatorname{diag}(6.40, 5.70, 0.36, 4.75) \qquad \boldsymbol{D}_{2,7} = \operatorname{diag}(5.26, 2.50, 8.01, 7.72) \tag{D.18}$$

$$\boldsymbol{D}_{1,8} = \operatorname{diag}(0.76, 1.56, 0.35, 1.96) \qquad \boldsymbol{D}_{2,8} = \operatorname{diag}(4.79, 3.38, 5.10, 2.16) \tag{D.19}$$

$$\boldsymbol{D}_{1,9} = \operatorname{diag}(1.89, 1.73, 4.63, 2.89) \qquad \boldsymbol{D}_{2,9} = \operatorname{diag}(6.96, 7.57, 5.87, 7.87) \tag{D.20}$$

$$\boldsymbol{D}_{1,10} = \operatorname{diag}(3.04, 6.66, 6.26, 2.69) \qquad \boldsymbol{D}_{2,10} = \operatorname{diag}(3.36, 9.62, 8.63, 9.07) \tag{D.21}$$

$$\boldsymbol{D}_{1,11} = \operatorname{diag}(7.77, 9.32, 3.21, 2.90) \qquad \boldsymbol{D}_{2,11} = \operatorname{diag}(3.95, 9.37, 0.80, 9.63) \tag{D.22}$$

$$\boldsymbol{D}_{1,12} = \operatorname{diag}(0.44, 5.33, 9.04, 9.08) \qquad \boldsymbol{D}_{2,12} = \operatorname{diag}(9.23, 3.41, 2.91, 5.36) \tag{D.23}$$

$$\boldsymbol{D}_{1,13} = \operatorname{diag}(1.44, 2.76, 6.12, 3.21) \qquad \boldsymbol{D}_{2,13} = \operatorname{diag}(8.71, 6.02, 5.85, 5.27) \tag{D.24}$$

$$\boldsymbol{D}_{1,14} = \operatorname{diag}(6.38, 9.15, 8.65, 2.26) \qquad \boldsymbol{D}_{2,14} = \operatorname{diag}(2.70, 2.86, 4.49, 4.29) \tag{D.25}$$

$$\boldsymbol{D}_{1,15} = \operatorname{diag}(8.66, 7.43, 8.19, 7.80) \qquad \boldsymbol{D}_{2,15} = \operatorname{diag}(1.10, 3.83, 0.30, 7.96) \tag{D.26}$$

$$\boldsymbol{Q}_{1,1} = \mathrm{diag}(8.40, 9.47, 1.29, 4.71) \qquad \boldsymbol{Q}_{2,1} = \mathrm{diag}(6.03, 4.90, 8.19, 0.48) \tag{D.27}$$

$$\boldsymbol{Q}_{1,2} = \mathrm{diag}(4.56, 8.81, 4.65, 3.66) \qquad \boldsymbol{Q}_{2,2} = \mathrm{diag}(7.51, 2.04, 2.90, 7.59) \tag{D.28}$$

$$\boldsymbol{Q}_{1,3} = \mathrm{diag}(5.09, 8.76, 2.71, 1.96) \qquad \boldsymbol{Q}_{2,3} = \mathrm{diag}(7.41, 6.37, 4.99, 2.71) \tag{D.29}$$

$$\boldsymbol{Q}_{1,4} = \mathrm{diag}(5.12, 6.05, 0.77, 8.14) \qquad \boldsymbol{Q}_{2,4} = \mathrm{diag}(8.27, 7.79, 7.36, 0.17) \tag{D.30}$$

$$\boldsymbol{Q}_{1,5} = \mathrm{diag}(5.58, 8.58, 3.80, 3.32) \qquad \boldsymbol{Q}_{2,5} = \mathrm{diag}(2.11, 8.80, 9.62, 2.54) \tag{D.31}$$

$$\boldsymbol{Q}_{1,6} = \mathrm{diag}(1.32, 9.49, 5.95, 5.71) \qquad \boldsymbol{Q}_{2,6} = \mathrm{diag}(7.11, 1.95, 5.09, 8.45) \tag{D.32}$$

$$\boldsymbol{Q}_{1,7} = \mathrm{diag}(1.57, 6.96, 6.55, 3.16) \qquad \boldsymbol{Q}_{2,7} = \mathrm{diag}(3.16, 6.88, 5.65, 0.67) \tag{D.33}$$

$$\boldsymbol{Q}_{1,8} = \mathrm{diag}(4.61, 7.88, 0.98, 5.70) \qquad \boldsymbol{Q}_{2,8} = \mathrm{diag}(2.88, 8.12, 0.96, 0.49) \tag{D.34}$$

$$\boldsymbol{Q}_{1,9} = \mathrm{diag}(6.74, 5.13, 8.26, 8.24) \qquad \boldsymbol{Q}_{2,9} = \mathrm{diag}(7.34, 2.84, 3.05, 8.56) \tag{D.35}$$

$$\boldsymbol{Q}_{1,10} = \mathrm{diag}(7.00, 9.44, 5.36, 9.15) \qquad \boldsymbol{Q}_{2,10} = \mathrm{diag}(4.83, 0.24, 8.11, 6.61) \tag{D.36}$$

$$\boldsymbol{Q}_{1,11} = \mathrm{diag}(9.58, 7.78, 5.43, 7.21) \qquad \boldsymbol{Q}_{2,11} = \mathrm{diag}(1.40, 1.80, 0.45, 4.35) \tag{D.37}$$

$$\boldsymbol{Q}_{1,12} = \mathrm{diag}(3.07, 4.88, 2.45, 0.32) \qquad \boldsymbol{Q}_{2,12} = \mathrm{diag}(0.97, 1.61, 6.02, 5.80) \tag{D.38}$$

$$\boldsymbol{Q}_{1,13} = \mathrm{diag}(8.85, 6.92, 5.37, 1.14) \qquad \boldsymbol{Q}_{2,13} = \mathrm{diag}(5.18, 9.53, 0.91, 9.05) \tag{D.39}$$

$$\boldsymbol{Q}_{1,14} = \mathrm{diag}(1.13, 6.20, 2.94, 4.21) \qquad \boldsymbol{Q}_{2,14} = \mathrm{diag}(1.48, 3.29, 4.65, 9.90) \tag{D.40}$$

$$\boldsymbol{Q}_{1,15} = \mathrm{diag}(3.11, 6.26, 5.56, 1.13) \qquad \boldsymbol{Q}_{2,15} = \mathrm{diag}(8.03, 8.87, 8.83, 1.09) \tag{D.41}$$

D.5 Numeric Data for the Simulation Results from Section 5.3

In Section 5.3, we use the time $t_{\delta,i}$ as a measure for the convergence of a reconstruction towards the variable to reconstruct. Table D.5 depicts the values of $t_{\delta,i}$ with $\delta = 0.01$ for the reconstructions obtained from the centralized observer (5.12) and the distributed observer (5.14) for the 120 states of the nonlinear system (5.7).

Table D.5: Time $t_{0.01,i}$ for the centralized observer (5.12) and the distributed observer (5.14)

State i	(5.12): $t_{0.01,i}$ in s	(5.14): $t_{0.01,i}$ in s	State i	(5.12): $t_{0.01,i}$ in s	(5.14): $t_{0.01,i}$ in s
1	0.34	0.46	61	0.35	0.34
2	0.41	0.46	62	0.15	0.17
3	0.35	0.46	63	0.89	0.95
4	0.40	0.46	64	3.25	4.34
5	0.11	0.11	65	0.48	0.46
6	0.29	0.32	66	0.43	0.46
7	0.08	0.10	67	0.40	0.46
8	1.19	1.07	68	0.37	0.46
9	0.63	0.46	69	0.10	0.09
10	0.58	0.46	70	0.20	0.22
11	0.70	0.46	71	0.21	0.26
12	0.75	0.46	72	0.07	0.07
13	0.31	0.14	73	0.58	0.46
14	0.55	0.44	74	0.37	0.46
15	0.32	0.19	75	1.26	0.46
16	0.23	0.10	76	0.50	0.46
17	0.53	0.46	77	0.42	0.28
18	0.14	0.46	78	1.63	1.98
19	0.48	0.46	79	0.25	0.07
20	0.37	0.46	80	0.11	0.08
21	0.07	0.06	81	0.30	0.46
22	0.05	0.13	82	0.80	0.46
23	0.13	0.16	83	0.28	0.46
24	0.15	0.20	84	0.50	0.46
25	0.53	0.46	85	0.68	0.83
26	0.14	0.46	86	0.59	0.27
27	0.48	0.46	87	3.41	12.82
28	0.37	0.46	88	0.15	0.11
29	0.07	0.07	89	0.50	0.46
30	0.58	2.21	90	0.25	0.46
31	0.22	0.24	91	0.90	0.46
32	2.85	3.17	92	0.48	0.46
33	0.53	0.46	93	0.55	0.51
34	0.14	0.46	94	0.72	0.84
35	0.48	0.46	95	0.54	0.26
36	0.37	0.46	96	0.18	0.15
37	0.46	0.39	97	0.60	0.46
38	0.03	0.07	98	0.64	0.46
39	0.10	0.13	99	0.32	0.46
40	0.29	0.37	100	0.48	0.46
41	0.48	0.46	101	0.23	0.10
42	0.43	0.46	102	0.18	0.08
43	0.40	0.46	103	1.55	0.87
44	0.37	0.46	104	0.12	0.10
45	0.07	0.07	105	0.23	0.46
46	0.61	0.65	106	0.51	0.46
47	0.30	0.36	107	1.10	0.46
48	0.06	0.06	108	0.46	0.46
49	0.28	0.46	109	0.70	1.15
50	0.42	0.46	110	0.54	0.49
51	0.44	0.46	111	0.60	0.22
52	0.41	0.46	112	0.12	0.11
53	0.15	0.28	113	0.36	0.46
54	0.26	0.27	114	0.67	0.46
55	0.10	0.10	115	0.07	0.46
56	0.98	0.89	116	0.30	0.46
57	0.48	0.46	117	0.41	0.52
58	0.43	0.46	118	0.30	0.14
59	0.40	0.46	119	0.47	1.74
60	0.37	0.46	120	0.39	0.54

D.6 Noise Study for the Observers from Section 5.3

At the end of Section 5.3, it was commented on the behavior of the centralized observer (5.12) and the distributed observer (5.14) in the presence of measurement noise. The section at hand provides the simulation details that build the basis for the comments from Section 5.3.

To analyze the behavior of the observers under measurement noise, the measurement equations (5.11) and (5.13) are extended by Gaussian random processes. The extended measurement equations for the centralized observer (5.12) and the distributed observer (5.14) read

$$\boldsymbol{m} = \boldsymbol{C}\,\boldsymbol{Q}\,\boldsymbol{x} + \boldsymbol{\varepsilon}\,, \tag{D.42a}$$

$$\boldsymbol{m}_j = \boldsymbol{C}_j\boldsymbol{Q}_j\boldsymbol{x}_j + \boldsymbol{\varepsilon}_j, \quad \forall j \in \mathbb{V}_s, \tag{D.42b}$$

respectively. In (D.42), $\boldsymbol{\varepsilon}$ and $\boldsymbol{\varepsilon}_j$ are vector-valued Gaussian random processes with zero mean and covariance matrices $\sigma\boldsymbol{I}_{60}$ and $\sigma\boldsymbol{I}_4$, respectively, where $\sigma \in \mathbb{R}_{\geq 0}$. The simulations of the centralized observer (5.12) and the distributed observer (5.14) are then rerun on the basis of the noisy measurements (D.42). Thereby, the noise variance is set to $\sigma^2 = 10^{-4}$ which leads to a mean SNR over all measurement signals of 51.8 dB. To ensure comparability, the same realization of the noise process is applied to the centralized and distributed observer. The results of the simulations under noise are presented in the sequel.

The bar diagram in Figure D.3 depicts the number of states with $t_{\delta,i} \leq 10\,\mathrm{s}$ for different values of the tolerance δ. The red and brown bars correspond the centralized and the distributed observer, respectively.

For the centralized observer, there are no states with $t_{\delta,i} \leq 10\,\mathrm{s}$ if $\delta \leq 10^{-3}$. For $\delta = 10^{-2}$ and $\delta = 10^{-1}$, we have 24 and 108 states with $t_{\delta,i} \leq 10\,\mathrm{s}$, respectively. Only for $\delta = 10^{0}$, all states achieve a $t_{\delta,i} \leq 10\,\mathrm{s}$. For the distributed observer, there are 59 states with $t_{\delta,i} \leq 10\,\mathrm{s}$ for $\delta \in [10^{-4}, 10^{-3}]$. For $\delta = 10^{-2}$, 119 states have a time $t_{\delta,i} \leq 10\,\mathrm{s}$;[1] for $\delta \geq 10^{-1}$ all 120 states have a $t_{\delta,i} \leq 10\,\mathrm{s}$.

Figure D.3 reveals the centralized observer to be negatively influenced by the measurement noise. Thereby, the noise deteriorates the convergence behavior of

[1] Note that in the absence of noise there is also one state in the distributed observer which has a $t_{\delta,i} > 10\,\mathrm{s}$ for $\delta = 10^{-2}$, see Figure 5.14.

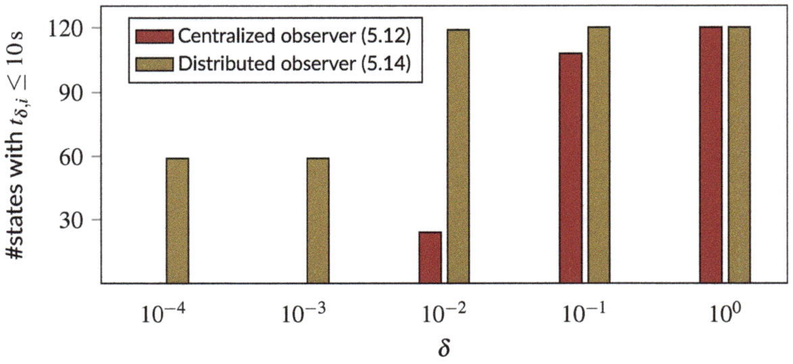

Figure D.3: Number of observer states with $t_{\delta,i} \leq 10\,\mathrm{s}$ for different values of δ for the centralized observer (5.12) and the distributed observer (5.14) in the presence of noise

all reconstructions. In the noise-free case, each of the 120 states has a $t_{\delta,i} \leq 10\,\mathrm{s}$ for $\delta = 10^{-2}$ (cf. Figure 5.14). In contrast, under noisy measurements this number reduces to 24 states. On the other hand, the figure shows that for $t \geq 10\,\mathrm{s}$ all reconstructions remain in a tolerance band of width $\delta = 10^{0}$ around the values of the true states.

For the distributed observer, the situation is different. Half of all reconstructions remain unaffected by the measurement noise. The explanation of this can be found in the subsystem observer dynamics (5.14a) and the structure of the matrices $\boldsymbol{R}_j$ in (5.10a). The error injection, and therewith the noise, acts on the first four elements of the state vector of each subsystem state. The corresponding reconstructions converge to and remain in a tolerance band of width $\delta = 10^{-1}$ around the value to be observed. In contrast, the last four elements of the observer state converge due to a natural damping contained in the matrix $\boldsymbol{R}_j$. Thereby, the matrix $\boldsymbol{R}_j$ is a 4×4 block diagonal matrix. Hence, there is no coupling from the first four to the second four observer states (and vice versa) and the latter remain unaffected by the noise.

In conclusion, the above results show that those observer states converging only due to the natural damping of the system remain unaffected by the measurement noise. In contrast, in observer states influenced by the error injection the property of asymptotic convergence reduces to the convergence into a tolerance band around the value to be observed. The width δ of this tolerance band depends on the noise level.

Abbreviations

AMOTO	automatic model generation and observer design tool
DAE	differential-algebraic equation
GUI	graphical user interface
IDA	interconnection and damping assignment
LMI	linear matrix inequality
ODE	ordinary differential equation
PBC	passivity-based control
PDE	partial differential equation
PHS	port-Hamiltonian system
PSDSE	power system dynamic state estimation
RESP	relative error signal power
SNR	signal-to-noise ratio
WLS	weighted least squares

List of Figures

List of Tables

References

Ackermann, J. (1972). Der Entwurf linearer Regelungssysteme im Zustandsraum [the design of linear control systems in the state-space]. *Regelungstechnik*, 20:297–300.

Alhelou, H. H., Golshan, M. E. H., and Hatziargyriou, N. D. (2019). Deterministic dynamic state estimation-based optimal LFC for interconnected power systems using unknown input observer. *IEEE Transactions on Smart Grid*, 11(2):1582–1592.

Arnold, V. I. (1989). *Mathematical methods of classical mechanics*. Springer, New York, 2nd edition.

Atitallah, M., El Harabi, R., and Abdelkrim, M. N. (2015). Fault detection and estimation based on full order unknown input Hamiltonian observers. In *12th IEEE International Multi-Conference on Systems, Signals Devices (SSD)*, pages 1–7.

Bachovchin, K. D. and Ilić, M. D. (2015). Automated modeling of power system dynamics using the Lagrangian formulation. *International Transactions on Electrical Energy Systems*, 25(10):2087–2108.

Baran, M. E. and Wu, F. F. (1989). Network reconfiguration in distribution systems for loss reduction and load balancing. *IEEE Power Engineering Review*, 9(4):101–102.

Batlle, C., Massana, I., and Simó, E. (2011). Representation of a general composition of Dirac structures. In *50th IEEE Conference on Decision and Control (CDC) and European Control Conference (ECC)*, pages 5199–5204.

Beattie, C., Mehrmann, V., Xu, H., and Zwart, H. (2018). Linear port-Hamiltonian descriptor systems. *Mathematics of Control, Signals, and Systems*, 30(4):17.

Biedermann, B. and Meurer, T. (2021). Observer design for a class of nonlinear systems combining dissipativity with interconnection and damping assignment. *International Journal of Robust and Nonlinear Control.*

Biedermann, B., Rosenzweig, P., and Meurer, T. (2018). Passivity-based observer design for state affine systems using interconnection and damping assignment. In *2018 IEEE Conference on Decision and Control (CDC)*, pages 4662–4667.

Bloch, A. and Crouch, P. E. (1999). Representations of Dirac structures on vector spaces and nonlinear L-C circuits. In *Differential Geometry and Control*, volume 64, pages 103–117.

Borutzky, W. (2010). *Bond graph methodology: development and analysis of multidisciplinary dynamic system models*. Springer, London.

Boyd, S., El Ghaoui, L., Feron, E., and Balakrishnan, V. (1994). *Linear matrix inequalities in system and control theory*. Society for Industrial and Applied Mathematics.

Brayton, R. K. and Moser, J. K. (1964). A theory of nonlinear networks, ii. *Quarterly of Applied Mathematics*, 22(2):81–104.

Breedveld, P. (2009). Port-based modeling of dynamic systems. In Duindam, V., Macchelli, A., Stramigioli, S., and Bruyninckx, H., editors, *Modeling and control of complex physical systems: the port-Hamiltonian approach*, chapter 1, pages 1–52. Springer, Berlin Heidelberg.

Cardoso Ribeiro, F. L. (2016). *Port-Hamiltonian modeling and control of a fluid-structure system: Application to sloshing phenomena in a moving container coupled to a flexible structure.* Doctoral thesis, Université de Toulouse, Toulouse.

Castanedo, F. (2013). A review of data fusion techniques. *The Scientific World Journal*, 2013.

Cederbaum, I. (1984). Some applications of graph theory to network analysis and synthesis. *IEEE Transactions on Circuits and Systems*, 31(1):64–68.

Chen, W. and Saif, M. (2006). Unknown input observer design for a class of nonlinear systems: an LMI approach. In *25th IEEE American Control Conference (ACC)*, pages 834–838.

Cheng, C.-F., Wang, W.-J., and Lin, Y.-P. (1994). Robust observer synthesis for nonlinear large-scale systems. *International Journal of Systems Science*, 25(6):1053–1066.

Cho, Y. M. and Rajamani, R. (1997). A systematic approach to adaptive observer synthesis for nonlinear systems. *IEEE Transactions on Automatic Control*, 42(4):534–537.

Chua, L. and Rohrer, R. (1965). On the dynamic equations of a class of nonlinear rlc networks. *IEEE Transactions on Circuit Theory*, 12(4):475–489.

Dai, S. (2016). *Compositional modeling and design of cyber-physical systems using port-Hamiltonian systems.* Doctoral thesis, Vanderbilt University, Nashville (TS), USA.

Darouach, M., Zasadzinski, M., and Xu, S. J. (1994). Full-order observers for linear systems with unknown inputs. *IEEE Transactions on Automatic Control*, 39(3):606–609.

Deroo, F. (2016). *Control of interconnected systems with distributed model knowledge.* Doctoral thesis, Technical University of Munich, Munich, Germany.

Deuflhard, P. and Hohmann, A. (1995). *Numerical analysis: a first course in scientific computation.* de Gruyter, Berlin and others.

Ding, S. X. (2008). *Model-based fault diagnosis techniques: design schemes, algorithms and tools.* Springer-Verlag, Berlin Heidelberg.

Donaire, A. and Junco, S. (2009). Derivation of input-state-output port-Hamiltonian systems from bond graphs. *Simulation Modelling Practice and Theory*, 17(1):137–151.

Dörfler, F., Simpson-Porco, J. W., and Bullo, F. (2018). Electrical networks and algebraic graph theory: models, properties, and applications. *Proceedings of the IEEE*, 106(5):977–1005.

Duindam, V., Macchelli, A., Stramigioli, S., and Bruyninckx, H., editors (2009). *Modeling and control of complex physical systems: the port-Hamiltonian approach.* Springer, Berlin Heidelberg.

Edwards, C. and Spurgeon, S. K. (1994). On the development of discontinuous observers. *International Journal of Control*, 59(5):1211–1229.

Falaize, A. and Hélie, T. (2016). Passive guaranteed simulation of analog audio circuits: a port-Hamiltonian approach. *Applied Sciences*, 6(10):273.

Falaize, A. and Hélie, T. (2019). PyPHS. `https://github.com/pyphs/pyphs`. Accessed: 2019-07-03.

Fiaz, S., Zonetti, D., Ortega, R., Scherpen, J. M. A., and van der Schaft, A. (2013). A port-Hamiltonian approach to power network modeling and analysis. *European Journal of Control*, 19(6):477–485.

Franceschini, G., Piazzi, A., and Tassoni, C. (1994). A genetic algorithm approach to design flux observer for induction servo motors. In *20th Annual Conference of IEEE Industrial Electronics (IECON)*, volume 3, pages 2132–2136.

Gao, Z., Nguang, S. K., and Kong, D.-X. (2019). Advances in modelling, monitoring, and control for complex industrial systems. *Complexity*, 2019.

Garziad, M. and Saka, A. (2017). An overview of the main topics and applications of graph theoretic problems in multiphysics engineering. *Journal of Applied Mechanical Engineering*, 6(5):284.

Gernandt, H., Haller, F., Reis, T., and van der Schaft, A. (2021). Port-Hamiltonian formulation of nonlinear electrical circuits. *Journal of Geometry and Physics*, 159:103959.

Golo, G., Breedveld, P., Maschke, B., and van der Schaft, A. (2000). Geometric formulation of generalized bond graph models - part i: generalized junction structures. Technical Report, University of Twente, Faculty of Mathematical Sciences, Enschede, the Netherlands.

Golo, G., van der Schaft, A., Breedveld, P., and Maschke, B. (2003). Hamiltonian formulation of bond graphs. In *Nonlinear and hybrid systems in automotive control*, pages 351–372. Springer, London.

Hautus, M. L. J. (1969). Controllability and observability conditions of linear autonomous systems. *Indagationes Mathematicae (Proceedings)*, 72(5):443–448.

Hautus, M. L. J. (1983). Strong detectability and observers. *Linear Algebra and its Applications*, 50:353–368.

Hou, M. and Patton, R. J. (1998). Input observability and input reconstruction. *Automatica*, 34(6):789–794.

Howell, A. and Hedrick, J. K. (2002). Nonlinear observer design via convex optimization. In *21st IEEE American Control Conference (ACC)*, volume 3, pages 2088–2093.

Ichihara, H. (2007). Observer design for polynomial systems using convex optimization. In *46th IEEE Conference on Decision and Control (CDC)*, pages 5347–5352.

IEEE Standards Association (2014). *IEEE Std 519-2014. Recommended practice and requirements for harmonic control in electric power systems.* IEEE Power and Energy Society, New York.

Irle, J. P. (2016). *Analyse und Entwurf von Beobachtern mit unbekannten Signalen und Parametern [Analysis and design of observers with unknown signals and parameters].* Doctoral thesis, Karlsruhe Institute of Technology (KIT), Karlsruhe, Germany.

Jeltsema, D. and Scherpen, J. M. A. (2009). Multidomain modeling of nonlinear networks and systems. *IEEE Control Systems Magazine*, 29(4):28–59.

Karnopp, D. (1969). Power-conserving transformations: physical interpretations and applications using bond graphs. *Journal of the Franklin Institute*, 288(3):175–201.

Karnopp, D. C., Margolis, D. L., and Rosenberg, R. C. (2012). *System dynamics: modeling, simulation, and control of mechatronic systems.* John Wiley & Sons, Incorporated, Hoboken (NJ), USA, 5th edition.

Kautsky, J., Nichols, N. K., and van Dooren, P. (1985). Robust pole assignment in linear state feedback. *International Journal of control*, 41(5):1129–1155.

Kersting, W. H. (2017). *Distribution System Modeling and Analysis.* Taylor & Francis, CRC Press, Boca Raton (FL), USA, 4th edition.

Khalil, H. K. (2002). *Nonlinear systems.* Prentice-Hall, Upper Saddle River (NJ), USA, 3rd edition.

Khalil, I. S. M., Sabanovic, A., and Misra, S. (2012). An energy-based state observer for dynamical subsystems with inaccessible state variables. In *2012 IEEE/RSJ International Conference on Intelligent Robots and Systems*, pages 3734–3740.

Kirchhoff, G. (1847). Über die Auflösung der Gleichungen, auf welche man bei der Untersuchung der linearen Vertheilung galvanischer Ströme geführt wird [On the solution of the equations to which we are led in the investigation of the linear distribution of galvanic currents]. *Annalen der Physik*, 148(12):497–508.

Kotyczka, P., Joos, H., Wu, Y., and Le Gorrec, Y. (2019). Finite-dimensional observers for port-Hamiltonian systems of conservation laws. In *58th IEEE Conference on Decision and Control (CDC)*, pages 6875–6880.

Kotyczka, P. and Wang, M. (2015). Dual observer-based compensator design for linear port-Hamiltonian systems. In *14th IEEE European Control Conference (ECC)*, pages 2908–2913.

Kugi, A. (2001). *Non-linear control based on physical models: electrical, mechanical and hydraulic systems*. Springer-Verlag, London.

Kupper, M. (2019). *Verteilte Zustandsschätzung fraktionaler Systeme und ihre Anwendung auf Lithium-Ionen-Batteriesysteme [Distributed state estimation in fractional systems with application to lithium-ion battery systems]*. Doctoral thesis, Karlsruhe Institute of Technology (KIT), Karlsruhe, Germany.

Laila, D. S., Lovera, M., and Astolfi, A. (2011). A discrete-time observer design for spacecraft attitude determination using an orthogonality-preserving algorithm. *Automatica*, 47(5):975–980.

Leu, M.-C. and Hemati, N. (1986). Automated symbolic derivation of dynamic equations of motion for robotic manipulators. *Journal of Dynamic Systems, Measurement and Control*, 108(3):172–179.

Lopes, N. (2016). *Approche passive pour la modélisation, la simulation et l'étude d'un banc de test robotisé pour les instruments de type cuivre [A passivity-based approach to the modeling, simulation, and analysis of a test bench for brass instruments]*. Doctoral thesis, Université Pierre et Marie Curie, Paris, France.

Lu, T.-T. and Shiou, S.-H. (2002). Inverses of 2×2 block matrices. *Computers & Mathematics with Applications*, 43(1-2):119–129.

Ludyk, G. (1995). *Theoretische Regelungstechnik 2: Zustandsrekonstruktion, optimale und nichtlineare Regelungssysteme [Theoretical control engineering 2: state reconstruction, optimal and nonlinear control systems]*. Springer-Verlag, Berlin Heidelberg.

Luenberger, D. G. (1964). Observing the state of a linear system. *IEEE Transactions on Military Electronics*, 8(2):74–80.

Luenberger, D. G. (1966). Observers for multivariable systems. *IEEE Transactions on Automatic Control*, 11(2):190–197.

Luenberger, D. G. (1971). An introduction to observers. *IEEE Transactions on Automatic Control*, 16(6):596–602.

Lyche, T. (2020). *Numerical linear algebra and matrix factorizations*, volume 22. Springer Nature, Cham, Switzerland.

Mantilla-Pérez, P., Pérez-Rúa, J.-A., Millán, M. A. D., Domínguez, X., and Arboleya, P. (2020). Power flow simulation in the product development process of modern vehicular DC distribution systems. *IEEE Transactions on Vehicular Technology*, 69(5):5025–5040.

Maschke, B. and van der Schaft, A. (1992). Port-controlled Hamiltonian systems: modelling origins and systemtheoretic properties. *IFAC Proceedings Volumes*, 25(13):359–365.

Maschke, B., van der Schaft, A., and Breedveld, P. (1992). An intrinsic Hamiltonian formulation of network dynamics: non-standard poisson structures and gyrators. *Journal of the Franklin Institute*, 329(5):923–966.

McPhee, J. J. (1998). Automatic generation of motion equations for planar mechanical systems using the new set of branch coordinates. *Mechanism and Machine Theory*, 33(6):805–823.

Merkert, L., Listmann, K., and Hohmann, S. (2019). Optimization of thermo-hydraulic systems using multiparametric delay modeling. *Energy*, 189:116125.

Milano, F., Dörfler, F., Hug, G., Hill, D. J., and Verbič, G. (2018). Foundations and challenges of low-inertia systems. In *IEEE Power Systems Computation Conference (PSCC)*.

Moreno, J. A. (2004). Observer design for nonlinear systems: a dissipative approach. *IFAC Proceedings Volumes*, 37(21):681–686.

Muller, C. (2018). *Port-Hamiltonian modelling of a wind turbine system with doubly fed induction generator*. Bachelor's thesis, Institute of Control Systems (IRS), Karlsruhe Institute of Technology (KIT), Karlsruhe, Germany.

Najnudel, J., Hélie, T., Boutin, H., Roze, D., Maniguet, T., and Vaiedelich, S. (2018). Analog circuits and port-Hamiltonian realizability issues: a resolution method for simulations via equivalent components. In *145th Audio Engineering Society Convention*.

O'Malley, M. J., Anwar, M. B., Heinen, S., Kober, T., McCalley, J., McPherson, M., Muratori, M., Orths, A., Ruth, M., Schmidt, T. J., and Tuohy, A. (2020). Multicarrier energy systems: shaping our energy future. *Proceedings of the IEEE*.

Ortega, R., van der Schaft, A., Castanos, F., and Astolfi, A. (2008). Control by interconnection and standard passivity-based control of port-Hamiltonian systems. *IEEE Transactions on Automatic Control*, 53(11):2527–2542.

Paynter, H. M. (1961). *Analysis and design of engineering systems (Course 2.751)*. Class Notes for M.I.T. The M.I.T. Press, Cambridge (MA), USA.

Pfeifer, M., Caspart, S., Hampel, S., Muller, C., Krebs, S., and Hohmann, S. (2020a). Explicit port-Hamiltonian formulation of multi-bond graphs for an automated model generation. *Automatica*, 120:109121.

Pfeifer, M., Caspart, S., Muller, C., Pfeiffer, S., Krebs, S., and Hohmann, S. (2020b). Explicit port-Hamiltonian formulation of bond graphs with dependent storages. In *Proceedings of the 21st IFAC World Congress*.

Pfeifer, M., Krebs, S., Hofmann, F., Kupper, M., and Hohmann, S. (2019a). Interval input-state-output estimation for linear port-Hamiltonian systems with application to power distribution systems. In *58th IEEE Conference on Decision and Control (CDC)*, pages 3176–3183.

Pfeifer, M., Pfeiffer, S., Caspart, S., Muller, C., Krebs, S., and Hohmann, S. (2019b). Explicit port-Hamiltonian formulation of bond graphs. In *GMA-Fachausschuss 1.30: Modellbildung, Identifikation und Simulation in der Automatisierungstechnik*, pages 391–404.

Pfeifer, M., Zimmerlin, M., Schwarzendorfer, M., Sauter, P., Krebs, S., Leibfried, T., and Hohmann, S. (2018). Weighted least squares state estimation for coupled power and gas distribution networks. In *53rd IEEE International Universities Power Engineering Conference (UPEC)*.

Pfeiffer, S., Muller, C., Pfeifer, M., Caspart, S., Krebs, S., and Hohmann, S. (2019). AMOTO: Automatic MOdeling TOol for the port-Hamiltonian formulation of bond graphs. In *GMA-Fachausschuss 1.30: Modellbildung, Identifikation und Simulation in der Automatisierungstechnik*, pages 363–367.

Primadianto, A. and Lu, C.-N. (2017). A review on distribution system state estimation. *IEEE Transactions on Power Systems*, 32(5):3875–3883.

Raghavan, S. and Hedrick, J. K. (1994). Observer design for a class of nonlinear systems. *International Journal of Control*, 59(2):515–528.

Rao, C. R. and Mitra, S. K. (1972). Generalized inverse of a matrix and its applications. In *Proceedings of the Sixth Berkeley Symposium on Mathematical Statistics and Probability, Volume 1: Theory of Statistics*, pages 601–620.

Roberson, R. E. and Wittenburg, J. (1968). A dynamical formalism for an arbitrary number of interconnected rigid bodies with reference to the problem of satellite attitude control. In *3rd Congress of the International Federation of Automatic Control 1966*.

Rosenberg, R. C. (1971). State-space formulation for bond graph models of multiport systems. *Journal of Dynamic Systems, Measurement, and Control*, 93(1):35–40.

Scheuermann, T., Kotyczka, P., Martens, C., Louati, H., Maschke, B., Zanota, M.-L., and Pitault, I. (2020). An object-oriented library for heat transfer modelling and simulation in open cell foams. In *Proceedings of the 21st IFAC World Congress*.

Schiffer, J., Zonetti, D., Ortega, R., Stankovic, A., Sezi, T., and Raisch, J. (2016). A survey on modeling of microgrids – from fundamental physics to phasors and voltage sources. *Automatica*, 74:135–150.

Schlacher, K., Kugi, A., and Scheidl, R. (1998). Tensor analysis based symbolic computation for mechatronic systems. *Mathematics and computers in simulation*, 46(5-6):517–525.

Schweers, C. (2017). *Adaptive Sigma-Punkte-Filter-Auslegung zur Zustands- und Parameterschätzung an Black-Box-Modellen [Adaptive design of sigma-point filters for a state- and parameter estimation with black-box models]*. Doctoral thesis, Universität Paderborn, Paderborn, Germany.

Shoukry, Y., Chong, M., Wakaiki, M., Nuzzo, P., Sangiovanni-Vincentelli, A., Seshia, S. A., Hespanha, J. P., and Tabuada, P. (2018). SMT-based observer design for cyber-physical systems under sensor attacks. *ACM Transactions on Cyber-Physical Systems*, 2(1):1–27.

Simpson-Porco, J. W., Dörfler, F., and Bullo, F. (2017). Voltage stabilization in microgrids via quadratic droop control. *IEEE Transactions on Automatic Control*, 3(62):1239 – 1253.

Singer, Y. (2019). *Automated input-state-output observer design for linear port-Hamiltonian systems*. Bachelor's thesis, Institute of Control Systems (IRS), Karlsruhe Institute of Technology (KIT), Karlsruhe, Germany.

Singh, A. K. and Pal, B. C. (2019). *Dynamic estimation and control of power systems*. Academic Press, London and others.

Sira-Ramírez, H. and Cruz-Hernández, C. (2001). Synchronization of chaotic systems: a generalized Hamiltonian systems approach. *International Journal of Bifurcation and Chaos*, 11(05):1381–1395.

Sontag, E. D. (1998). *Mathematical control theory: deterministic finite dimensional systems*. Springer, New York, 2nd edition.

Strehle, F., Malan, A. J., Krebs, S., and Hohmann, S. (2019). A port-Hamiltonian approach to plug-and-play voltage and frequency control in islanded inverter-based AC microgrids. In *58th IEEE Conference on Decision and Control (CDC)*, pages 4648–4655.

Strehle, F., Malan, A. J., Krebs, S., and Hohmann, S. (2020). Passivity conditions for plug-and-play operation of nonlinear static AC loads. In *Proceedings of the 21st IFAC World Congress*.

Strehle, F., Pfeifer, M., Kölsch, L., Degünther, C., Ruf, J., Andresen, L., and Hohmann, S. (2018). Towards port-Hamiltonian modeling of multi-carrier energy systems: a case study for a coupled electricity and gas distribution system. *IFAC-PapersOnLine*, 51(2):463–468.

Syrmos, V. L. (1993). Computational observer design techniques for linear systems with unknown inputs using the concept of transmission zeros. *IEEE Transactions on Automatic Control*, 38(5):790–794.

Tahir, A., Böling, J., Haghbayan, M.-H., Toivonen, H. T., and Plosila, J. (2019). Swarms of unmanned aerial vehicles — a survey. *Journal of Industrial Information Integration*, 16:100106.

Trumpf, J. (2013). On state observers. In Trumpf, J. and Hüper, K., editors, *Mathematical System Theory - Festschrift in Honor of Uwe Helmke on the Occasion of his Sixtieth Birthday*, pages 421–435. CreateSpace, Scotts Valley (CA), USA.

Trumpf, J., Trentelman, H. L., and Willems, J. C. (2011). An internal model principle for observers. In *50th IEEE Conference on Decision and Control (CDC) and European Control Conference (ECC)*, pages 3992–3999.

van der Schaft, A. (2009). Port-Hamiltonian systems. In Duindam, V., Macchelli, A., Stramigioli, S., and Bruyninckx, H., editors, *Modeling and control of complex physical systems: the port-Hamiltonian approach*, chapter 2, pages 53–130. Springer, Berlin Heidelberg.

van der Schaft, A. (2017). *L2-gain and passivity techniques in nonlinear control*. Springer, Cham, Switzerland, 3rd edition.

van der Schaft, A. and Jeltsema, D. (2014). Port-Hamiltonian systems theory: an introductory overview. *Foundations and Trends® in Systems and Control*, 1(2-3):173–378.

van der Schaft, A. and Maschke, B. (1995). The Hamiltonian formulation of energy conserving physical systems with external ports. *AEU - Archiv für Elektronik und Übertragungstechnik*, 49(5-6):362–371.

van der Schaft, A. and Maschke, B. (2013). Port-Hamiltonian systems on graphs. *SIAM Journal on Control and Optimization*, 51(2):906–937.

van der Schaft, A. and Maschke, B. (2018). Generalized port-Hamiltonian DAE systems. *Systems & Control Letters*, 121:31–37.

Venkatraman, A. and van der Schaft, A. (2010). Full-order observer design for a class of port-Hamiltonian systems. *Automatica*, 46(3):555–561.

Vidyasagar, M. (1993). *Nonlinear systems analysis*. Prentice-Hall, London and others, 2nd edition.

Vincent, B., Hudon, N., Lefèvre, L., and Dochain, D. (2016). Port-Hamiltonian observer design for plasma profile estimation in tokamaks. *IFAC-PapersOnLine*, 49(24):93–98.

Wang, H., Wang, J., Liu, J., and Lam, J. (2003). Iterative LMI approach for robust fault detection observer design. In *42nd IEEE International Conference on Decision and Control (CDC)*, pages 1974–1979.

Wang, Y., Ge, S. S., and Cheng, D. (2005). Observer and observer-based H_∞ control of generalized Hamiltonian systems. *Science in China Series F: Information Sciences*, 48(2):211–224.

Warsewa, A., Böhm, M., Sawodny, O., and Tarín, C. (2020). A port-Hamiltonian approach to modeling the structural dynamics of complex systems. *Applied Mathematical Modelling*, 89:1528–1546.

Watson, N. R. and Yu, K. K. C. (2008). Transient state estimation. In *2008 13th IEEE International Conference on Harmonics and Quality of Power*.

Wellstead, P. E. (1979). *Introduction to physical system modelling*. Acadamic Press, London.

Willems, J. C. (2007). The behavioral approach to open and interconnected systems. *IEEE Control Systems*, 27(6):46–99.

Wong, J. Y. (2020). IEEE 33 bus system. (`https://www.mathworks.com/matlabcentral/fileexchange/73127-ieee-33-bus-system`), MATLAB Central File Exchange. Retrieved August 25, 2020.

Yaghmaei, A. and Yazdanpanah, M. J. (2019a). Output control design and separation principle for a class of port-Hamiltonian systems. *International Journal of Robust and Nonlinear Control*, 29(4):867–881.

Yaghmaei, A. and Yazdanpanah, M. J. (2019b). Structure preserving observer design for port-Hamiltonian systems. *IEEE Transactions on Automatic Control*, 64(3):1214–1220.

Zhao, J., Gomez-Exposito, A., Netto, M., Mili, L., Abur, A., Terzija, V., Kamwa, I., Pal, B. C., Singh, A. K., Qi, J., Huang, Z., and Meliopoulos, A. P. S. (2019). Power system dynamic state estimation: motivations, definitions, methodologies and future work. *IEEE Transactions on Power Systems*, 34(4):3188 - 3198.

Zhao, J., Netto, M., and Mili, L. (2017). A robust iterated extended Kalman filter for power system dynamic state estimation. *IEEE Transactions on Power Systems*, 32(4):3205–3216.

List of Publications and Supervised Theses

Journal Articles

- **Martin Pfeifer**, Sven Caspart, Silja Hampel, Charles Muller, Stefan Krebs, Sören Hohmann, *Explicit port-Hamiltonian formulation of multi-bond graphs for an automated model generation*, Automatica 2020, 120:109121
- Christoph M. Hackl[2], Pol Jané-Soneira[2], **Martin Pfeifer**[2], Korbinian Schechner[2], Sören Hohmann, *Full- and reduced-order state-space modeling of wind turbine systems with permanent magnet synchronous generator*, Energies 2018, 11(7):1809
- Stefan Krebs, Christoph Schnurr, **Martin Pfeifer**, Jörg Weigold, Sören Hohmann, *Reduced-order hybrid interval observer for verified state estimation of an induction machine*, Control Engineering Practice 2016, 57:157–168

Refereed Conference Articles

- **Martin Pfeifer**, Sven Caspart, Felix Strehle, Sören Hohmann, *Full-order observer design for a class of nonlinear port-Hamiltonian systems*, 7th IFAC Workshop on Lagrangian and Hamiltonian Methods for Nonlinear Control, 2021, Berlin, Germany
- **Martin Pfeifer**, Sven Caspart, Charles Muller, Silja Pfeiffer, Stefan Krebs, Sören Hohmann, *Explicit port-Hamiltonian formulation of bond graphs with dependent storages*, 21st IFAC World Congress, 2020, Berlin, Germany
- Felix Strehle, **Martin Pfeifer**, Albertus Johannes Malan, Stefan Krebs, Sören Hohmann, *A scalable port-Hamiltonian approach to plug-and-play voltage*

[2]First four authors contributed equally to the paper and were listed in alphabetical order.

stabilization in DC microgrids, 4th IEEE Conference on Control Technology and Applications (CCTA), 2020, Montréal, Canada

- Oliver Stark, **Martin Pfeifer**, Stefan Krebs, Sören Hohmann, *Bias-free parameter identification for non-commensurable fractional systems*, 18th European Control Conference (ECC), 2020, Saint Petersburg, Russian Federation
- **Martin Pfeifer**, Stefan Krebs, Felix Hofmann, Martin Kupper, Sören Hohmann, *Interval input-state-output estimation for linear port-Hamiltonian systems with application to power distribution systems*, 58th IEEE Conference on Decision and Control (CDC), 2019, Nice, France
- **Martin Pfeifer**, Silja Pfeiffer, Sven Caspart, Charles Muller, Stefan Krebs, Sören Hohmann, *Explicit port-Hamiltonian formulation of bond graphs*, VDI/VDE GMA-Fachausschuss 1.30 "Modellbildung, Identifikation und Simulation in der Automatisierungstechnik", 2019, Anif, Austria
- Silja Pfeiffer, Charles Muller, Sven Caspart, **Martin Pfeifer**, Stefan Krebs, Sören Hohmann, *AMOTO: Automatic MOdeling TOol for the port-Hamiltonian formulation of bond graphs*, VDI/VDE GMA-Fachausschuss 1.30 "Modellbildung, Identifikation und Simulation in der Automatisierungstechnik", 2019, Anif, Austria
- Martin Kupper, **Martin Pfeifer**, Stefan Krebs, Sören Hohmann, *Unknown input Kalman filtering for linear discrete-time fractional order systems with direct feedthrough*, 18th European Control Conference (ECC), 2019, Napoli, Italy
- **Martin Pfeifer**, Martin Zimmerlin, Martin Schwarzendorfer, Patrick Sauter, Stefan Krebs, Sören Hohmann, *Weighted least squares state estimation for coupled power and gas distribution networks*, 53rd International Universities Power Engineering Conference (UPEC), 2018, Glasgow, Scotland
- Felix Stehle, **Martin Pfeifer**, Lukas Kölsch, Charlotte Degünther, Johannes Ruf, Lisa Andresen, Sören Hohmann, *Towards port-Hamiltonian modeling of multi-carrier energy systems: a case study for a coupled electricity and gas distribution system*, 10th Vienna International Conference on Mathematical Modelling (MATHMOD), 2018, Vienna, Austria
- Sören Hohmann[3], Heiko Maaß[3], Carina Mieth[3], **Martin Pfeifer**[3], Dorothea Wagner[3], Franziska Wegner[3], *Graph-theoretic model for observability in multi-carrier energy distribution networks*, 6th D-A-CH+ Conference on Energy Informatics, 2017, Lugano, Switzerland
- Patrick Sauter, Philipp Karg, **Martin Pfeifer**, Mathias Kluwe, Martin Zimmerlin, Thomas Leibfried, Sören Hohmann, *Neural network-based load forecasting in distribution grids for predictive energy management systems*, International ETG Congress, 2017, Bonn, Germany

- Martin Zimmerlin, Michael Suriyah, Thomas Leibfried, Patrick Sauter, Jona Maurer, **Martin Pfeifer**, Mathias Kluwe, Sören Hohmann, Wolfgang Köppel, Charlotte Degünther, Christian Müller, Johannes Ruf, Praseeth Prabhakaran, Peter Missal, Joachim Walter, Leoni Herold, Thomas Eichinger, *Das Verbundprojekt RegEnKibo: Regionalisierung der Energieversorgung auf Verteilnetzebene unter Anwendung eines multimodalen, zellularen Ansatzes*, International ETG Congress, 2017, Bonn, Germany
- Stefan Krebs, **Martin Pfeifer**, Sebastian Fugel, Jörg Weigold, Sören Hohmann, *Interval observer for LPV systems based on time-variant transformations*, 58th IEEE Conference on Decision and Control (CDC), 2016, Las Vegas (NV), USA
- **Martin Pfeifer**, Thomas Weickert, Silke Klose, *Rule-based PI controller autotuning for drive systems*, 14th European Control Conference (ECC), 2015, Linz, Austria
- **Martin Pfeifer**, Gustavo Lenis, Olaf Dössel, *A general approach for dynamic modeling of physiological time series*, Biomedizinische Technik (BMT), 2013, Graz, Austria

Invited Talks

- **Martin Pfeifer**, Lukas Kölsch, Sören Hohmann, *Automated design of an input-state-output observer for linear port-Hamiltonian systems with application to the state estimation in unbalanced distribution systems*, 2021 SIAM Conference on Computational Science and Engineering, 2021, Fort Worth (TX), USA
- **Martin Pfeifer**, Martin Zimmerlin, Stefan Krebs, Thomas Leibfried, Sören Hohmann, *State estimation for coupled electricity and gas distribution networks*, 4th Scientific SCI Meeting, 2018, Stuttgart, Germany
- Sören Hohmann[3], Heiko Maaß[3], Carina Mieth[3], **Martin Pfeifer**[3], Dorothea Wagner[3], Franziska Wegner[3], *Graph-theoretic model for observability in multi-carrier energy distribution networks*, 3rd Scientific SCI Meeting, 2017, Karlsruhe, Germany

[3]Authors were listed in alphabetical order.

Supervised Student Theses

- Syrine Khammari, *Automatisierter Entwurf von Unknown-Input-Beobachtern für asymmetrische Verteilnetze*, Bachelor's thesis, Institute of Control Systems (IRS), Karlsruhe Institute of Technology (KIT), 2020
- Yoel Singer, *Automated input-state-output observer design for linear port-Hamiltonian systems*, Bachelor's thesis, Institute of Control Systems (IRS), Karlsruhe Institute of Technology (KIT), 2019
- Charles Muller, *Port-Hamiltonian modelling of a wind turbine system with doubly fed induction generator*, Bachelor's thesis, Institute of Control Systems (IRS), Karlsruhe Institute of Technology (KIT), 2019
- Silja Pfeiffer, *Automated determination of explicit port-Hamiltonian models from bond graphs*, Bachelor's thesis, Institute of Control Systems (IRS), Karlsruhe Institute of Technology (KIT), 2018
- Felix Hofmann, *Interval input-state-output estimation for port-Hamiltonian systems with application to power systems*, Master's thesis, Institute of Control Systems (IRS), Karlsruhe Institute of Technology (KIT), 2018
- Julian Pflüger, *Structured port-Hamiltonian modeling and observer synthesis for linear dynamic systems with application to electrical distribution networks*, Master's thesis, Institute of Control Systems (IRS), Karlsruhe Institute of Technology (KIT), 2018
- Martin Schwarzendorfer, *Entwicklung eines multimodalen Zustandsschätzers für gekoppelte Elektrizitäts- und Gasverteilnetze*, Master's thesis, Institute of Electrical Energy Systems and High Voltage Technology (IEH), Karlsruhe Institute of Technology (KIT), 2017
- Carina Mieth, *Observability of multi-carrier energy distribution networks and optimal sensor placement*, Master's thesis, Institute of Control Systems (IRS), Karlsruhe Institute of Technology (KIT), 2017
- Hanns Jonathan Jonuschat, *Dynamic modeling of combined heat and power systems with internal combustion engine*, Bachelor's thesis, Institute of Control Systems (IRS), Karlsruhe Institute of Technology (KIT), 2017
- Felix Strehle, *Modeling of multi-carrier energy distribution systems with port-Hamiltonian systems*, Master's thesis, Institute of Control Systems (IRS), Karlsruhe Institute of Technology (KIT), 2017
- Tobias Weinkamm, *Differentialalgebraisches Modellierungskonzept für die Mittelspannungsebene elektrischer Verteilnetze*, Master's thesis, Institute of Control Systems (IRS), Karlsruhe Institute of Technology (KIT), 2016

- Pol Jané-Soneira, *Dynamic modelling of variable frequency synchronous generator wind power plants*, Bachelor's thesis, Institute of Control Systems (IRS), Karlsruhe Institute of Technology (KIT), 2016

Thanks

To Prof. Hohmann for being an inspiring teacher.

To Prof. Kugi for taking the time to assess this work.

To the IRS, the AES research group, to Jairo and Mathias. To Felicitas, Steven, and Frederik. To Sven for his dedication in lecturing the problem classes "Lineare Algebra II für die Fachrichtung Informatik". To Silja and Charles for excellent coding.

To Stefan, Felix, and Relax for a fruitful friendship and a critical review of this thesis.

To Umberto Eco for his book "How to Write a Thesis".

To Lena and Zoé for setting up my writing room.

Karlsruher Beiträge zur Regelungs- und Steuerungstechnik (ISSN 2511-6312)
Institut für Regelungs- und Steuerungssysteme

Band 01 Diehm, Gunter
Identifikation des menschlichen Bewegungsverhaltens auf der Basis von Primitiven.
ISBN 978-3-7315-0608-9

Band 02 Flad, Michael
Kooperative Regelungskonzepte auf Basis der Spieltheorie und deren Anwendung auf Fahrerassistenzsysteme.
ISBN 978-3-7315-0610-2

Band 03 Eckert, Marius
Modellbasierte Identifikation fraktionaler Systeme und ihre Anwendung auf die Lithium-Ionen-Zelle.
ISBN 978-3-7315-0690-4

Band 04 Krebs, Stefan
Intervallbeobachter für lineare parametervariante Systeme und deren Anwendung auf die Asynchronmaschine.
ISBN 978-3-7315-0857-1

Band 05 Kaspar, Stephan
Fahrdynamikuntersuchungen eines Elektrofahrzeugs mit Einzelrad-Hinterradantrieb.
ISBN 978-3-7315-0916-5

Band 06 Sauter, Patrick S.
Modellierung und zentrale prädiktive Regelung von multimodalen Energieverteilnetzen.
ISBN 978-3-7315-0963-9

Band 07 Kupper, Martin
Verteilte Zustandsschätzung fraktionaler Systeme und ihre Anwendung auf Lithium-Ionen-Batteriesysteme.
ISBN 978-3-7315-0971-4

Band 08 Merkert, Lennart
Optimal Scheduling of Combined Heat and Power Generation Considering Heating Grid Dynamics.
ISBN 978-3-7315-1056-7

Die Bände sind unter www.ksp.kit.edu als PDF frei verfügbar oder als Druckausgabe bestellbar.

Band 09 Ludwig, Julian
Automatisierte kooperative Transition einer Regelungsaufgabe zwischen Mensch und Maschine am Beispiel des hochautomatisierten Fahrens.
ISBN 978-3-7315-1069-7

Band 10 Inga Charaja, Juan Jairo
Inverse Dynamic Game Methods for Identification of Cooperative System Behavior.
ISBN 978-3-7315-1080-2

Band 11 Schnurr, Christoph Xaver
Ein Verfahren zur lexikographischen modellprädiktiven Regelung mit der Anwendung auf eine permanenterregte Synchronmaschine.
ISBN 978-3-7315-1095-6

Band 12 Schwab, Stefan
Guaranteed Verification of Dynamic Systems.
ISBN 978-3-7315-0965-3

Band 13 Pfeifer, Martin
Automated Model Generation and Observer Design for Interconnected Systems: A Port-Hamiltonian Approach.
ISBN 978-3-7315-1135-9